LONDON MATHEMATICAL SOCIETY STUDENT TEXTS

Managing Editor: Professor D. Benson,
Department of Mathematics, University of Aberdeen, UK

Joseph Liouville
1809–1882
By permission of the
Académie des Sciences, Institut de France.

London Mathematical Society Student Texts 76

Number Theory in the Spirit of Liouville

KENNETH S. WILLIAMS

Carleton University, Ottawa

CAMBRIDGE
UNIVERSITY PRESS

CAMBRIDGE UNIVERSITY PRESS
Cambridge, New York, Melbourne, Madrid, Cape Town, Singapore,
São Paulo, Delhi, Dubai, Tokyo, Mexico City

Cambridge University Press
The Edinburgh Building, Cambridge CB2 8RU, UK

Published in the United States of America by Cambridge University Press, New York

www.cambridge.org
Information on this title: www.cambridge.org/9781107002531

First published 2011

Printed in the United Kingdom at the University Press, Cambridge

A catalog record for this publication is available from the British Library

Library of Congress Cataloging in Publication data
Williams, Kenneth S.
Number theory in the spirit of Liouville / Kenneth S. Williams.
p.   cm. – (London Mathematical Society student texts ; 76)
Includes bibliographical references and index.
ISBN 978-1-107-00253-1 (hardback)
1. Number theory.   2. Liouville, Joseph, 1809–1882.   I. Title.
QA241.W625   2010
512.7′2 – dc22   2010030399

ISBN 978-1-107-00253-1 Hardback
ISBN 978-0-521-17562-3 Paperback

Dedicated to my granddaughter
ISABELLE SOFIE OLSEN
born January 11, 2009

and

to the memory of
JOSEPH LIOUVILLE
March 24, 1809–September 8, 1882

Contents

Preface

In a series of eighteen papers published between the years 1858 and 1865 the French mathematician Joseph Liouville (1809–1882) introduced a powerful new method into elementary number theory. Liouville's idea was to give a number of elementary (but not simple to prove) identities from which flowed many number-theoretic results by specializing the functions involved in the formulae.

Although Liouville's ideas are now 150 years old, they still do not usually form part of a standard course in elementary number theory. Moreover there is no book in English devoted entirely to Liouville's method, and, although some elementary number theory texts devote a chapter to Liouville's ideas, most do not. In this book we hope to remedy this situation by providing a gentle introduction to Liouville's method. We will not give a comprehensive treatment of all of Liouville's identities but rather give a sufficient number of his identities in order to provide elementary arithmetic proofs of such number-theoretic results as the Girard-Fermat theorem, a recurrence relation for the sum of divisors function, Lagrange's theorem, Legendre's formula for the number of representations of a nonnegative integer as the sum of four triangular numbers, Jacobi's formula for the number of representations of a positive integer as the sum of eight squares, and many others. We will also treat some of the more recent results that have been obtained using Liouville's ideas.

Liouville's method, although beautiful and arithmetic, is still an elementary one and as such has its limitations. As it is based on a number of identities, in order to obtain a particular number-theoretic result using it, the right identity has to be chosen as well as the right choice of the function occurring in it. And this is not always easy to do! Also, as with any elementary method, there are boundaries to what it can achieve. Indeed there are number-theoretic formulae which cannot be proved by Liouville's method and other tools are required to prove them. However, on the other hand, although we do not know the limitations of Liouville's approach, there are still new number-theoretic formulae waiting to be discovered and proved by Liouville's method. Hopefully, after reading this book, the reader will find some.

The prerequisites for this book include the basics of elementary number theory such as divisibility, primes, the fundamental theorem of arithmetic, quadratic reciprocity, the Legendre-Jacobi-Kronecker symbol, and a little about the representation of integers by binary quadratic forms such as $x^2 + xy + y^2$, $x^2 + y^2$ and $x^2 + 2y^2$. Hopefully in reading this book, the reader will enjoy and appreciate the elegant arithmetic proofs that Liouville's method enables us to give. After reading this book the interested reader is encouraged to study the theory of modular forms, where formulae similar to but deeper than the ones given in this book can be found.

The author is grateful to his colleagues A. Alaca and S. Alaca for their comments on the draft of this book, and to M. Huband for her help with Chapter 1. The author is also grateful for the suggestions and corrections that he received from B. C. Berndt of the University of Illinois. He also acknowledges the kindness of Professor Berndt in allowing him to name this book in a similar fashion to Berndt's excellent book "Number Theory in the Spirit of Ramanujan." He also thanks his wife Carole for her help with the references and index.

Kenneth S. Williams
Ottawa, Ontario, Canada
March 2010

Notation

$\mathbb{N}$ = set of positive integers = $\{1, 2, 3, \ldots\}$

$\mathbb{N}_0$ = set of nonnegative integers = $\{0, 1, 2, 3, \ldots\}$

$\mathbb{Z}$ = set of all integers = $\{\ldots, -3, -2, -1, 0, 1, 2, 3, \ldots\}$

$\mathbb{Q}$ = set of all rational numbers

$\mathbb{R}$ = set of all real numbers

$\mathbb{C}$ = set of all complex numbers

$\mathrm{Re}(z)$ = real part of $z \in \mathbb{C}$, that is $\mathrm{Re}(z) = x$, where $z = x + iy$, $x, y \in \mathbb{R}$

$\mathrm{Im}(z)$ = imaginary part of $z \in \mathbb{C}$, that is $\mathrm{Im}(z) = y$, where $z = x + iy$, $x, y \in \mathbb{R}$

B_n = n-th Bernoulli number ($B_0 = 1$, $B_1 = -\frac{1}{2}$, $B_2 = \frac{1}{6}$, $\ldots$)

$d \mid n$ the integer d divides the integer n

$d \nmid n$ the integer d does not divide the integer n

$p^a \parallel n$ the prime p is such that $p^a \mid n$ and $p^{a+1} \nmid n$

$\gcd(m, n)$ = greatest common divisor of the integers m and n (not both zero), which we abbreviate to (m, n) if space requires

$[x]$ = the greatest integer less than or equal to the real number x

$\emptyset$ = the empty set

$$F_k(n) = \begin{cases} 1, & \text{if } k \mid n, \\ 0, & \text{if } k \nmid n. \end{cases}$$

$$G_2(\ell) = \begin{cases} 0, & \text{if } 2 \mid \ell, \\ 1, & \text{if } 2 \nmid \ell. \end{cases}$$

$$s(n) = \begin{cases} 1, & \text{if } n \text{ is a perfect square}, \\ 0, & \text{otherwise}. \end{cases}$$

$$\sigma_k(n) = \begin{cases} \displaystyle\sum_{\substack{d \in \mathbb{N} \\ d \mid n}} d^k, & \text{if } n \in \mathbb{N}, \\ 0, & \text{if } n \notin \mathbb{N}. \end{cases}$$

$$d(n) = \sigma_0(n) = \sum_{\substack{d \in \mathbb{N} \\ d \mid n}} 1 = \text{number of positive divisors of } n \in \mathbb{N}$$

$$\sigma(n) = \sigma_1(n) = \sum_{\substack{d \in \mathbb{N} \\ d \mid n}} d = \text{sum of positive divisors of } n \in \mathbb{N}$$

$$\sigma_k^*(n) = \sum_{\substack{d \in \mathbb{N} \\ d \mid n \\ n/d \text{ odd}}} d^k$$

$$d^*(n) = \sigma_0^*(n) = \sum_{\substack{d \in \mathbb{N} \\ d \mid n \\ n/d \text{ odd}}} 1$$

$$\sigma^*(n) = \sigma_1^*(n) = \sum_{\substack{d \in \mathbb{N} \\ d \mid n \\ n/d \text{ odd}}} d$$

$$d_{k,m}(n) = \sum_{\substack{d \in \mathbb{N} \\ d \mid n \\ d \equiv k \,(\mathrm{mod}\, m)}} 1$$

$$A(n) = \{(i, j, k) \in \mathbb{Z} \times \mathbb{N} \times \mathbb{N} \mid i^2 + jk = n, \ k \text{ odd}\}$$

$$A_k(n) = \sum_{\substack{m \in \mathbb{N} \\ 1 \le m < n/k}} \sigma(m)\sigma(n - km)$$

$$r_k(n) = \mathrm{card}\{(x_1, \ldots, x_k) \in \mathbb{Z}^k \mid n = x_1^2 + \cdots + x_k^2\}$$

$$p_k(n) = \mathrm{card}\left\{(x_1, \ldots, x_k) \in \mathbb{Z}^k \mid n = x_1^2 + \cdots + x_k^2, \gcd(x_1, \ldots, x_k) = 1\right\}$$

$$s_{2k}(n) = \mathrm{card}\left\{(x_1, \ldots, x_{2k}) \in \mathbb{Z}^{2k} \mid n = x_1^2 + x_1 x_2 + x_2^2 \right.$$
$$\left. + \cdots + x_{2k-1}^2 + x_{2k-1}x_{2k} + x_{2k}^2\right\}$$

$$t_k(n) = \mathrm{card}\left\{(x_1, \ldots, x_k) \in \mathbb{N}_0^k \mid n = \tfrac{1}{2}x_1(x_1 + 1) + \cdots + \tfrac{1}{2}x_k(x_k + 1)\right\}$$

$$r(n) = \mathrm{card}\left\{(x, y) \in \mathbb{N}^2 \mid n = x^2 + y^2\right\}$$

$$R_1(n) = \mathrm{card}\left\{(x, y, z, t) \in \mathbb{N}^4 \mid n = x^2 + y^2 + z^2 + 2t^2\right\}$$

$$R_2(n) = \mathrm{card}\left\{(x, y, z, t) \in \mathbb{N}^4 \mid n = x^2 + y^2 + 2z^2 + 2t^2\right\}$$

$$R_3(n) = \mathrm{card}\left\{(x, y, z, t) \in \mathbb{N}^4 \mid n = x^2 + 2y^2 + 2z^2 + 2t^2\right\}$$

$$\left(\frac{d}{n}\right) = \text{Legendre-Jacobi-Kronecker symbol, which is}$$

defined for $d \in \mathbb{Z}$ with $d \equiv 0, 1 \pmod 4$ and $n \in \mathbb{N}$ (d is called the discriminant)

$$\left(\frac{-3}{n}\right) = \begin{cases} +1, & \text{if } n \equiv 1 \pmod 3, \\ -1, & \text{if } n \equiv 2 \pmod 3, \\ 0, & \text{if } n \equiv 0 \pmod 3. \end{cases}$$

$$\left(\frac{-4}{n}\right) = \begin{cases} +1, & \text{if } n \equiv 1 \pmod 4, \\ -1, & \text{if } n \equiv 3 \pmod 4, \\ 0, & \text{if } n \equiv 0 \pmod 2. \end{cases}$$

$$\left(\frac{-7}{n}\right) = \begin{cases} +1, & \text{if } n \equiv 1, 2, 3 \pmod 7, \\ -1, & \text{if } n \equiv 3, 5, 6 \pmod 7, \\ 0, & \text{if } n \equiv 0 \pmod 7. \end{cases}$$

$$\left(\frac{-8}{n}\right) = \begin{cases} +1, & \text{if } n \equiv 1 \text{ or } 3 \pmod 8, \\ -1, & \text{if } n \equiv 5 \text{ or } 7 \pmod 8, \\ 0, & \text{if } n \equiv 0 \pmod 2. \end{cases}$$

$$\left(\frac{8}{n}\right) = \begin{cases} +1, & \text{if } n \equiv \pm 1 \pmod 8, \\ -1, & \text{if } n \equiv \pm 3 \pmod 8, \\ 0, & \text{if } n \equiv 0 \pmod 2. \end{cases}$$

$$s(j, k) = \begin{cases} +1, & \text{if } j \equiv k \equiv 0 \pmod 2, \\ -1, & \text{otherwise.} \end{cases}$$

$$S_{e,f}(n) = \sum_{m=1}^{n-1} \sigma_e(m)\sigma_f(n - m)$$

$$T_{e,f,g}(n) = \sum_{\substack{m \in \mathbb{N} \\ m < n/g}} \sigma_e(m)\sigma_f(n - gm)$$

$$W_{a,b}(n) := \sum_{\substack{m \in \mathbb{N} \\ m < n \\ m \equiv a(\mathrm{mod}\ b)}} \sigma(m)\sigma(n - m)$$

$$S(A, B, C, D, f; n) = \sum_{\substack{(a, b, x, y) \in \mathbb{N}^4 \\ Cax + Dby = n}} (f(Aa - Bb) - f(Aa + Bb))$$

$\mu(n) =$ Möbius function

$$= \begin{cases} 1, & \text{if } n = 1, \\ (-1)^k, & \text{if } n = p_1 p_2 \ldots p_k, \text{ where } p_1, \ldots, p_k \text{ are distinct primes,} \\ 0, & \text{otherwise.} \end{cases}$$

$$\tilde{\sigma}_s(n) := \sum_{\substack{d \in \mathbb{N} \\ d \mid n}} (-1)^{d-1} d^s = \sigma_s(n) - 2^{s+1}\sigma_s(n/2)$$

$$\hat{\sigma}_s(n) := \sum_{\substack{d \in \mathbb{N} \\ d \mid n}} (-1)^{n/d-1} d^s = \sigma_s(n) - 2\sigma_s(n/2)$$

$$\tilde{\sigma}(n) := \tilde{\sigma}_1(n) = \sigma(n) - 4\sigma(n/2)$$

$$\hat{\sigma}(n) := \hat{\sigma}_1(n) = \sigma(n) - 2\sigma(n/2)$$

$$d_1(n) := \sum_{\substack{d \in \mathbb{N} \\ d \mid n}} d = \sigma(n)$$

$$d_2(n) := \sum_{\substack{d \in \mathbb{N} \\ d \mid n \\ 2 \nmid n}} d = \sigma(n) - 2\sigma(n/2)$$

$$d_3(n) := \sum_{\substack{d \in \mathbb{N} \\ d \mid n \\ 2 \mid n}} d = 2\sigma(n/2)$$

$$d_4(n) := \sum_{\substack{d \in \mathbb{N} \\ d \mid n \\ 2 \nmid n/d}} d = \sigma(n) - \sigma(n/2)$$

$$d_5(n) := \sum_{\substack{d \in \mathbb{N} \\ d \mid n \\ 2 \mid n/d}} d = \sigma(n/2)$$

$$d_6(n) := \sum_{\substack{d \in \mathbb{N} \\ d \mid n}} (-1)^{d-1} d = \sigma(n) - 4\sigma(n/2)$$

$$d_7(n) := \sum_{\substack{d \in \mathbb{N} \\ d \mid n}} (-1)^{n/d-1} d = \sigma(n) - 2\sigma(n/2)$$

$$D(r, s; n) := \sum_{m=1}^{n-1} d_r(m) d_s(n - m)$$

$\Delta :=$ set of triangular numbers
$= \{0, 1, 3, 6, 10, 15, \ldots\}$

$R(n) := \mathrm{card}\{(t_1, t_2, t_3, t_4) \in \Delta^4 \mid n = t_1 + t_2 + 2t_3 + 2t_4\}$

$\mathbb{H} := \{z \in \mathbb{C} \mid \mathrm{Im}(z) > 0\}$

$$SL_2(\mathbb{Z}) := \left\{ \begin{bmatrix} a & b \\ c & d \end{bmatrix} \mid a, b, c, d \in \mathbb{Z}, \quad ad - bc = 1 \right\}$$

$$E_k(q) := 1 - \frac{2k}{B_k} \sum_{n=1}^{\infty} \sigma_{k-1}(n)q^n, \quad q = e^{2\pi i z}, \ z \in \mathbb{H}, \ k(\text{even}) \geq 2$$

$M_k(SL_2(\mathbb{Z})) :=$ space of modular forms of weight k for $SL_2(\mathbb{Z})$

1

Joseph Liouville (1809–1882)

In 1809, the French mathematician Jean Fourier discovered that a complex wave is the sum of several simple waves; Napoléon Bonaparte was engaged against the British forces in the Peninsular War; and in the little town of Saint Omer, Pas de Calais, France, Joseph Liouville was born on March 24. His father, Claude Joseph, was a captain in Napoléon's army; his mother, Thérèse Balland, was a maternal cousin of Claude's. They came from a distinguished, upper middle-class Lorraine family. The Liouvilles had an impeccable reputation, many of the family serving France honorably in the military, government and the law. The baby Joseph had a brother, Felix, born six years earlier at the family home in Toul, near Nancy.

Because of their father's profession, the family could not always be together and the two Liouville boys spent some of their childhood with an uncle in Vignot, near Commercy, where they attended their first school. Joseph, according to C.-E. Dumont [97], writing in 1843, was no child prodigy. Indeed, his first teacher predicted he would not go far because he played too much. Though this early assessment proved to be very wrong, young Joseph demonstrated his early inclination for mathematics through games, most notably chess.

When Captain Liouville retired the family moved back to Toul. There, Joseph entered the local *collège* where he studied ancient languages. He continued his studies in mathematics at the *Collège St. Louis* in Paris and began sketching his own mathematical ideas, transcribing them into notebooks, a practice he maintained throughout his life. When he graduated from the *Collège St. Louis* at the age of sixteen, he enrolled next in the famous engineering institute *École Polytechnique*, established in 1794 and considered the premier institution of its kind in Europe. His teachers were of the highest calibre and the course of study was rigorous. Liouville studied applied analysis, geometry, mechanics, physics, chemistry, history, topographic drawing, architecture, geodesy, and literature. He was taught mechanical analysis by André Marie Ampère (1775–1836), a distinguished professor and mentor of Augustin Cauchy (1789–1857), who also taught at the school and was considered the bright light of the mathematical community in Paris.

In 1827, Liouville transferred to the engineering college, *École des Ponts et Chaussées*. During his three years at that institution he distinguished himself by preparing seven memoirs, mainly on the theory of electricity and heat, which he presented to the *Académie des Sciences* where they were well received. Some were published in the current journals of the day and Liouville enjoyed an early reputation as a rising star in the scientific community. It was at this time (1830) that four important things occurred in his life. He was ordered by the school to train "in the field" as an engineer in Normandy; he became ill – perhaps with rheumatism – and returned to Toul on a special leave of absence; his mother died; and he decided to marry his cousin, Marie Louise Balland. The result was that after many letters to and from the school, several delays, and a brief honeymoon following his June 15 wedding, Liouville did report for engineering duty near Grenoble. His interest in engineering, however, had ended and in October he resigned his post to follow a career in mathematics. He was fortunate to be appointed as *répétiteur* (or substitute) for Claude Louis Mathieu (1783–1875) at the *École Polytechnique* in 1831.

In 1833, through the good graces of his friend Jean Colladon (1802–1893), Liouville was appointed professor of rational mechanics at the *École Centrale des Arts et Manufactures*, a position he held until 1838 in spite of reports that his teaching was too theoretical and strayed from the curriculum. During this period, in order to make ends meet at home, Liouville also held positions at various private schools, at a *collège*, as well as at the *École Polytechnique*, teaching from 35 to 40 hours a week. The salaries paid to instructors were low and there was strenuous competition among aspiring academics to secure work. In Liouville's case, his father continued to support him and Marie Louise until 1833. The young family settled down in Toul where Liouville spent the long summer vacations from June until November. He also maintained an apartment in Paris. In 1836 he earned his doctorate, a necessary step in eventually teaching at the university level.

Although teaching kept him active and involved, Liouville's interest in research required another forum. He therefore joined various discussion groups and in 1832 was elected to the *Société Philomatique*. The *Académie des Sciences*, however, offered the definitive judgement on original work and Liouville continued to submit works to that body, including his comprehensive study on fractional calculus and his theory of integration in finite terms. The latter was favorably reviewed by Siméon-Denis Poisson (1781–1840) of the academy and published in one of its official reports. The paper contained what would later be labeled "Liouville's Theorem."

The state of French learned journals, especially scientific journals, was in flux during the 1830s, but Liouville continued to find outlets for his work. The *Journal de l'École Polytechnique* accepted two of his papers and he turned next to a German publication edited by August Leopold Crelle (1780–1855). This publication was Germany's answer to the respected *Annales de Mathématiques Pures et Appliquées* edited by Joseph Diez Gergonne (1771–1859), which had printed its last number in 1831. Crelle's journal accepted mainly work by German authors, but Liouville was able to join a small group of French scholars who had papers published. This gave Liouville his first contact with, and exposure to, German mathematicians. In particular, some time in 1830 Liouville met and began a correspondence with a famous German mathematician, Peter Gustav Dirichlet (1805–1859), a few years his senior and on a level with such well-known German mathematicians as Carl Jacobi (1804–1851) and even the great Carl Friedrich Gauss (1777–1855).

By 1833, Liouville decided that France needed a new journal of mathematics, one that would be devoted entirely to the study of mathematics and would at last replace the *Annales*, which had finished in 1831. To that end, he set about assembling his first edition. As Jesper Lützen describes in his book [192]:

> Liouville intended to accept both original papers on advanced subjects and new approaches to more elementary mathematics. Liouville also made it clear that he would not allow his Journal to degenerate into a forum for the everlasting and often slanderous quarrels in the competitive Parisian academic circles.
>
> *(Lützen, 1990, p. 38)*

From the first edition in 1836, the journal was of an excellent and enduring quality and attracted the best and brightest mathematical minds in France, as well as from Germany and later, other parts of Europe. Liouville demonstrated a skill at selection and editing not usually seen in a scholar so young and inexperienced. In his first editions he published contributions from the likes of Peter Gustav Dirichlet, Charles Sturm (1803–1855), Ándre Ampère (1775–1836), Michel Chasles (1793–1880), Count Gugliemo Libri-Carrucci (1803–1869), Gabriel Lamé (1795–1870), Victor Lebesgue (1791–1875) and Julius Plücker (1801–1868). The journal was so respected, so popular, and so successful that it soon became known simply as *Journal de Liouville*. In addition to the works of established mathematicians, Liouville accepted papers by such promising young mathematicians as Joseph Serret (1819–1885), Urbain Leverrier (1811–1877) and Charles Hermite (1822–1901). He continued to edit the journal for the next 40 years. It provided him with an international reputation and a forum for his own work.

While continuing his exhaustive routine of teaching, research and editing, Liouville, during the 1830s and 1840s, was unrelenting in his quest for secure appointments at prestigious institutions, in part to further his career but also to provide an adequate income for his growing family. The Liouvilles produced four children, three daughters and a son. They seem to have been a close-knit group, enjoying the company of uncles, aunts, cousins and grandparents, with long sojourns in Toul tending the family vineyards and gardens and making occasional forays to visit relatives and friends.

Liouville lost a close contest to Jean Marie Constant Duhamel (1797–1872) for a permanent professorship at the *École Polytechnique* in 1835, but was soon in contention again, this time for a seat at the prestigious *Académie des Sciences*. Liouville chose to compete for the seat in geometry and submitted various results to the academy for consideration, among them new work he had completed with Charles Sturm, which later became known as Sturm-Liouville theory. According to Jesper Lützen, at the last moment, he

> devised an ingenious convergence proof for the Fourier series of the second-order Sturm-Liouville problems.
>
> *(Lützen, 1990, p. 45)*

In the end, however, it was not enough and the seat went to the more experienced Sturm. Finally, in 1837 Liouville obtained a position at the *Collège de France* as a substitute for Jean Baptiste Biot (1774–1862), teaching mathematical physics.

In 1838 he was named Professor of Analysis and Mechanics at the *École Polytechnique*. Liouville was an active member of the teaching staff at the *École Polytechnique*, and in addition to those duties he served on numerous administrative committees and was in charge of the library at least until 1847. He also served as editor of the school's journal, the *Journal de l'École Polytechnique* for many years.

In 1839, the astronomer Michel Lefrançois Lalande (1766–1839) died causing a vacancy at the *Académie des Sciences*, and, ever on the alert for well-paying academic positions, Liouville was again in competition for a seat at the Académie. Having earlier done work on celestial mechanics and planetary theory, Liouville presented three papers to the academy within two weeks of Lalande's death. Vigorously opposing his bid was his nemesis, Libri, whom he had accused publicly of plagiarism. Nonetheless, with the strong support of other members of the academy, principally Dominique Arago (1786–1853) and his close friend and colleague, Charles Sturm, Liouville succeeded to the post. The animosity between Libri and Liouville continued at the academy, each man missing no opportunity to snarl at the other. Liouville is quoted as saying in a letter to Dirichlet, that Libri was a man:

who is beginning to be despised almost as much as he deserves.

(Lützen, 1990, p. 64)

(Four years later, in 1843, Liouville resigned his temporary position at the *Collège de France* in protest when Libri was elected to a seat there).

Liouville had many contacts in the international mathematics community, bringing a wide perspective to his participation on the various committees on which he served during his many years at the Académie. He took a special interest in prize competitions and was often asked to select topics as well as to choose and serve on judging committees. In addition, he judged annual competitions at other academic institutions.

In 1840, Liouville was elected to the *Bureau des Longitudes*, a society founded during the French Revolution for the discussion of astronomical ideas. It was the pre-eminent society of its type in France and election to the Bureau was a high honor. It also paid handsomely. Liouville presented various research papers to the Bureau, among them his work on astronomy and geodesy; his discovery of transcendental numbers; and a series of papers on rational mechanics. Liouville served as president of the Bureau three times: in 1843, 1847 and 1872.

By the age of just 31, Liouville had managed to secure a good living from his various teaching positions, his membership at the Bureau and his income from the vineyards in Toul.

Eventually, in 1851, Liouville returned to the *Collège de France*, defeating Augustin Cauchy for a permanent teaching chair in mathematics. In so doing he took over the position his enemy Libri had held at that institution. Libri had fled France for England, in disgrace, having been accused of the theft of valuable books and manuscripts from French libraries. Libri was sentenced to 10 years in prison in absentia for his crime, but he never returned to France.

The *Collège de France* provided an atmosphere for teaching that delighted Liouville as it allowed him to teach topics of his own choosing and present his ideas and research to his students. The astronomer Hervé Faye, recalling Liouville's lectures at this time, had this to say about his former professor:

Monsieur Liouville was one of the most brilliant professors one has ever heard. His lectures impressed me so strongly in my youth that today I still have a vivid recollection of the startling clarity with which he was endowed.

(Lützen, 1990, p. 83)

Faye was not the only student influenced by Liouville. Over the course of his 50-year teaching career, he encouraged, supported and shared ideas with many fine young mathematicians, among them: Joseph Serret, Jacob Steiner (1796–1863), Charles Hermite, Joseph Bertrand (1822–1900), Paul Laurent (1841–1908) and William Thomson Kelvin (later Lord Kelvin) (1824–1907). He also took on the enormous challenge of trying to explain the work of Evariste Galois (1811–1832) on the theory of equations and elliptic functions. Galois died in 1832 at the age of 20 after fighting a duel and his mathematical ideas had gone largely ignored until Liouville published them in his Journal in 1846. He also gave a series of lectures on Galois' theories and influenced others to study Galois' work.

While it may seem that Liouville led a rather frantic existence – teaching at several institutions, attending meetings and assemblies, conducting his own varied research, editing his *Journal de Liouville* and contributing to others, tending to his family and travelling back and forth from Toul to Paris – nonetheless, as Lützen points out, by the end of the 1840s, mathematics appeared to him as one body of interconnected ideas. Liouville, he says:

> cultivated only limited areas at different places of the mathematical landscape, but there is no doubt that he had a great view over it.
>
> *(Lützen, 1990, p. 147)*

Liouville was an avid republican all his adult life. He had supported his colleague and friend Dominique Arago's proposal for universal suffrage which he made in the Chamber of Deputies in 1840. He took part in so-called *banquets réformistes* to protest government policies, and he acted as chairman (after getting his father's permission) of such an event in Toul in 1840 during which he gathered over 1,000 signatures on an anti-government petition. The result of the unrest of 1840 was a move from the liberal monarchist Louis Adolphe Thiers to the conservative monarchist François Guizot. Although Liouvilles's political ideas were considered radical, he still had no problem swearing an oath of loyalty to King Louis-Phillipe (unlike the Bourbonist Cauchy, who could not do so).

In 1848, a popular uprising, complete with barricades and banners, caused the King to flee Paris and a republican provisional government was formed. Encouraged by Arago, Liouville agreed to stand for election to the Chamber of Deputies on a moderate, republican platform. He was elected in a republican sweep that saw 500 of the 800 seats going to form the "Second Republic." It was a proud moment for the republicans, but it did not last. The workers rebelled against the bourgeoisie and, on December 10, 1848, the French people elected as President Louis-Napoléon Bonaparte, nephew of the former emperor. In the election of 1849, which Liouville again contested as a republican, he lost his seat in a conservative win. In 1850, Louis-Napoléon led a coup against the assembly, and in a bizarre turn of events, he extended his term as president to 10 years and ultimately was elected Emperor by the French people.

Liouville retreated. He would not meet with friends in Toul, nor would he attend social events in Paris. In a letter to M. Grébus, the principal of the *collège* in Toul he proclaimed:

> I live in peace with my conscience, and I can go with my head held high.
>
> *(Lützen, 1990, p. 161)*

In fact, the defeat left him feeling bitter and disillusioned. When his beloved father died in 1850, he experienced further misery, proclaiming in a letter to his old friend P. Vogin:

> It is for me an immense and irreparable emptiness and yet another reason for me to sink into solitude.
>
> *(Lützen, 1990, p. 162)*

The number of mathematics and other science courses declined during the Second Empire due to conservative influences, including the Roman Catholic Church. But there was progress in industry and agriculture and Louis-Napoléon re-built Paris and extended the French railway system. He was also victorious (with England) in the Crimean War. Meanwhile, Liouville stayed out of politics and after a period of inactivity, he again immersed himself in his work, published numerous papers on his new interest of number theory and continued the impressive run of his respected Journal. He was rewarded in 1857 by an appointment as professor of mechanics at the *Faculté des Sciences*. This last achievement assured him of a secure and generous income, for, since his father's death, he had inherited with his brother, Felix, the lucrative remuneration from the vineyards in Toul.

In his lifetime Liouville made many friends, especially among his mathematics colleagues in France and elsewhere. He also maintained the alliances of his youth and befriended young and aspiring scientists. He enjoyed the friendship and respect of some of the great men in public life and was honored in many ways by them. In contrast, there were enemies, notably Libri, but also his former student, Serret, who turned against him after acrimonious arguments over "possession rights" to Galois' work and, in 1857, the fierce competition between them for the chair in mechanics at the *Académie des Sciences* (which Liouville won, as mentioned previously). In 1854, however, the most bitter hostility arose between Liouville and his former student Urbain Jean Joseph Leverrier.

Leverrier was appointed director of the Observatory in Paris following the death of the great scientist and public figure, Dominique Arago (1786–1853).

The appointment of the conservative, Catholic, Leverrier was destined to cause problems, not only for Liouville but also for his son, Ernest, who had been appointed by Arago as an *élève astronome* at the Observatory in 1852. Whether deliberately or not, Liouville renewed old quarrels with Leverrier; he criticized his work, took issue with his research; and strenuously opposed Leverrier's increasing powers and interference with the *Bureau des Longitudes*. Liouville was not alone in opposing Leverrier, however he was probably the only member of the scientific community who had a son working directly for his enemy. Ernest was dismissed along with many others who had worked at the Observatory with Arago. In a letter to Dirichlet, recorded in its entirety in Jesper Lützen's biography, Liouville described the painful circumstances surrounding this episode. He wrote:

> You are probably aware of the painful emotions I have had to face during the six months which have passed since your nice visit. The death of our excellent Arago, the expulsion of his family and his friends from their formerly peaceful stay at the Observatory, the degradation of the Bureau des Longitudes, the suicide of poor Mauvais, whose life was made unbearable by too much trouble, this is but a small part of what we have had to suffer.
>
> *(Lützen, 1990, p. 183)*

Ernest soon found a new apartment, gave up his career as an astronomer and turned to the profession of his Uncle Felix, the law. He continued his interest in astronomy, however, and his father published several of his papers on the subject in his Journal. Ernest was also of great assistance to his father with the editing of the Journal during Liouville's various illnesses.

The one true and enduring friendship in Liouville's life was with Gustav Dirichlet, the most authoritative expert of his day on the theory of numbers. From the time of their first meeting in Paris in 1830, until Dirichlet's death in 1859, the two corresponded, visited each other frequently, involved their families in the friendship, and perhaps, most important, collaborated in the field of number theory. Liouville had been instrumental in getting Dirichlet, who by 1855 was successor to Gauss at Göttingen, appointed "foreign associate" at the *Académie des Sciences*. Dirichlet returned the favor by arranging to have Liouville receive the Gauss Medal and securing his appointment as foreign member of the Academy at Göttingen.

By 1856, Dirichlet's friendship and influence had led Liouville to an examination of analytic number theory. It was a new field for him and he embraced it with enthusiasm. Indeed, he promised to send his ideas on the theory of numbers to Dirichlet, and in 1857 he lectured on the subject at the *Collège de France*. In addition, Liouville presented "notes" to the Académie, wrote numerous papers, published several in his Journal, and corresponded with a

Russian number theorist, Viktor Bouniakowsky (1804–1889). In a letter dated August, 1858, Dirichlet praises his friend's efforts:

> The theorem you prove and the theorem of Mr. Bouniakowsky (whose memoir I have not seen yet) seem to me the more remarkable because they are of an entirely new type.
>
> *(Lützen, 1990, p. 193)*

Jesper Lützen concludes that the years 1856 and 1857 were two of the most productive years in Liouville's career. He states:

> Not only did he publish notable works on rational mechanics, definite integrals and number theory, he also laid the foundation for all his subsequent works in the latter field.
>
> *(Lützen, 1990, p. 193)*

At the *Collège de France* where he was free to choose the subjects of his lectures, he devoted 11 courses over the next 20 years on "Theory of Numbers."

There is much warmth in the personal correspondence between Dirichlet and Liouville, attesting to the respect and affection they held for each other. The Dirichlet family spent time in Toul; the wives were fond of one another and also corresponded; the Liouvilles were invited to visit in Germany (although they were never able to go). Liouville was therefore devastated when he heard, in rapid succession, of Dirichlet's heart attack in Switzerland in 1858, his wife's unexpected demise in 1859, and finally, the death of Dirichlet himself on May 5, 1859. With the passing of Dirichlet, Liouville lost a faithful friend, colleague and supporter, and his most prolific correspondent.

As early as the 1830s Liouville complained of ill health. At that time it was stomach problems and possibly rheumatism. Later, he developed gout, a painful condition affecting the toes and feet and caused by a surfeit of uric acid in the blood. There was no treatment for gout at that time except to elevate the affected limb. Liouville endured many bouts of pain and immobility due to this ailment. In 1858, Liouville wrote wistfully to his brother Felix, who was on holiday in Italy after a bout of cancer:

> I myself, who only like my hole, cannot think without enthusiasm of the pleasure I would enjoy by taking a walk in this beautiful country full of great places. But if God has denied me this pleasure forever, he has at least granted me in the study of mathematics, a great consolation within my reach.
>
> *(Lützen, 1990, p. 217)*

Felix died in 1860, missing the opportunity to see his brother Joseph named an Officer of the *Légion d'Honneur*. Liouville then became the head of the family.

From 1860, Liouville immersed himself in number theory, using his Journal to pour out notes (more than 50 in two years) about number-theoretical functions. As Lützen explains:

> ... he continued to publish rather uninteresting applications of his general ideas instead of revealing the core of these ideas and his methods of proof.
>
> *(Lützen, 1990, p. 229)*

Lützen goes on to speculate that Liouville may have, as Hermite stated in a letter 20 years later:

> ... kept for himself the entire harvest of all the consequences of his original discovery.
>
> *(Lützen, 1990, p. 229)*

because he never was able to find the time or energy for the task. Although he promised himself and others that there would be published results, not only on number-theoretical functions, but also on celestial mechanics, rational mechanics, quadratic forms and other subjects, none were forthcoming. For one so prolific in his notebooks, it was a strange omission.

Through the 1860s and 1870s, in spite of ill health, Liouville kept to his rigorous teaching schedule, attended meetings of the *Bureau des Longitudes* and the *Académie*, encouraged and promoted young talent, and made copious entries in his personal notebooks. He also remained at the helm of the *Journal de Liouville*, which he referred to fondly as "his child." It was not until 1875 that he finally turned it over to Henri Resal, a member of the *Académie des Sciences*. But it was a bad choice. Resal did not attract top-rank mathematicians and by 1885 the journal was in serious trouble. It was finally rescued by Camille Jordan, who brought it back to its former glory, and the Journal has continued to this day with the title page still bearing the name of its originator.

It must have given Liouville considerable satisfaction to record in his notebook that in 1870, with a new liberal government encouraging the Emperor to adopt reforms, his enemy Leverrier was dismissed from the Observatory. Liouville himself served on a committee to select a replacement, however the choice of the committee, Charles Delaunay, died within two years and Leverrier returned to the position. Four years later Leverrier died. Liouville, who had written of him:

> such a stupid tyrant cannot last
> *(Lützen, 1990, p. 240)*

had the satisfaction of gluing the death notice in his notebook.

By 1870, France was plunged into the Franco-Prussian War and the Liouville family home in Toul was under siege. Liouville escaped to Paris, the Emperor

Napoléon surrendered, and a republican government was proclaimed. The new government, however, decided to continue the war. Paris was besieged and the conditions in the capital became critical as the Prussians bombarded the city with artillery, causing considerable damage and loss of life. Liouville had several relatives serving in the French army, one of whom died in the conflict. After the defeat the German army paraded through Paris on March 1, then abruptly left. What succeeded, however, was much worse. The radical mobs of the Paris Commune plundered the city, burning, looting and killing and Liouville was again forced to flee. By June, however, with the situation calmed, he was back lecturing at the *Collège de France*.

In 1876, Liouville was elected a member of the Berlin Academy. Karl Weierstrass (1815–1897) wrote a glowing letter of recommendation and in his acceptance, Liouville commented, somewhat wryly:

> The most ardent patriotism does not dispense with personal recognition.
>
> *(Lützen, 1990, p. 245)*

He was, by then, greatly honored. He was a member of the British Royal Society and 14 other learned academies, in addition to being raised to the rank of *Commandeur de la Légion d'Honneur* in France. He also had a moon crater and a street in his home town of Toul named after him.

Although Liouville continued to lecture at the *Collège de France*, where the topics were always of his choosing, he was forced to turn his other teaching assignments over to substitutes from the mid-1870s until his death. Even at the Collège there were many cancelled lectures and shortened courses.

The notebooks of his last six years bear poignant testimony to his ill health, pain and depression. In addition to gout, he now suffered from arthritis and insomnia, as well as a seemingly unfounded bitterness against contemporary mathematicians, many of whom were his friends. In 1876, he wrote in his notebook:

> When one crushes me from one side it is only fair that one crushes me from the other side. This is what my life has become. This is what my life is composed of. Pillaged, insulted, or rather insulted then pillaged. Ah! Death is gentle when it relieves us from such a life! There is no more justice on Earth.
>
> *(Lützen, 1990, p. 256)*

Later the insults against his friends grew worse:

> ... vile thieves ... those scoundrels have no doubt been my assassins ... unworthy animals, but Jesuits ...
>
> *(Lützen, 1990, p. 245)*

The accusations were irrational and unworthy of him, but as Lützen points out:

> ... they show how desperate Liouville had become because of his miserable life.
>   *(Lützen, 1990, p. 257)*

He continued to spend the summer months in Toul where he both endured and found some pleasure in visits from family members. But in 1880 he suffered a double tragedy. His beloved wife of more than 50 years, Marie Louise, died, followed soon after by his only son, Ernest. Liouville went into a final decline, though as Hervé Faye (1814–1902) recorded, he attended a meeting of the *Bureau des Longitudes* just five days before his death, which occurred on September 8, 1882.

At his funeral two days later, attended by his three surviving daughters, their husbands and children, as well as by many colleagues and friends, speeches were made by representatives of all of the institutions to which Liouville had been associated during his 73 years. The *Académie des Sciences* met the following day and after a tribute to their dead colleague, the meeting was closed, a rare honor. Similarly, a few days later, the *Bureau des Longitudes* honored Liouville by declaring its meeting closed in his memory.

His legacy is contained in the more than 400 memoirs he left, 200 of them devoted to his last great interest, number theory. The memoirs include over 100 papers on mathematical analysis; studies on mathematical physics, including electrodynamics and the theory of heat and sound; work on celestial mechanics; on rational mechanics; papers on the theory of algebraic equations and group theory; and 20 papers on geometry. In addition, he is remembered for saving the work of Galois from oblivion, and finally, for the enduring contribution he made to mathematics through his *Journal de Liouville*, now over 170 years old.

Notes on Chapter 1

This brief biography of Joseph Liouville could not have been written without consulting the excellent scholarly biography of Liouville written by J. Lützen [192]. This is a very readable account of Liouville's life and work and is a "must read" for anyone interested in Liouville. Articles by Chrystal [74], Stander [247] and Taton [251] were also consulted. The website http://www-groups.dcs.st-and.ac.uk/history/mathematicians/Liouville.html contains a biography of Liouville, as well as further references.

2

Liouville's Ideas in Number Theory

In 1856, at the age of 47, Liouville began his first profound research in the field that was to become his favorite for the rest of his life, the theory of numbers, particularly the representation of integers by quadratic forms. In the years 1856 and 1857, probably the two most productive years of his career, Liouville discovered the underlying number theoretic principles from which flowed numerous results. Unfortunately Liouville never published the proofs of his formulae. The formulae themselves were stated in a series of eighteen articles and the application of them to quadratic forms in a series of ninety papers. Although he indicated he would do so, the proofs never appeared. This was perhaps due to Liouville's declining health and the heavy commitments on his time due to committee work and teaching. Later other mathematicians proved his formulae although not always in the elementary way that Liouville intended. Liouville's point of view was that many of the arithmetic formulae proved by his colleagues Jacobi, Kronecker and others by analytic methods should follow from a few basic elementary arithmetic principles. Thus, for example, the arithmetic formula for the number of representations of a positive integer as the sum of four squares, which follows from Jacobi's monumental work on elliptic functions, should be provable by entirely elementary arithmetic arguments. This in no way downgrades the use of analysis, complex variable theory, modular forms, elliptic functions and theta functions in proving arithmetic formulae but rather recognizes these formulae as elementary formulae.

If k and n are positive integers, the arithmetic function that counts the sum of the k-th powers of the positive integers dividing n is denoted by $\sigma_k(n)$. When $k = 1$ we write $\sigma(n)$ for $\sigma_1(n)$. Thus, for example, $\sigma(8) = 1 + 2 + 4 + 8 = 15$, $\sigma_3(9) = 1^3 + 3^3 + 9^3 = 757$ and $\sigma_5(2) = 1^5 + 2^5 = 33$. In his beautiful article on lattices and linear codes, Elkies [102] describes the identity

$$\sum_{m=1}^{n-1} \sigma_3(m)\sigma_3(n-m) = \frac{1}{120}(\sigma_7(n) - \sigma_3(n)) \tag{2.1}$$

as "mysterious." However formula (2.1) is a purely arithmetic one since it only involves finite sums of divisors of positive integers. Huard, Ou, Spearman and Williams [137] have shown in the spirit of Liouville that (2.1) can be deduced from an elementary identity of the type given by Liouville. Thus (2.1) may be regarded as "elementary" rather than "mysterious."

In addition to not giving proofs of his formulae, Liouville also did not give any idea how he came upon his formulae. We know that Liouville was familiar with the path-breaking work of Jacobi on elliptic functions and theta functions since he gave lectures on them using the powerful theorems of Cauchy in the theory of functions of a complex variable. (Jacobi had used purely algebraic methods to obtain his results.) It is therefore possible that formulae from the theory of theta functions led Liouville to the discovery of his elementary arithmetic formulae.

The four basic theta functions are defined in the notation of Whittaker and Watson [258, p. 464] as follows. Let $\tau \in \mathbb{C}$ be such that $\text{Im}(\tau) > 0$. Set $q = e^{\pi i \tau}$ so that $|q| < 1$. For $z \in \mathbb{C}$

$$\theta_1(z, q) := 2\sum_{n=1}^{\infty} (-1)^{n-1} q^{(2n-1)^2/4} \sin(2n-1)z,$$

$$\theta_2(z, q) := 2\sum_{n=1}^{\infty} q^{(2n-1)^2/4} \cos(2n-1)z,$$

$$\theta_3(z, q) := 1 + 2\sum_{n=1}^{\infty} q^{n^2} \cos 2nz,$$

$$\theta_4(z, q) := 1 + 2\sum_{n=1}^{\infty} (-1)^n q^{n^2} \cos 2nz.$$

The two formulae

$$\frac{\theta_2(0, q)\theta_3(0, q)\theta_1(z, q)}{\theta_4(z, q)} = 4\sum_{m=0}^{\infty} \frac{q^{(2m+1)/2}}{1 - q^{2m+1}} \sin(2m+1)z \tag{2.2}$$

and

$$\frac{\theta_2(0, q)^2\theta_3(0, q)^2\theta_1(z, q)^2}{\theta_4(z, q)^2} = 8\sum_{m=1}^{\infty} \frac{mq^m}{1 - q^{2m}}(1 - \cos 2mz) \tag{2.3}$$

are implicit in the work of Jacobi [146], [148, Vol. I, pp. 49–239], see the notes at the end of the chapter. These formulae would have been known to Liouville. The left hand side of (2.3) is the square of the left hand side of (2.2) so that

$$\sum_{m=1}^{\infty} \frac{mq^m}{1-q^{2m}}(1-\cos 2mz) = 2\left(\sum_{m=0}^{\infty} \frac{q^{(2m+1)/2}}{1-q^{2m+1}}\sin(2m+1)z\right)^2. \quad (2.4)$$

Now let us speculate on Liouville's thoughts. Perhaps he wondered what identity would result from (2.4) by equating coefficients of powers of q. Let us do this and see what we obtain. The left hand side of (2.4) is

$$\sum_{m=1}^{\infty} m(1-\cos 2mz)\sum_{t=0}^{\infty} q^{(2t+1)m} = \sum_{n=1}^{\infty} q^n \sum_{\substack{m=1}}^{\infty} \sum_{\substack{t=0 \\ (2t+1)m=n}}^{\infty} m(1-\cos 2mz)$$

$$= \sum_{n=1}^{\infty} q^n \sum_{\substack{m \in \mathbb{N} \\ m \mid n \\ n/m \text{ odd}}} m(1-\cos 2mz).$$

Also

$$\sum_{m=0}^{\infty} \frac{q^{(2m+1)/2}}{1-q^{2m+1}}\sin(2m+1)z = \sum_{m=0}^{\infty} q^{(2m+1)/2}\sin(2m+1)z \sum_{t=0}^{\infty} q^{(2m+1)t}$$

$$= \sum_{m=0}^{\infty}\sum_{t=0}^{\infty} q^{(2m+1)(2t+1)/2}\sin(2m+1)z$$

$$= \sum_{\substack{r=1 \\ r \text{ odd}}}^{\infty} q^{r/2} \sum_{\substack{(a,b) \in \mathbb{N}^2 \\ ab=r}} \sin az$$

$$= \sum_{\substack{r=1 \\ r \text{ odd}}}^{\infty} q^{r/2} \sum_{\substack{a \in \mathbb{N} \\ a \mid r}} \sin az$$

so the right hand side of (2.4) is

$$2\sum_{\substack{r,s=1 \\ r,s \text{ odd}}}^{\infty} q^{(r+s)/2} \sum_{\substack{(a,b) \in \mathbb{N}^2 \\ a \mid r, b \mid s}} \sin az \sin bz = 2\sum_{n=1}^{\infty} q^n \sum_{\substack{(r,s) \in \mathbb{N}^2 \\ r+s=2n \\ r,s \text{ odd}}} \sum_{\substack{(a,b) \in \mathbb{N}^2 \\ a \mid r, b \mid s}} \sin az \sin bz.$$

Equating coefficients of q^n ($n \in \mathbb{N}$) on the left and right hand sides, we obtain the following arithmetical property of the trigonometric functions

$$\sum_{\substack{m \in \mathbb{N} \\ m \mid n \\ n/m \text{ odd}}} m(1 - \cos 2mz) = 2 \sum_{\substack{(r, s) \in \mathbb{N}^2 \\ r + s = 2n \\ r, s \text{ odd}}} \sum_{\substack{(a, b) \in \mathbb{N}^2 \\ a \mid r, b \mid s}} \sin az \sin bz, \quad n \in \mathbb{N}.$$

Writing r as ax and s as by, and using the trigonometric identity

$$2 \sin az \sin bz = \cos(a - b)z - \cos(a + b)z,$$

we obtain

$$\sum_{\substack{m \in \mathbb{N} \\ m \mid n \\ n/m \text{ odd}}} m(1 - \cos 2mz) = \sum_{\substack{(a, b, x, y) \in \mathbb{N}^4 \\ ax + by = 2n \\ a, b, x, y \text{ odd}}} (\cos(a - b)z - \cos(a + b)z). \quad (2.5)$$

Identity (2.5) is not a truly arithmetic formula since it involves the cosine function. This suggests that we expand each cosine in powers of z using

$$1 - \cos z = \sum_{k=1}^{\infty} (-1)^{k-1} \frac{z^{2k}}{2k!}$$

and equate coefficients of $(-1)^{k-1} z^{2k}/(2k)!$ ($k \in \mathbb{N}$). We obtain

$$\sum_{\substack{m \in \mathbb{N} \\ m \mid n \\ n/m \text{ odd}}} 2^{2k} m^{2k+1} = \sum_{\substack{(a, b, x, y) \in \mathbb{N}^4 \\ ax + by = 2n \\ a, b, x, y \text{ odd}}} \left((a + b)^{2k} - (a - b)^{2k} \right), \quad k, n \in \mathbb{N}. \quad (2.6)$$

This is a genuine arithmetic formula. However we can make formula (2.6) even more general (as Liouville did).

Let $f : \mathbb{Z} \to \mathbb{C}$ be an even function, that is $f(x) = f(-x)$ for all $x \in \mathbb{Z}$. Then for each $n \in \mathbb{N}$ there exists a polynomial

$$g_n(x) = \sum_{k=0}^{n} g(n, k) x^{2k}$$

such that

$$f(2j) = g_n(2j), \quad j \in \{0, \pm 1, \pm 2, \ldots, \pm n\}.$$

We note that

$$f(0) = g_n(0) = g(n, 0).$$

Then

$$\sum_{\substack{(a,b,x,y) \in \mathbb{N}^4 \\ ax+by=2n \\ a,b,x,y \text{ odd}}} \left(f(a+b) - f(a-b) \right)$$

$$= \sum_{\substack{(a,b,x,y) \in \mathbb{N}^4 \\ ax+by=2n \\ a,b,x,y \text{ odd}}} \left(g_n(a+b) - g_n(a-b) \right)$$

$$= \sum_{\substack{(a,b,x,y) \in \mathbb{N}^4 \\ ax+by=2n \\ a,b,x,y \text{ odd}}} \sum_{k=0}^{n} g(n,k)((a+b)^{2k} - (a-b)^{2k})$$

$$= \sum_{\substack{(a,b,x,y) \in \mathbb{N}^4 \\ ax+by=2n \\ a,b,x,y \text{ odd}}} \sum_{k=1}^{n} g(n,k)((a+b)^{2k} - (a-b)^{2k})$$

$$= \sum_{k=1}^{n} g(n,k) \sum_{\substack{(a,b,x,y) \in \mathbb{N}^4 \\ ax+by=2n \\ a,b,x,y \text{ odd}}} ((a+b)^{2k} - (a-b)^{2k})$$

$$= \sum_{k=1}^{n} g(n,k) \sum_{\substack{m \in \mathbb{N} \\ m \mid n \\ n/m \text{ odd}}} 2^{2k} m^{2k+1}$$

$$= \sum_{\substack{m \in \mathbb{N} \\ m \mid n \\ n/m \text{ odd}}} m \sum_{k=1}^{n} g(n,k) 2^{2k} m^{2k}$$

$$= \sum_{\substack{m \in \mathbb{N} \\ m \mid n \\ n/m \text{ odd}}} m(g_n(2m) - g(n,0))$$

$$= \sum_{\substack{m \in \mathbb{N} \\ m \mid n \\ n/m \text{ odd}}} m(f(2m) - f(0)),$$

which is the first of Liouville's formulae [170, 1st article, formula (A), p. 144], namely:

Let $n \in \mathbb{N}$. Let $f : \mathbb{Z} \to \mathbb{C}$ be an even function. Then

$$\sum_{\substack{(a,b,x,y) \in \mathbb{N}^4 \\ ax+by=2n \\ a,b,x,y \text{ odd}}} (f(a+b) - f(a-b)) = \sum_{\substack{m \in \mathbb{N} \\ m \mid n \\ n/m \text{ odd}}} m(f(2m) - f(0)). \qquad (2.7)$$

In suggesting this as a possible route by which Liouville was led to his formula (2.7), we were guided by ideas of Humbert [143].

Formulae such as (2.7) form the basis of Liouville's method. Suitable choices of the even function f lead to a wealth of arithmetic formulae. We content ourselves here with just one example. We take $f(x) = x^2$ ($x \in \mathbb{Z}$). Then the left hand side of (2.7) becomes

$$\sum_{\substack{(a,b,x,y) \in \mathbb{N}^4 \\ ax+by=2n \\ a,b,x,y \text{ odd}}} ((a+b)^2 - (a-b)^2) = 4 \sum_{\substack{(a,b,x,y) \in \mathbb{N}^4 \\ ax+by=2n \\ a,b,x,y \text{ odd}}} ab$$

$$= 4 \sum_{\substack{(\ell,m) \in \mathbb{N}^2 \\ \ell+m=2n \\ \ell,m \text{ odd}}} \sum_{\substack{(a,x) \in \mathbb{N}^2 \\ ax=\ell}} a \sum_{\substack{(b,y) \in \mathbb{N}^2 \\ by=m}} b$$

$$= 4 \sum_{\substack{\ell=1 \\ \ell \text{ odd}}}^{2n-1} \sum_{\substack{a \in \mathbb{N} \\ a \mid \ell}} a \sum_{\substack{b \in \mathbb{N} \\ b \mid 2n-\ell}} b$$

$$= 4 \sum_{\substack{\ell=1 \\ \ell \text{ odd}}}^{2n-1} \sigma(\ell)\sigma(2n - \ell)$$

and Liouville's formula (2.7) yields the arithmetic identity

$$\sum_{\substack{\ell=1 \\ \ell \text{ odd}}}^{2n-1} \sigma(\ell)\sigma(2n - \ell) = \sum_{\substack{m \in \mathbb{N} \\ m \mid n \\ n/m \text{ odd}}} m^3, \quad n \in \mathbb{N}, \qquad (2.8)$$

which was stated by Liouville [170, 1st article, p. 146] in the case when n is odd.

Now let us ask how could Liouville have proved (2.7) in an elementary way avoiding any use of analysis. Looking at the formula (2.7) it is easy to see how the terms involving $f(0)$ occur on the left hand side. They can only arise from those terms with $a = b$. Thus $f(0)$ occurs on the left hand side with multiplicity

$$\sum_{\substack{(a,b,x,y) \in \mathbb{N}^4 \\ ax+by = 2n \\ a,b,x,y \text{ odd} \\ a=b}} 1 = \sum_{\substack{(a,x,y) \in \mathbb{N}^3 \\ a(x+y) = 2n \\ a,x,y \text{ odd}}} 1 = \sum_{\substack{a \in \mathbb{N} \\ a \mid 2n \\ a \text{ odd}}} \sum_{\substack{(x,y) \in \mathbb{N}^2 \\ x+y = 2n/a \\ x,y \text{ odd}}} 1 = \sum_{\substack{a \in \mathbb{N} \\ a \mid n \\ a \text{ odd}}} n/a = \sum_{\substack{m \in \mathbb{N} \\ m \mid n \\ n/m \text{ odd}}} m$$

in agreement with the right hand side.

Now let us look at the terms $f(m)$ with $m \neq 0$. Since a and b in the sum on the left hand side of (2.7) are both odd, $a + b$ and $a - b$ are both even so these terms occur as $f(2k)$ with $k \in \mathbb{Z}$ and $k \neq 0$. Let $k \in \mathbb{N}$. The term $f(2k)$ occurs on the left hand side, remembering that as f is even it also occurs as $f(-2k)(= f(2k))$, with multiplicity

$$\sum_{\substack{(a,b,x,y) \in \mathbb{N}^4 \\ ax+by = 2n \\ a,b,x,y \text{ odd} \\ a+b = 2k}} 1 - \sum_{\substack{(a,b,x,y) \in \mathbb{N}^4 \\ ax+by = 2n \\ a,b,x,y \text{ odd} \\ a-b = 2k}} 1 - \sum_{\substack{(a,b,x,y) \in \mathbb{N}^4 \\ ax+by = 2n \\ a,b,x,y \text{ odd} \\ a-b = -2k}} 1.$$

In the last sum the change of variable $(a, b, x, y) \to (b, a, y, x)$ shows that it is equal to the second sum. Thus the multiplicity of $f(2k)$ ($k \in \mathbb{N}$) is

$$\sum_{\substack{(a,b,x,y) \in \mathbb{N}^4 \\ ax+by = 2n \\ a,b,x,y \text{ odd} \\ a+b = 2k}} 1 - 2 \sum_{\substack{(a,b,x,y) \in \mathbb{N}^4 \\ ax+by = 2n \\ a,b,x,y \text{ odd} \\ a-b = 2k}} 1.$$

On the right hand side of (2.7) the term $f(2k)$ occurs with multiplicity

$$\sum_{\substack{m \in \mathbb{N} \\ m \mid n \\ n/m \text{ odd} \\ m = k}} m = \begin{cases} k, & \text{if } k \mid n \text{ and } n/k \text{ odd}, \\ 0, & \text{otherwise}. \end{cases}$$

Hence Liouville's formula (2.7) follows in an elementary fashion from the underlying identity

$$\sum_{\substack{(a,b,x,y) \in \mathbb{N}^4 \\ ax+by = 2n \\ a,b,x,y \text{ odd} \\ a+b = 2k}} 1 - 2 \sum_{\substack{(a,b,x,y) \in \mathbb{N}^4 \\ ax+by = 2n \\ a,b,x,y \text{ odd} \\ a-b = 2k}} 1 = \begin{cases} k, & \text{if } k \mid n \text{ and } n/k \text{ odd}, \\ 0, & \text{otherwise}, \end{cases} \qquad (2.9)$$

by multiplying this identity by $f(2k)$ and summing over $k \in \mathbb{N}$ with $k \leq n$. We show how to prove (2.9) and similar formulae in Chapter 10.

Throughout his eighteen articles [170] on his method, Liouville gave a great many arithmetic formulae, some quite similar to (2.7), others quite different. From these formulae he deduced many surprising results. We just mention one of them.

In [170, 11th article, p. 286], Liouville gave the following formula. Let n be a positive odd integer. Then

$$\text{card} \left\{ (x_1, x_2, x_3, x_4, x_5) \in \mathbb{Z}^5 \mid n = x_1^2 + x_2^2 + x_3^2 + x_4^2 + 2x_5^2, \ x_5 \text{ even} \right\}$$
$$-\text{card} \left\{ (x_1, x_2, x_3, x_4, x_5) \in \mathbb{Z}^5 \mid n = x_1^2 + x_2^2 + x_3^2 + x_4^2 + 2x_5^2, \ x_5 \text{ odd} \right\}$$
$$= 8 \sum (-1)^{(y_1-1)/2} y_1,$$

where the sum is taken over all $(y_1, y_2, y_3) \in \mathbb{N} \times \mathbb{N} \times \mathbb{Z}$ such that

$$2n = y_1^2 + y_2^2 + y_3^2, \ y_1 \equiv 1 \ (\text{mod } 2), \ y_2 \equiv 1 \ (\text{mod } 2).$$

For example with $n = 29$ we have

$$\text{card} \left\{ (x_1, x_2, x_3, x_4, x_5) \in \mathbb{Z}^5 \mid 29 = x_1^2 + x_2^2 + x_3^2 + x_4^2 + 2x_5^2, \ x_5 \text{ even} \right\} = 752,$$
$$\text{card} \left\{ (x_1, x_2, x_3, x_4, x_5) \in \mathbb{Z}^5 \mid 29 = x_1^2 + x_2^2 + x_3^2 + x_4^2 + 2x_5^2, \ x_5 \text{ odd} \right\} = 832,$$

and

$$8 \sum_{\substack{(y_1, y_2, y_3) \in \mathbb{N} \times \mathbb{N} \times \mathbb{Z} \\ y_1^2 + y_2^2 + y_3^2 = 58 \\ y_1, y_2 \text{ odd}}} (-1)^{(y_1-1)/2} y_1 = -80 = 752 - 832.$$

It is not our purpose in this book to give all of Liouville's formulae. The interested reader can find an excellent summary of Liouville's formulae in Chapter 11 of Volume 2 of Dickson's *History of the Theory of Numbers* [93, pp. 329–339] as well as references to proofs of his formulae by other authors. Rather we present a selection of his formulae and use them to prove many standard results in elementary number theory. We tend to stress those of Liouville's formulae which have not been treated in textbooks on number theory such as Uspensky and Heaslet [254] and Nathanson [209].

Notes on Chapter 2

We can only truly speculate as to the proofs that Liouville had in mind for his arithmetic formulae. His published papers contain only statements and numerical examples. Perhaps he revealed his ideas in the course of his lectures at the *Collège de France*. There are 340 notebooks of Liouville's in the *Bibliotheque de l'Institut de France* in Paris that need to be examined in detail to see if they contain clues to Liouville's methods. A rapid examination of some of these notebooks suggests that they consist mainly of outlines and final drafts of his published works, see Taton [251].

We now show how formula (2.2) follows from the work of Jacobi [146], [148, Vol. I, pp. 49-239]. We make use of Jacobi's quantities k, K, H and Θ and relate them to the theta functions θ_1, θ_2, θ_3 and θ_4. From Jacobi [148, Vol. I, eqs. (6), (7), p. 235] we have

$$\sqrt{\frac{2K}{\pi}} = 1 + 2q + 2q^4 + 2q^9 + \cdots$$

and

$$\sqrt{\frac{2kK}{\pi}} = 2q^{1/4} + 2q^{9/4} + 2q^{25/4} + \cdots$$

so that

$$\sqrt{\frac{2K}{\pi}} = \theta_3(0, q), \qquad \sqrt{\frac{2kK}{\pi}} = \theta_2(0, q).$$

Hence

$$\theta_2(0, q)\theta_3(0, q) = \frac{2\sqrt{k}K}{\pi}.$$

From Jacobi [148, Vol. I, eqs. (1), (2), p. 231] we have

$$\Theta\left(\frac{2Kz}{\pi}\right) = 1 - 2q\cos 2z + 2q^4\cos 4z - 2q^9\cos 6z + \cdots$$

and

$$H\left(\frac{2Kz}{\pi}\right) = 2q^{1/4}\sin z - 2q^{9/4}\sin 3z + 2q^{25/4}\sin 5z - \cdots.$$

Hence

$$\Theta\left(\frac{2Kz}{\pi}\right) = \theta_4(z, q), \qquad H\left(\frac{2Kz}{\pi}\right) = \theta_1(z, q).$$

Thus

$$\frac{\theta_2(0, q)\theta_3(0, q)\theta_1(z, q)}{\theta_4(z, q)} = \frac{2\sqrt{k}K}{\pi}\frac{H\left(\dfrac{2Kz}{\pi}\right)}{\Theta\left(\dfrac{2Kz}{\pi}\right)}.$$

Now by Jacobi [148, Vol. I, p. 224] we have

$$\sin \text{am} \frac{2Kz}{\pi} = \frac{1}{\sqrt{k}} \frac{H\left(\dfrac{2Kz}{\pi}\right)}{\Theta\left(\dfrac{2Kz}{\pi}\right)}.$$

Thus

$$\frac{\theta_2(0, q)\theta_3(0, q)\theta_1(z, q)}{\theta_4(z, q)} = \frac{2kK}{\pi} \sin \text{am} \frac{2Kz}{\pi}.$$

From Jacobi [148, Vol. I, p. 165] we have

$$\frac{2kK}{\pi} \sin \text{am} \frac{2Kz}{\pi} = \frac{4q^{1/2} \sin z}{1 - q} + \frac{4q^{3/2} \sin 3z}{1 - q^3} + \frac{4q^{5/2} \sin 5z}{1 - q^5} + \cdots$$

so

$$\frac{\theta_2(0, q)\theta_3(0, q)\theta_1(z, q)}{\theta_4(z, q)} = 4 \sum_{m=0}^{\infty} \frac{q^{(2m+1)/2} \sin(2m + 1)z}{1 - q^{2m+1}},$$

which is (2.2). Formula (2.3) can be proved in a similar manner.

Finally we justify that if $f : \mathbb{Z} \to \mathbb{C}$ is an even function then for each $n \in \mathbb{N}$ there exists a polynomial

$$g_n(x) = \sum_{k=0}^{n} g(n, k)x^{2k}$$

such that

$$f(2j) = g_n(2j), \quad j \in \{0, \pm 1, \pm 2, \ldots, \pm n\}.$$

Let $a_1, \ldots, a_n \in \mathbb{C}$. The Vandermonde determinant is

$$\begin{vmatrix} 1 & a_1 & a_1^2 & \cdots & a_1^{n-1} \\ 1 & a_2 & a_2^2 & \cdots & a_2^{n-1} \\ 1 & a_3 & a_3^2 & \cdots & a_3^{n-1} \\ \cdot & \cdot & \cdot & \cdots & \cdot \\ 1 & a_n & a_n^2 & \cdots & a_n^{n-1} \end{vmatrix} = \prod_{1 \leq j < i \leq n} (a_i - a_j),$$

see for example [239, p. 114]. Replacing n by $n + 1$ and taking $a_i = (2i - 2)^2$ ($i = 1, 2, \ldots, n + 1$), we deduce

$$\begin{vmatrix} 1 & 0 & 0 & \cdots & 0 \\ 1 & 2^2 & 2^4 & \cdots & 2^{2n} \\ 1 & 4^2 & 4^4 & \cdots & 4^{2n} \\ \cdot & \cdot & \cdot & \cdots & \cdot \\ 1 & (2n)^2 & (2n)^4 & \cdots & (2n)^{2n} \end{vmatrix} = \prod_{1 \leq j < i \leq n+1} ((2i - 2)^2 - (2j - 2)^2) \neq 0.$$

Thus the system of linear equations

$$
\begin{bmatrix}
1 & 0 & 0 & \cdots & 0 \\
1 & 2^2 & 2^4 & \cdots & 2^{2n} \\
1 & 4^2 & 4^4 & \cdots & 4^{2n} \\
\cdot & \cdot & \cdot & \cdots & \cdot \\
1 & (2n)^2 & (2n)^4 & \cdots & (2n)^{2n}
\end{bmatrix}
\begin{bmatrix}
x_0 \\
x_1 \\
x_2 \\
\cdot \\
x_n
\end{bmatrix}
=
\begin{bmatrix}
f(0) \\
f(2) \\
f(4) \\
\cdot \\
f(2n)
\end{bmatrix}
$$

has a unique solution $(x_0, x_1, x_2, \ldots, x_n) = (g(n, 0), g(n, 1), g(n, 2), \ldots, g(n, n))$. Set

$$
g_n(x) = \sum_{k=0}^{n} g(n, k) x^{2k}.
$$

Then for $j = 0, 1, 2, \ldots, n$ we have

$$
g_n(2j) = \sum_{k=0}^{n} g(n, k)(2j)^{2k} = f(2j).
$$

As g_n and f are both even functions this result also holds for $j = -1, -2, \ldots, -n$.

3

The Arithmetic Functions
$\sigma_k(n)$, $\sigma_k^*(n)$, $d_{k,m}(n)$ and $F_k(n)$

Let n be a positive integer. An integer d is said to divide n or be a divisor of n if there exists an integer e such that $n = de$. If d divides n we write $d \mid n$. If d does not divide n we write $d \nmid n$. Clearly $0 \nmid n$ and $1 \mid n$.

A function $f : \mathbb{N} \longrightarrow \mathbb{C}$ is called an arithmetic function. If f is an arithmetic function and $n \notin \mathbb{N}$ we set $f(n) = 0$. The arithmetic function f is said to be multiplicative if $f(n_1 n_2) = f(n_1)f(n_2)$ for all $n_1, n_2 \in \mathbb{N}$ with $\gcd(n_1, n_2) = 1$. If f is a multiplicative arithmetic function and $n = p_1^{m_1} p_2^{m_2} \cdots p_r^{m_r}$ is the factorization of $n \in \mathbb{N}$ into distinct prime powers then $f(n) = f(p_1^{m_1})f(p_2^{m_2}) \cdots f(p_r^{m_r})$.

Example 3.1. Let $n \in \mathbb{N}$. As d runs through the positive integers dividing n, n/d runs through the positive divisors of n in the reverse order. Thus, if f is an arithmetic function, we have

$$\sum_{\substack{d \in \mathbb{N} \\ d \mid n}} f(d) = \sum_{\substack{d \in \mathbb{N} \\ d \mid n}} f(n/d).$$

Example 3.2. Let $n \in \mathbb{N}$. We show that

$$\{d \in \mathbb{N} \mid d \mid n, \ n/d \text{ even}\} = \{d \in \mathbb{N} \mid d \mid n/2\}$$

and

$$\{d \in \mathbb{N} \mid d \mid n, \ n/d \text{ odd}\} = \{d \in \mathbb{N} \mid d \mid n\} - \{d \in \mathbb{N} \mid d \mid n/2\}.$$

For both odd and even n, we have

$$\begin{aligned}
\{d \in \mathbb{N} \mid d \mid n, \ n/d \text{ even}\} &= \{d \in \mathbb{N} \mid d \mid n, \ 2 \mid n/d\} \\
&= \{d \in \mathbb{N} \mid d \mid n, \ d \mid n/2\} \\
&= \{d \in \mathbb{N} \mid d \mid n/2\},
\end{aligned}$$

which is the first equality. Then

$$\{d \in \mathbb{N} \mid d \mid n, \; n/d \text{ odd}\} = \{d \in \mathbb{N} \mid d \mid n\} - \{d \in \mathbb{N} \mid d \mid n, \; n/d \text{ even}\}$$
$$= \{d \in \mathbb{N} \mid d \mid n\} - \{d \in \mathbb{N} \mid d \mid n/2\},$$

which is the second equality.

Example 3.3. Let $n \in \mathbb{N}$. Let f be an arithmetic function. It follows immediately from the first equality in Example 3.2 that

$$\sum_{\substack{d \in \mathbb{N} \\ d \mid n \\ n/d \text{ even}}} f(d) = \sum_{\substack{d \in \mathbb{N} \\ d \mid n/2}} f(d).$$

Example 3.4. Let $n \in \mathbb{N}$. Let f be an arithmetic function. It follows immediately from the second equality in Example 3.2 that

$$\sum_{\substack{d \in \mathbb{N} \\ d \mid n \\ n/d \text{ odd}}} f(d) = \sum_{\substack{d \in \mathbb{N} \\ d \mid n}} f(d) - \sum_{\substack{d \in \mathbb{N} \\ d \mid n/2}} f(d).$$

In this chapter we give the properties of four arithmetic functions that we use throughout the book. The first function is the arithmetic function σ_k defined for all $k \in \mathbb{Z}$ by

$$\sigma_k(n) := \sum_{\substack{d \in \mathbb{N} \\ d \mid n}} d^k, \quad n \in \mathbb{N}, \tag{3.1}$$

where d runs through the positive integers dividing n. We set

$$d(n) := \sigma_0(n) = \sum_{\substack{d \in \mathbb{N} \\ d \mid n}} 1 \tag{3.2}$$

and

$$\sigma(n) := \sigma_1(n) = \sum_{\substack{d \in \mathbb{N} \\ d \mid n}} d. \tag{3.3}$$

If p is a prime and a is a nonnegative integer, we have

$$\sigma_k(p^a) = 1 + p^k + p^{2k} + \cdots + p^{ak} = \frac{p^{(a+1)k} - 1}{p^k - 1}.$$

Theorem 3.1. (i) σ_k *is multiplicative.*

(ii) *Let p be a prime. Let $k, n \in \mathbb{N}$. Then*

$$\sigma_k(pn) - (p^k + 1)\sigma_k(n) + p^k\sigma_k(n/p) = 0.$$

Proof. (i) Let $n_1, n_2 \in \mathbb{N}$ be such that $\gcd(n_1, n_2) = 1$. Let $d \in \mathbb{N}$ be a divisor of $n_1 n_2$. Then there exist unique positive integers d_1 and d_2 such that $d = d_1 d_2$ with $d_1 \mid n_1$ and $d_2 \mid n_2$. Conversely, if $d_1 \in \mathbb{N}$ and $d_2 \in \mathbb{N}$ are such that $d_1 \mid n_1$ and $d_2 \mid n_2$ then $d = d_1 d_2 \mid n_1 n_2$. Hence we have

$$\sigma_k(n_1 n_2) = \sum_{\substack{d \in \mathbb{N} \\ d \mid n_1 n_2}} d^k = \sum_{\substack{d_1 \in \mathbb{N} \\ d_1 \mid n_1}} \sum_{\substack{d_2 \in \mathbb{N} \\ d_2 \mid n_2}} (d_1 d_2)^k$$

$$= \sum_{\substack{d_1 \in \mathbb{N} \\ d_1 \mid n_1}} d_1^k \sum_{\substack{d_2 \in \mathbb{N} \\ d_2 \mid n_2}} d_2^k = \sigma_k(n_1)\sigma_k(n_2),$$

showing that σ_k is multiplicative.

(ii) Let $n = p^s N$, where $s \in \mathbb{N}_0$, $N \in \mathbb{N}$ and $\gcd(N, p) = 1$. Then

$$\sigma_k(p^{s+1}) - (p^k + 1)\sigma_k(p^s) + p^k\sigma_k(p^{s-1})$$

$$= \frac{p^{(s+2)k} - 1}{p^k - 1} - (p^k + 1)\frac{p^{(s+1)k} - 1}{p^k - 1} + p^k\frac{p^{sk} - 1}{p^k - 1} = 0.$$

Multiplying this equality by $\sigma_k(N)$, and appealing to part (i), we obtain the assertion of part (ii). $\qquad\square$

In particular we have

$$\sigma(2n) = 3\sigma(n) - 2\sigma(n/2),$$
$$\sigma(3n) = 4\sigma(n) - 3\sigma(n/3),$$
$$\sigma_3(2n) = 9\sigma_3(n) - 8\sigma_3(n/2),$$
$$\sigma_3(3n) = 28\sigma_3(n) - 27\sigma_3(n/3),$$

for all $n \in \mathbb{N}$. These identities will be used on many occasions throughout this book. We also note that $\sigma(n) \equiv 0 \pmod{4}$ if $n \equiv 3 \pmod{4}$.

The second arithmetic function that we are interested in is σ_k^*, which is defined for all $k \in \mathbb{Z}$ by

$$\sigma_k^*(n) := \sum_{\substack{d \in \mathbb{N} \\ d \mid n \\ n/d \text{ odd}}} d^k, \quad n \in \mathbb{N}. \qquad (3.4)$$

Clearly if n is odd we have $\sigma_k^*(n) = \sigma_k(n)$. We set

$$d^*(n) := \sigma_0^*(n) = \sum_{\substack{d \in \mathbb{N} \\ d \mid n \\ n/d \text{ odd}}} 1 \tag{3.5}$$

and

$$\sigma^*(n) := \sigma_1^*(n) = \sum_{\substack{d \in \mathbb{N} \\ d \mid n \\ n/d \text{ odd}}} d. \tag{3.6}$$

The first five values are $\sigma^*(1) = 1$, $\sigma^*(2) = 2$, $\sigma^*(3) = 4$, $\sigma^*(4) = 4$ and $\sigma^*(5) = 6$. If a is a nonnegative integer, we have

$$\sigma_k^*(2^a) = 2^{ak}.$$

Theorem 3.2. σ_k^* *is multiplicative.*

Proof. Let $n_1, n_2 \in \mathbb{N}$ satisfy $\gcd(n_1, n_2) = 1$. Then, as in the proof of Theorem 3.1(i), we have

$$\sigma_k^*(n_1 n_2) = \sum_{\substack{d \in \mathbb{N} \\ d \mid n_1 n_2 \\ n_1 n_2/d \text{ odd}}} d^k = \sum_{\substack{d_1 \in \mathbb{N} \\ d_1 \mid n_1 \\ n_1/d_1 \text{ odd}}} d_1^k \sum_{\substack{d_2 \in \mathbb{N} \\ d_2 \mid n_2 \\ n_2/d_2 \text{ odd}}} d_2^k = \sigma_k^*(n_1)\sigma_k^*(n_2),$$

showing that σ_k^* is multiplicative. $\qquad\square$

Our next theorem relates the functions σ_k and σ_k^*.

Theorem 3.3. *Let $n, k \in \mathbb{N}$. Then*

$$\sigma_k^*(n) = \sigma_k(n) - \sigma_k(n/2).$$

Proof. This theorem follows immediately from the identity of Example 3.4 by taking $f(x) = x^k$. We give another proof, which uses the fact that σ_k is a multiplicative arithmetic function.

If n is odd we have

$$\sigma_k^*(n) = \sum_{\substack{d \in \mathbb{N} \\ d \mid n \\ n/d \text{ odd}}} d^k = \sum_{\substack{d \in \mathbb{N} \\ d \mid n}} d^k = \sigma_k(n) = \sigma_k(n) - \sigma_k(n/2).$$

If n is even then we have $n = 2^a n_1$, where $a \in \mathbb{N}$, $n_1 \in \mathbb{N}$, and n_1 is odd. As $d \mid n$ if and only if $d = ef$ with $e \mid 2^a$ and $f \mid n_1$, we have

$$\sigma_k^*(n) = \sum_{\substack{e \in \mathbb{N} \\ e \mid 2^a \\ 2^a/e \text{ odd}}} e^k \sum_{\substack{f \in \mathbb{N} \\ f \mid n_1}} f^k = 2^{ak} \sum_{\substack{f \in \mathbb{N} \\ f \mid n_1}} f^k = 2^{ak} \sigma_k(n_1).$$

Now as σ_k is a multiplicative function, we have

$$\sigma_k(n) = \sigma_k(2^a n_1) = \sigma_k(2^a)\sigma_k(n_1) = \frac{2^{(a+1)k} - 1}{2^k - 1} \sigma_k(n_1)$$

and

$$\sigma_k(n/2) = \sigma_k(2^{a-1} n_1) = \sigma_k(2^{a-1})\sigma_k(n_1) = \frac{2^{ak} - 1}{2^k - 1} \sigma_k(n_1)$$

so that

$$\sigma_k(n) - \sigma_k(n/2) = 2^{ak} \sigma_k(n_1) = \sigma_k^*(n).$$

This completes the proof of the theorem. $\qquad\qquad\qquad\qquad\qquad$ $\square$

Taking $k = 0$ and $k = 1$ in Theorem 3.3 we obtain the following theorem.

Theorem 3.4. *For all $n \in \mathbb{N}$ we have*

$$d^*(n) = d(n) - d(n/2)$$

and

$$\sigma^*(n) = \sigma(n) - \sigma(n/2).$$

We now use Theorem 3.2 to prove a result about the parity of $\sigma^*(n)$.

Theorem 3.5. *Let n be a positive integer. If n is even then so is $\sigma^*(n)$.*

Proof. Let n be an even positive integer. Then $n = 2^a N$, where $a \in \mathbb{N}$, $N \in \mathbb{N}$, and N is odd. Hence, by Theorem 3.2, we have

$$\sigma^*(n) = \sigma^*(2^a N) = \sigma^*(2^a)\sigma^*(N) = 2^a \sigma^*(N) \equiv 0 \ (\mathrm{mod}\ 2),$$

completing the proof. $\qquad\qquad\qquad\qquad\qquad\qquad\qquad\qquad\qquad$ $\square$

Our next theorem gives a necessary and sufficient condition for $\sigma(n)$ to be odd.

Theorem 3.6. *Let $n \in \mathbb{N}$. Let N denote the largest odd integer dividing n. Then $\sigma(n)$ is odd if and only if N is a perfect square.*

Proof. For $m \in \mathbb{N}_0$ we have

$$\sigma(2^m) = 2^{m+1} - 1 \equiv 1 \ (\mathrm{mod}\ 2).$$

For $m \in \mathbb{N}_0$ and p an odd prime, we have

$$\sigma(p^m) = 1 + p + p^2 + \cdots + p^m \equiv \begin{cases} 0 \ (\mathrm{mod}\ 2), & \text{if } m \equiv 1 \ (\mathrm{mod}\ 2), \\ 1 \ (\mathrm{mod}\ 2), & \text{if } m \equiv 0 \ (\mathrm{mod}\ 2). \end{cases}$$

The asserted result now follows as σ is multiplicative. □

From Theorems 3.4, 3.5 and 3.6 we deduce the following result, which is used in the proof of the Girard-Fermat theorem (Theorem 7.1).

Theorem 3.7. *Let $n \in \mathbb{N}$. Then $\sigma^*(n)$ is odd if and only if n is an odd perfect square.*

Proof. Suppose first that $\sigma^*(n)$ is odd. Then, by Theorem 3.5, we deduce that n is odd. Hence, by Theorem 3.4, $\sigma(n)$ is odd and thus, by Theorem 3.6, n is a perfect square.

Now suppose that n is an odd perfect square. Hence, by Theorem 3.6, $\sigma(n)$ is odd. Then, by Theorem 3.4, $\sigma^*(n)$ is odd. □

The third arithmetic function we make use of is the arithmetic function $d_{k,m}$, which is defined for $m \in \mathbb{N}$ and $k \in \mathbb{Z}$ by

$$d_{k,m}(n) := \sum_{\substack{d \in \mathbb{N} \\ d \mid n \\ d \equiv k \ (\mathrm{mod}\ m)}} 1, \quad n \in \mathbb{N}.$$

Clearly if $k_1 \in \mathbb{Z}$ and $k_2 \in \mathbb{Z}$ are such that $k_1 \equiv k_2 \ (\mathrm{mod}\ m)$ then $d_{k_1,m}(n) = d_{k_2,m}(n)$. Hence we usually restrict k to the set $\{0, 1, 2, \ldots, m-1\}$. In particular we have

$$d_{0,1}(n) = d(n).$$

The basic properties of $d_{k,m}(n)$ are given in the next theorem.

Theorem 3.8. *Let $m, n \in \mathbb{N}$ and $k \in \{0, 1, 2, \ldots, m-1\}$. Then*

$$d_{0,m}(n) = d(n/m), \tag{3.7}$$

$$\sum_{k=0}^{m-1} d_{k,m}(n) = d(n), \tag{3.8}$$

and if $e \in \mathbb{N}$ is such that $e \mid k$ and $e \mid m$ then

$$d_{k,m}(n) = d_{k/e,m/e}(n/e). \tag{3.9}$$

Proof. We have

$$d_{0,m}(n) = \sum_{\substack{d \in \mathbb{N} \\ d \mid n \\ d \equiv 0 \,(\mathrm{mod}\, m)}} 1 = \sum_{\substack{e \in \mathbb{N} \\ me \mid n}} 1 = \sum_{\substack{e \in \mathbb{N} \\ e \mid n/m}} 1 = d(n/m),$$

which is the first assertion of the theorem.

Next we have

$$\sum_{k=0}^{m-1} d_{k,m}(n) = \sum_{k=0}^{m-1} \sum_{\substack{d \in \mathbb{N} \\ d \mid n \\ d \equiv k \,(\mathrm{mod}\, m)}} 1 = \sum_{\substack{d \in \mathbb{N} \\ d \mid n}} 1 = d(n),$$

which is the second assertion.

Set $k_1 = k/e \in \mathbb{N}_0$ and $m_1 = m/e \in \mathbb{N}$. Then we have

$$d_{k,m}(n) = \sum_{\substack{d \in \mathbb{N} \\ d \mid n \\ d \equiv ek_1 \,(\mathrm{mod}\, em_1)}} 1.$$

Now $d \equiv ek_1 \ (\mathrm{mod}\ em_1)$ implies $d \equiv 0 \ (\mathrm{mod}\ e)$ so $e \mid d$. Changing the summation variable from d to $d_1 = d/e \in \mathbb{N}$, we obtain

$$d_{k,m}(n) = \sum_{\substack{d_1 \in \mathbb{N} \\ d_1 \mid n/e \\ d_1 \equiv k_1 \,(\mathrm{mod}\, m_1)}} 1 = d_{k_1,m_1}(n/e) = d_{k/e,m/e}(n/e),$$

which is the final assertion of the theorem. $\square$

From Theorems 3.8 and 3.4, we deduce that

$$d_{0,2}(n) = d(n/2), \quad d_{1,2}(n) = d(n) - d(n/2) = d^*(n). \tag{3.10}$$

In Chapter 9 we determine $d_{i,4}(n)$ for $i \in \{0, 1, 2, 3\}$ (see Theorem 9.4) and in Chapter 17 we determine $d_{i,3}(n)$ for $i \in \{0, 1, 2\}$ (see Theorem 17.2).

If x is a real number we denote the greatest integer less than or equal to x by $[x]$, that is, $[x] = m$, where m is the unique integer satisfying $m \le x < m + 1$. Our penultimate theorem of this chapter evaluates the sum

$$\sum_{\substack{d \in \mathbb{N} \\ d \mid n}} \left[\frac{ad + b}{m} \right]$$

for integers $a(\ne 0)$ and b and positive integers m and n.

Theorem 3.9. *Let $a \in \mathbb{Z}\backslash\{0\}$, $b \in \mathbb{Z}$, $m \in \mathbb{N}$ and $n \in \mathbb{N}$. Set*

$$g := \gcd(a, m), \quad a_1 := a/g, \quad m_1 := m/g,$$

so that $\gcd(a_1, m_1) = 1$. Define $a_2 \in \{1, 2, \ldots, m_1 - 1\}$ by

$$a_1 a_2 \equiv 1 \pmod{m_1}.$$

Then

$$\sum_{\substack{d \in \mathbb{N} \\ d \mid n}} \left[\frac{ad+b}{m} \right] = \frac{a}{m}\sigma(n) + \frac{b}{m}d(n) - \frac{1}{m} \sum_{\substack{\ell = 1 \\ \ell \equiv b \,(\mathrm{mod}\, g)}}^{m-1} \ell d_{a_2(\ell-b)/g, m_1}(n).$$

Proof. Let $t \in \mathbb{Z}$. Let ℓ be the unique integer in $\{0, 1, 2, \ldots, m - 1\}$ such that $t \equiv \ell \pmod{m}$. Then $\frac{t-\ell}{m} \in \mathbb{Z}$ and

$$\frac{t-\ell}{m} \leq \frac{t}{m} < \frac{t-\ell}{m} + 1$$

so that

$$\left[\frac{t}{m} \right] = \frac{t-\ell}{m}.$$

Hence

$$\sum_{\substack{d \in \mathbb{N} \\ d \mid n}} \left[\frac{ad+b}{m} \right] = \sum_{\ell=0}^{m-1} \sum_{\substack{d \in \mathbb{N} \\ d \mid n \\ ad+b \equiv \ell \,(\mathrm{mod}\, m)}} \left[\frac{ad+b}{m} \right]$$

$$= \sum_{\ell=0}^{m-1} \sum_{\substack{d \in \mathbb{N} \\ d \mid n \\ ad+b \equiv \ell \,(\mathrm{mod}\, m)}} \frac{ad+b-\ell}{m}$$

$$= \frac{1}{m} \sum_{\ell=0}^{m-1} \sum_{\substack{d \in \mathbb{N} \\ d \mid n \\ ad+b \equiv \ell \,(\mathrm{mod}\, m)}} (ad+b) - \frac{1}{m} \sum_{\ell=0}^{m-1} \ell \sum_{\substack{d \in \mathbb{N} \\ d \mid n \\ ad+b \equiv \ell \,(\mathrm{mod}\, m)}} 1$$

$$= \frac{1}{m} \sum_{\substack{d \in \mathbb{N} \\ d \mid n}} (ad+b) - \frac{1}{m} \sum_{\ell=1}^{m-1} \ell \sum_{\substack{d \in \mathbb{N} \\ d \mid n \\ ad \equiv \ell - b \,(\mathrm{mod}\, m)}} 1.$$

Now the congruence

$$ad \equiv \ell - b \pmod{m}$$

is solvable for d if and only if $\gcd(a, m) \mid \ell - b$, that is, if and only if

$$\ell \equiv b \pmod{g}$$

in which case the solutions d are given by

$$d \equiv a_2(\ell - b)/g \pmod{m_1}.$$

Hence

$$\sum_{\substack{d \in \mathbb{N} \\ d \mid n}} \left[\frac{ad + b}{m}\right] = \frac{a}{m}\sigma(n) + \frac{b}{m}d(n) - \frac{1}{m} \sum_{\substack{\ell = 1 \\ \ell \equiv b \pmod{g}}}^{m-1} \ell \sum_{\substack{d \in \mathbb{N} \\ d \mid n \\ d \equiv a_2(\ell - b)/g \pmod{m_1}}} 1$$

as asserted. $\qquad\qquad\qquad\qquad\qquad\qquad\qquad\qquad\qquad\qquad\qquad$ □

Example 3.5. When $\gcd(a, m) = 1$ we have $g = 1$, $a_1 = a$ and $m_1 = m$ so that Theorem 3.9 gives

$$\sum_{\substack{d \in \mathbb{N} \\ d \mid n}} \left[\frac{ad + b}{m}\right] = \frac{a}{m}\sigma(n) + \frac{b}{m}d(n) - \frac{1}{m} \sum_{\ell=1}^{m-1} \ell d_{a^{-1}(\ell-b), m}(n), \qquad (3.11)$$

where a^{-1} denotes any integer x such that $ax \equiv 1 \pmod{m}$. In particular when $a = 1$ we can take $a^{-1} = 1$ and (3.11) becomes

$$\sum_{\substack{d \in \mathbb{N} \\ d \mid n}} \left[\frac{d + b}{m}\right] = \frac{1}{m}\sigma(n) + \frac{b}{m}d(n) - \frac{1}{m} \sum_{\ell=1}^{m-1} \ell d_{\ell-b, m}(n). \qquad (3.12)$$

When $b = 0$ this formula becomes

$$\sum_{\substack{d \in \mathbb{N} \\ d \mid n}} \left[\frac{d}{m}\right] = \frac{1}{m}\sigma(n) - \frac{1}{m} \sum_{\ell=1}^{m-1} \ell d_{\ell, m}(n). \qquad (3.13)$$

Finally in this chapter we define for $k, n \in \mathbb{N}$ the arithmetic function $F_k(n)$ given by

$$F_k(n) = \begin{cases} 1, & \text{if } k \mid n, \\ 0, & \text{if } k \nmid n. \end{cases}$$

In this case we extend the definition of $F_k(n)$ to all $n \in \mathbb{Z}$ by $F_k(-n) = F_k(n)$, $n \in \mathbb{N}$. Clearly

$$F_1(n) = 1, \quad F_2(n) = \begin{cases} 1, & \text{if } 2 \mid n \\ 0, & \text{if } 2 \nmid n \end{cases} = \frac{1}{2}(1 + (-1)^n)$$

for all $n \in \mathbb{N}$. The following properties of $F_k(n)$ are easily proved.

$$F_k(n) = F_k(n + k\ell), \quad \ell, n \in \mathbb{Z}, \quad k \in \mathbb{N}, \tag{3.14}$$

$$F_k(n) = F_k(-n), \quad n \in \mathbb{Z}, \quad k \in \mathbb{N}, \tag{3.15}$$

and

$$F_k(an) = F_{k/(a,k)}(n), \quad a, k \in \mathbb{N}, \quad n \in \mathbb{Z}, \tag{3.16}$$

where we have abbreviated $\gcd(a, k)$ to (a, k).

Example 3.6. Let $d, k \in \mathbb{N}$. We evaluate the sum

$$\sum_{\ell=1}^{d} F_k(\ell).$$

We have

$$\sum_{\ell=1}^{d} F_k(\ell) = \sum_{\substack{\ell \in \mathbb{N} \\ 1 \le \ell \le d \\ k \mid \ell}} 1 = \sum_{\substack{m \in \mathbb{N} \\ 1 \le km \le d}} 1 = \sum_{\substack{m \in \mathbb{N} \\ 1 \le m \le d/k}} 1 = \left[\frac{d}{k} \right].$$

Example 3.7. Let $d, k, n \in \mathbb{N}$. We evaluate the sum

$$\sum_{\substack{d \in \mathbb{N} \\ d \mid n}} \sum_{\ell=1}^{d} F_k(\ell).$$

Appealing to Example 3.6 and (3.13), we obtain

$$\sum_{\substack{d \in \mathbb{N} \\ d \mid n}} \sum_{\ell=1}^{d} F_k(\ell) = \sum_{\substack{d \in \mathbb{N} \\ d \mid n}} \left[\frac{d}{k} \right] = \frac{1}{k} \sigma(n) - \frac{1}{k} \sum_{\ell=1}^{k-1} \ell d_{\ell,k}(n).$$

Our final theorem of this chapter relates the function F_4 to the Legendre-Jacobi-Kronecker symbol for discriminant -4. This identity is used in the proof of Jacobi's four squares theorem given in Theorem 11.1.

Theorem 3.10. *For all $a, b \in \mathbb{N}$*

$$F_4(a - b) - F_4(a + b) = \left(\frac{-4}{ab} \right).$$

Proof. If $a \equiv 0 \pmod 2$ or $b \equiv 0 \pmod 2$ then

$$\text{either } 4 \nmid a - b, \ 4 \nmid a + b \text{ or } 4 \mid a - b, \ 4 \mid a + b$$

so

$$F_4(a - b) - F_4(a + b) = 0 - 0 \text{ or } 1 - 1 = 0 = \left(\frac{-4}{ab} \right).$$

If $a \equiv b \equiv 1 \pmod 2$ and $a \equiv b \pmod 4$ then $4 \mid a - b$, $4 \nmid a + b$ and $ab \equiv a^2 \equiv 1 \pmod 4$ so

$$F_4(a - b) - F_4(a + b) = 1 - 0 = 1 = \left(\frac{-4}{ab} \right).$$

If $a \equiv b \equiv 1 \pmod 2$ and $a \equiv -b \pmod 4$ then $4 \nmid a - b$, $4 \mid a + b$ and $ab \equiv -a^2 \equiv 3 \pmod 4$ so

$$F_4(a - b) - F_4(a + b) = 0 - 1 = -1 = \left(\frac{-4}{ab} \right).$$

This completes the proof. □

Many of the results in the problems of Exercises 3 are used in the remaining chapters of the book.

Exercises 3

1. Let $k, n \in \mathbb{N}$. Prove that

$$\sum_{\substack{d \in \mathbb{N} \\ d \mid n \\ n/d \text{ even}}} d^k = \sigma_k(n/2)$$

and

$$\sum_{\substack{d \in \mathbb{N} \\ d \mid n \\ n/d \text{ odd}}} d^k = \sigma_k(n) - \sigma_k(n/2).$$

Deduce Theorem 3.3.

2. Let $k, n \in \mathbb{N}$. Prove that

$$\sum_{\substack{d \in \mathbb{N} \\ d \mid n \\ d \text{ even}}} d^k = 2^k \sigma_k(n/2)$$

and

$$\sum_{\substack{d \in \mathbb{N} \\ d \mid n \\ d \text{ odd}}} d^k = \sigma_k(n) - 2^k \sigma_k(n/2).$$

3. Let $k, n \in \mathbb{N}$. Prove the following four formulae:

$$\sum_{\substack{d \in \mathbb{N} \\ d \mid n \\ d \text{ even}, \, n/d \text{ even}}} d^k = 2^k \sigma_k(n/4),$$

$$\sum_{\substack{d \in \mathbb{N} \\ d \mid n \\ d \text{ odd}, \, n/d \text{ even}}} d^k = \sigma_k(n/2) - 2^k \sigma_k(n/4),$$

$$\sum_{\substack{d \in \mathbb{N} \\ d \mid n \\ d \text{ even}, \, n/d \text{ odd}}} d^k = 2^k \sigma_k(n/2) - 2^k \sigma_k(n/4),$$

$$\sum_{\substack{d \in \mathbb{N} \\ d \mid n \\ d \text{ odd}, \, n/d \text{ odd}}} d^k = \sigma_k(n) - (2^k + 1)\sigma_k(n/2) + 2^k \sigma_k(n/4).$$

4. Prove the assertions of (3.14), (3.15) and (3.16).
5. Let $a, b \in \mathbb{N}$. Prove that

$$F_3(a - b) - F_3(a + b) = \left(\frac{-3}{ab} \right)$$

and

$$(F_7(a - b) - F_7(a + b)) + (F_7(a - 2b) - F_7(a + 2b))$$
$$+ (F_7(2a - b) - F_7(2a + b)) = \left(\frac{-7}{ab} \right).$$

(These results are used in the proofs of Theorems 14.5 and 17.3.)
6. Let $k, n \in \mathbb{N}$. Prove that

$$\sum_{\substack{d \in \mathbb{N} \\ d \mid n}} F_k(d) = d(n/k).$$

7. Let $k, n \in \mathbb{N}$. Prove that

$$\sum_{\substack{d \in \mathbb{N} \\ d \mid n}} d F_k(d) = k\sigma(n/k).$$

8. Let $k, n \in \mathbb{N}$. Prove that

$$\sum_{\substack{d \in \mathbb{N} \\ d \mid n}} \frac{n}{d} F_k(d) = \sigma(n/k).$$

9. For $x \in \mathbb{R}$ prove that

$$\left[\frac{x}{2}\right] + \left[\frac{x+1}{2}\right] = [x].$$

10. Let $k, n \in \mathbb{N}$. Prove that

$$\sum_{\substack{\ell = 1 \\ 2 \mid \ell}}^{n} F_k(\ell) = \begin{cases} \left[\dfrac{n}{k}\right], & \text{if } 2 \mid k, \\[2mm] \left[\dfrac{n}{2k}\right], & \text{if } 2 \nmid k. \end{cases}$$

11. Let $k, n \in \mathbb{N}$. Use the results of Example 3.6 and Problems 9 and 10 to prove that

$$\sum_{\substack{\ell = 1 \\ 2 \nmid \ell}}^{n} F_k(\ell) = \begin{cases} 0, & \text{if } 2 \mid k, \\[2mm] \left[\dfrac{n+k}{2k}\right], & \text{if } 2 \nmid k. \end{cases}$$

12. Let $k, n \in \mathbb{N}$. Prove that

$$\sum_{\substack{d \in \mathbb{N} \ell = 1 \\ d \mid n \quad 2 \mid \ell \\ 2 \mid d}}^{d} F_k(\ell) = \begin{cases} \displaystyle\sum_{\substack{d \in \mathbb{N} \\ d \mid n/2}} \left[\dfrac{2d}{k}\right], & \text{if } 2 \mid k, \\[4mm] \displaystyle\sum_{\substack{d \in \mathbb{N} \\ d \mid n/2}} \left[\dfrac{d}{k}\right], & \text{if } 2 \nmid k. \end{cases}$$

13. Let $k, n \in \mathbb{N}$. Prove that

$$\sum_{\substack{d \in \mathbb{N} \ell = 1 \\ d \mid n \quad 2 \nmid \ell \\ 2 \mid d}}^{d} F_k(\ell) = \begin{cases} 0, & \text{if } 2 \mid k, \\[4mm] \displaystyle\sum_{\substack{d \in \mathbb{N} \\ d \mid n/2}} \left[\dfrac{2d+k}{2k}\right], & \text{if } 2 \nmid k. \end{cases}$$

14. Let $k, n \in \mathbb{N}$. Prove that

$$\sum_{\substack{d \in \mathbb{N} \\ d \mid n \\ 2 \nmid d}} \sum_{\substack{\ell = 1 \\ 2 \mid \ell}}^{d} F_k(\ell) = \begin{cases} \displaystyle\sum_{\substack{d \in \mathbb{N} \\ d \mid n}} \left[\frac{d}{k}\right] - \sum_{\substack{d \in \mathbb{N} \\ d \mid n/2}} \left[\frac{2d}{k}\right], & \text{if } 2 \mid k, \\[6mm] \displaystyle\sum_{\substack{d \in \mathbb{N} \\ d \mid n}} \left[\frac{d}{2k}\right] - \sum_{\substack{d \in \mathbb{N} \\ d \mid n/2}} \left[\frac{d}{k}\right], & \text{if } 2 \nmid k. \end{cases}$$

15. Let $k, n \in \mathbb{N}$. Prove that

$$\sum_{\substack{d \in \mathbb{N} \\ d \mid n \\ 2 \nmid d}} \sum_{\substack{\ell = 1 \\ 2 \nmid \ell}}^{d} F_k(\ell) = \begin{cases} 0, & \text{if } 2 \mid k, \\[4mm] \displaystyle\sum_{\substack{d \in \mathbb{N} \\ d \mid n}} \left[\frac{d+k}{2k}\right] - \sum_{\substack{d \in \mathbb{N} \\ d \mid n/2}} \left[\frac{2d+k}{2k}\right], & \text{if } 2 \nmid k. \end{cases}$$

16. Let $k, n \in \mathbb{N}$. Prove that

$$\sum_{\substack{d \in \mathbb{N} \\ d \mid n}} \sum_{\substack{\ell = 1 \\ \ell \equiv d \,(\mathrm{mod}\ 2)}}^{d} F_k(\ell)$$

$$= \begin{cases} \displaystyle\sum_{\substack{d \in \mathbb{N} \\ d \mid n/2}} \left[\frac{2d}{k}\right], & \text{if } 2 \mid k, \\[6mm] \displaystyle\sum_{\substack{d \in \mathbb{N} \\ d \mid n}} \left[\frac{d+k}{2k}\right] + \sum_{\substack{d \in \mathbb{N} \\ d \mid n/2}} \left[\frac{d}{k}\right] - \sum_{\substack{d \in \mathbb{N} \\ d \mid n/2}} \left[\frac{2d+k}{2k}\right], & \text{if } 2 \nmid k. \end{cases}$$

17. Let $k, n \in \mathbb{N}$. Prove that

$$\sum_{\substack{d \in \mathbb{N} \\ d \mid n}} F_k(2d) = \begin{cases} d(2n/k), & \text{if } 2 \mid k, \\[2mm] d(n/k), & \text{if } 2 \nmid k. \end{cases}$$

18. Let $k, n \in \mathbb{N}$. Prove that

$$\sum_{\substack{d \in \mathbb{N} \\ d \mid n}} \left[\frac{d+k}{2k}\right] = \frac{1}{2k}\sigma(n) - \frac{1}{2k}\sum_{\ell=-k}^{k-1} \ell d_{\ell, 2k}(n).$$

19. Let $k, n \in \mathbb{N}$. If $2 \nmid k$ prove that

$$\sum_{\substack{d \in \mathbb{N} \\ d \mid n}} \left[\frac{2d + k}{2k} \right] = \frac{1}{k} \sigma(n) - \frac{1}{k} \sum_{\ell = -(k-1)/2}^{(k-1)/2} \ell d_{\ell, k}(n).$$

20. Let $k, n \in \mathbb{N}$. Prove that

$$\sum_{\substack{d \in \mathbb{N} \\ d \mid n}} \left[\frac{d + 2k}{2k} \right] = \frac{1}{2k} \sigma(n) + d(n) - \frac{1}{2k} \sum_{\ell = 1}^{2k-1} \ell d_{\ell, 2k}(n).$$

21. Let $k, n \in \mathbb{N}$. Deduce from Problem 12 and (3.13) that

$$\sum_{\substack{d \in \mathbb{N} \\ d \mid n \\ 2 \mid d}} \sum_{\substack{\ell = 1 \\ 2 \mid \ell}}^{d} F_k(\ell) = \begin{cases} \dfrac{2}{k} \sigma(n/2) - \dfrac{2}{k} \displaystyle\sum_{\ell = 1}^{k/2 - 1} \ell d_{\ell, k/2}(n/2), & \text{if } 2 \mid k, \\[4mm] \dfrac{1}{k} \sigma(n/2) - \dfrac{1}{k} \displaystyle\sum_{\ell = 1}^{k-1} \ell d_{\ell, k}(n/2), & \text{if } 2 \nmid k. \end{cases}$$

22. Let $k, n \in \mathbb{N}$. Deduce from Problems 13 and 19 that

$$\sum_{\substack{d \in \mathbb{N} \\ d \mid n \\ 2 \mid d}} \sum_{\substack{\ell = 1 \\ 2 \nmid \ell}}^{d} F_k(\ell) = \begin{cases} 0, & \text{if } 2 \mid k, \\[4mm] \dfrac{1}{k} \sigma(n/2) - \dfrac{1}{k} \displaystyle\sum_{\ell = -(k-1)/2}^{(k-1)/2} \ell d_{\ell, k}(n/2), & \text{if } 2 \nmid k. \end{cases}$$

23. Let $k, n \in \mathbb{N}$. Deduce from Problem 14, (3.9) and (3.13) that

$$\sum_{\substack{d \in \mathbb{N} \\ d \mid n \\ 2 \nmid d}} \sum_{\substack{\ell = 1 \\ 2 \mid \ell}}^{d} F_k(\ell) = \begin{cases} \dfrac{1}{k} \sigma(n) - \dfrac{2}{k} \sigma(n/2) - \dfrac{1}{k} \displaystyle\sum_{\substack{\ell = 1 \\ 2 \nmid \ell}}^{k-1} \ell d_{\ell, k}(n), & \text{if } 2 \mid k, \\[6mm] \dfrac{1}{2k} \sigma(n) - \dfrac{1}{k} \sigma(n/2) - \dfrac{1}{2k} \displaystyle\sum_{\substack{\ell = 1 \\ 2 \nmid \ell}}^{2k-1} \ell d_{\ell, 2k}(n), & \text{if } 2 \nmid k. \end{cases}$$

24. Let $k, n \in \mathbb{N}$. Deduce from Problems 15, 18 and 19 that

$$\sum_{\substack{d \in \mathbb{N} \\ d \mid n \\ 2 \nmid d}} \sum_{\substack{\ell = 1 \\ 2 \nmid \ell}}^{d} F_k(\ell) = \begin{cases} 0, & \text{if } 2 \mid k, \\[4mm] \dfrac{1}{2k} \sigma(n) - \dfrac{1}{k} \sigma(n/2) - \dfrac{1}{2k} \displaystyle\sum_{\substack{\ell = -k \\ 2 \nmid \ell}}^{k-1} \ell d_{\ell, 2k}(n), & \text{if } 2 \nmid k. \end{cases}$$

25. Let $k, n \in \mathbb{N}$. Deduce from Problems 21 and 24 that

$$\sum_{\substack{d \in \mathbb{N} \\ d \mid n}} \sum_{\substack{\ell = 1 \\ \ell \equiv d \, (\text{mod } 2)}}^{d} F_k(\ell)$$

$$= \begin{cases} \dfrac{2}{k}\sigma(n/2) - \dfrac{2}{k}\displaystyle\sum_{\ell=1}^{k/2-1} \ell d_{\ell,k/2}(n/2), & \text{if } 2 \mid k, \\[3ex] \dfrac{1}{2k}\sigma(n) - \dfrac{1}{k}\displaystyle\sum_{\ell=1}^{k-1} \ell d_{\ell,k}(n/2) - \dfrac{1}{2k}\displaystyle\sum_{\substack{\ell=-k \\ 2 \nmid \ell}}^{k-1} \ell d_{\ell,2k}(n), & \text{if } 2 \nmid k. \end{cases}$$

26. Let $n \in \mathbb{N}_0$. Prove that

$$\sum_{\substack{d \in \mathbb{N} \\ d \mid 4n+3}} (-1)^{(d-1)/2} = 0.$$

27. Deduce from Problem 26 that

$$d_{1,4}(4n+3) = d_{3,4}(4n+3), \quad n \in \mathbb{N}_0.$$

28. Let $k, n \in \mathbb{N}$. Use Bernoulli's identity for the sum $\sum_{v=1}^{n-1} v^k$ (see the notes at the end of this chapter) to prove that

$$\sum_{\substack{d \in \mathbb{N} \\ d \mid n}} \sum_{v=1}^{d-1} v^k = \frac{1}{k+1} \sum_{j=0}^{k} \binom{k+1}{j} B_j \sigma_{k+1-j}(n),$$

where B_j is the j-th Bernoulli number. Deduce that

$$\sum_{\substack{d \in \mathbb{N} \\ d \mid n}} \sum_{v=1}^{d} v = \frac{1}{2}\sigma_2(n) + \frac{1}{2}\sigma(n),$$

$$\sum_{\substack{d \in \mathbb{N} \\ d \mid n}} \sum_{v=1}^{d} v^2 = \frac{1}{3}\sigma_3(n) + \frac{1}{2}\sigma_2(n) + \frac{1}{6}\sigma(n),$$

$$\sum_{\substack{d \in \mathbb{N} \\ d \mid n}} \sum_{v=1}^{d} v^3 = \frac{1}{4}\sigma_4(n) + \frac{1}{2}\sigma_3(n) + \frac{1}{4}\sigma_2(n).$$

29. Let $n \in \mathbb{N}$. Prove that

$$\sum_{\substack{d \in \mathbb{N} \\ d \mid n}} \sum_{\substack{k=1 \\ k \equiv d \,(\mathrm{mod}\,2)}}^{d} k^2 = \frac{1}{6}\sigma_3(n) + \frac{1}{2}\sigma_2(n) + \frac{1}{3}\sigma(n).$$

30. Let $k, n \in \mathbb{N}$. Prove that

$$\sum_{\substack{d \in \mathbb{N} \\ d \mid n}} \sum_{\ell=1}^{2d} F_k(\ell) = \begin{cases} \dfrac{2}{k}\sigma(n) - \dfrac{1}{k}\displaystyle\sum_{\ell=1}^{k-1} \ell d_{(k+1)\ell/2,k}(n), & \text{if } 2 \nmid k, \\[4mm] \dfrac{2}{k}\sigma(n) - \dfrac{2}{k}\displaystyle\sum_{\ell=1}^{k/2-1} \ell d_{\ell,k/2}(n), & \text{if } 2 \mid k. \end{cases}$$

31. Let $n \in \mathbb{N}$. Prove that

$$\sum_{\substack{d \in \mathbb{N} \\ d \mid n}} \sum_{\substack{\ell \in \mathbb{N} \\ \ell < d}} F_2(\ell) = \frac{1}{2}(\sigma(n) - d(n) - d(n/2)).$$

32. Let $n \in \mathbb{N}$. Let f be an arithmetic function. Prove that

$$\sum_{\substack{d \in \mathbb{N} \\ d \mid n}} \sum_{\substack{\ell \in \mathbb{N} \\ \ell < d}} f(n/d)F_2(d) = \sum_{\substack{d \in \mathbb{N} \\ d \mid n/2}} \left(\frac{n}{d} - 1\right) f(d).$$

33. Let $n \in \mathbb{N}$. Let f be an arithmetic function. Prove that

$$\sum_{\substack{d \in \mathbb{N} \\ d \mid n}} \sum_{\substack{x \in \mathbb{N} \\ x < d}} f(d)F_2(n/d) = \sum_{\substack{d \in \mathbb{N} \\ d \mid n/2}} (d - 1)f(d).$$

34. Let $n \in \mathbb{N}$. Let f be an arithmetic function. Prove that

$$F_2(n) \sum_{\substack{d \in \mathbb{N} \\ d \mid n}} f(d) = \sum_{\substack{d \in \mathbb{N} \\ d \mid n/2}} f(d) + \sum_{\substack{d \in \mathbb{N} \\ d \mid n/2}} f(2d) - \sum_{\substack{d \in \mathbb{N} \\ d \mid n/4}} f(2d).$$

35. Let $n \in \mathbb{N}$. Prove that

$$\sum_{\substack{d \in \mathbb{N} \\ d \mid n}} \sum_{k=1}^{2d} k^2 = \frac{8}{3}\sigma_3(n) + 2\sigma_2(n) + \frac{1}{3}\sigma(n).$$

36. Let $n \in \mathbb{N}$. Let f be an arithmetic function. Prove that

$$\sum_{\substack{d \in \mathbb{N} \\ d \mid n}} \sum_{\substack{x \in \mathbb{N} \\ x < d}} f(x)F_2(n/d) = \sum_{\substack{d \in \mathbb{N} \\ d \mid n/2}} \left(\sum_{x=1}^{d} f(x)\right) - \sum_{\substack{d \in \mathbb{N} \\ d \mid n/2}} f(d).$$

37. Let $n \in \mathbb{N}$. Let f be an arithmetic function. Prove that

$$\sum_{\substack{d \in \mathbb{N} \\ d \mid n}} \sum_{\substack{x \in \mathbb{N} \\ x < d}} f(n/d) F_2(d - x) = \frac{1}{2} \sum_{\substack{d \in \mathbb{N} \\ d \mid n}} \left(\frac{n}{d} - 1\right) f(d) - \frac{1}{2} \sum_{\substack{d \in \mathbb{N} \\ d \mid n/2}} f(d).$$

38. Let $n \in \mathbb{N}$. Prove that if $\sigma(n) \equiv 1 \pmod{2}$ then $n = m^2$ or $2m^2$ for some $m \in \mathbb{N}$.

39. Let $n \in \mathbb{N}$. Prove that if $\sigma(n) \equiv 2 \pmod 4$ then there exist $t, e \in \mathbb{N}$ and a prime $p \equiv 1 \pmod 4$ such that $p \nmid t$, $e \equiv 1 \pmod 4$ and $n = p^e t^2$ or $2 p^e t^2$.

Notes on Chapter 3

If d is a (positive) divisor of $n (\in \mathbb{N})$ then n/d is also a divisor of n called the conjugate divisor of d. The quantity $\sigma^*(n)$ counts the sum of the divisors of n whose conjugate divisors are odd. The result that $\sigma^*(n) \equiv 1 \pmod 2$ implies that n is the square of an odd integer (Theorem 3.7) is proved in the books by Uspensky and Heaslet [254, pp. 447–8] and Nathanson [209, Lemma 13.2, p. 406].

The function $F_k(n)$ was defined in Huard, Ou, Spearman and Williams [137, eq. (2.6), p. 233]. The identity of Theorem 3.10 was used in Huard, Ou, Spearman and Williams [137, p. 261] to prove Legendre's formula for the number $t_4(n)$ of representations of a nonnegative integer n as the sum of four triangular numbers, namely $t_4(n) = \sigma(2n + 1)$; see Theorem 16.7. The first identity of Problem 5 is implicit in Huard, Ou, Spearman and Williams [137, p. 268]. The second identity of Problem 5 is stated and proved in Williams [269, Theorem 4.3, p. 800]. Example 3.6 is equation (2.6) of Williams [269]. Problems 6, 7, 8, 10, 11, 12, 13, 14, 15 and 16 are equations (2.3), (2.4), (2.5), (2.7), (2.8), (2.9), (2.10), (2.11), (2.12) and (2.13) of Williams [269] respectively.

Problem 9 is a well-known property of the greatest integer function. It is a special case of the more general result

$$[x] + \left[x + \frac{1}{n}\right] + \cdots + \left[x + \frac{n-1}{n}\right] = [nx], \quad x \in \mathbb{R}, \; n \in \mathbb{N},$$

see for example the book by Niven, Zuckerman and Montgomery [211, p. 186].

The Bernoulli numbers B_n ($n \in \mathbb{N}_0$) are defined by the recurrence relation

$$B_n = -\frac{1}{n+1} \sum_{j=0}^{n-1} \binom{n+1}{j} B_j, \quad n \in \mathbb{N}, \tag{3.17}$$

where $B_0 = 1$, or equivalently by

$$\frac{x}{e^x - 1} = \sum_{m=0}^{\infty} B_m \frac{x^m}{m!}, \quad x \in \mathbb{R}, \ |x| < 2\pi. \tag{3.18}$$

The first few Bernoulli numbers are $B_1 = -1/2$, $B_2 = 1/6$, $B_3 = 0$, $B_4 = -1/30$, $B_5 = 0$, $B_6 = 1/42$, $B_7 = 0$, $B_8 = -1/30$, $B_9 = 0$, $B_{10} = 5/66$, $B_{11} = 0$, $B_{12} = -691/2730$. Indeed $B_{2n+1} = 0, n \in \mathbb{N}$, see for example Berndt's book [45, Theorem 4.1.2, p. 85]. Bernoulli's identity states that

$$\sum_{r=1}^{n-1} r^k = \frac{1}{k+1} \sum_{j=0}^{k} \binom{k+1}{j} B_j n^{k+1-j}, \quad n, k \in \mathbb{N}.$$

Simple proofs of this identity have been given by Nunemacher and Young [212] and Williams [263]. Problem 28 is a simple consequence of Bernoulli's identity.

4

The Equation $i^2 + jk = n$

For a positive integer n we consider the solutions in integers i, j, k of the equation

$$i^2 + jk = n. \tag{4.1}$$

If no restrictions are placed on the integers i, j and k, the equation has infinitely many solutions as

$$(i, j, k) = (i, 1, n - i^2), \quad i = 0, 1, 2, \ldots,$$

are distinct solutions of the equation. In order to ensure that (4.1) has only a finite number of solutions we require j and k to be positive integers. Then each solution (i, j, k) of (4.1) satisfies

$$0 \le |i| < \sqrt{n}, \quad 1 \le j \le n, \quad 1 \le k \le n,$$

so that (4.1) has at most

$$(2[\sqrt{n}] + 1)n^2$$

solutions (i, j, k) with $i \in \mathbb{Z}$ and $j, k \in \mathbb{N}$. For our purposes it is convenient to further restrict k to be odd. The set $A(n)$ of solutions of (4.1) that we are interested in is

$$A(n) := \{(i, j, k) \in \mathbb{Z} \times \mathbb{N} \times \mathbb{N} \mid i^2 + jk = n, \ k \text{ odd}\}. \tag{4.2}$$

As $0^2 + n \cdot 1 = n$ we see that

$$(0, n, 1) \in A(n). \tag{4.3}$$

43

Thus the set $A(n)$ is nonempty for all $n \in \mathbb{N}$. Further $i^2 + jk = n$ if and only if $(-i)^2 + jk = n$ so that

$$(i, j, k) \in A(n) \iff (-i, j, k) \in A(n). \qquad (4.4)$$

We have

$$A(1) = \{(0, 1, 1)\},$$
$$A(2) = \{(-1, 1, 1), (0, 2, 1), (1, 1, 1)\},$$
$$A(3) = \{(-1, 2, 1), (0, 1, 3), (0, 3, 1), (1, 2, 1)\},$$
$$A(4) = \{(-1, 1, 3), (-1, 3, 1), (0, 4, 1), (1, 1, 3), (1, 3, 1)\},$$
$$A(5) = \{(-2, 1, 1), (-1, 4, 1), (0, 1, 5), (0, 5, 1), (1, 4, 1), (2, 1, 1)\}.$$

We are interested in alternating sums over the set $A(n)$ such as

$$\sum_{(i,j,k)\in A(n)} (-1)^i i. \qquad (4.5)$$

Clearly

$$\sum_{(i,j,k)\in A(1)} (-1)^i i = (-1)^0 0 = 0,$$

$$\sum_{(i,j,k)\in A(2)} (-1)^i i = (-1)^{-1}(-1) + (-1)^0 0 + (-1)^1 1 = 1 - 1 = 0,$$

$$\sum_{(i,j,k)\in A(3)} (-1)^i i = (-1)^{-1}(-1) + (-1)^0 0 + (-1)^0 0 + (-1)^1 1 = 1 - 1 = 0.$$

We prove that the sum (4.5) vanishes for all $n \in \mathbb{N}$.

Theorem 4.1. *Let n be a positive integer. Then*

$$\sum_{(i,j,k)\in A(n)} (-1)^i i = 0.$$

Proof. For each $(i, j, k) \in A(n)$ with $i = 0$ the summand $(-1)^i i = 0$. For $(i, j, k) \in A(n)$ with $i \neq 0$ we have $(-i, j, k)(\neq (i, j, k)) \in A(n)$ by (4.4). Their corresponding summands $(-1)^i i$ and $(-1)^{-i}(-i)$ have sum $(-1)^i i + (-1)^{-i}(-i) = 0$. Hence

$$\sum_{(i, j, k) \in A(n)} (-1)^i i = 0,$$

as asserted. $\qquad\qquad\qquad\qquad\qquad\qquad\qquad\qquad\qquad\qquad\square$

We next consider the sum formed by replacing the summand $(-1)^i i$ in the sum in Theorem 4.1 by $(-1)^i j$. We evaluate this new sum in terms of the function σ^* discussed in Chapter 3.

Theorem 4.2. *Let n be a positive integer. Then*

$$\sum_{(i,j,k) \in A(n)} (-1)^i j = \sigma^*(n) + 2 \sum_{\substack{i \in \mathbb{N} \\ 1 \le i < \sqrt{n}}} (-1)^i \sigma^*(n - i^2).$$

Proof. We have

$$\sum_{(i,j,k) \in A(n)} (-1)^i j = \sum_{\substack{i \in \mathbb{Z}, j, k \in \mathbb{N} \\ k \text{ odd} \\ i^2 + jk = n}} (-1)^i j. \tag{4.6}$$

The terms with $i = 0$ in the sum on the right hand side of (4.6) contribute

$$\sum_{\substack{j, k \in \mathbb{N} \\ k \text{ odd} \\ jk = n}} j = \sum_{\substack{j \in \mathbb{N} \\ j \mid n \\ n/j \text{ odd}}} j = \sigma^*(n).$$

The terms with $i \neq 0$ in the sum on the right hand side of (4.6) contribute

$$\sum_{\substack{i \in \mathbb{Z} \\ 0 < i^2 < n}} (-1)^i \sum_{\substack{j, k \in \mathbb{N} \\ k \text{ odd} \\ jk = n - i^2}} j = 2 \sum_{\substack{i \in \mathbb{N} \\ 0 < i < \sqrt{n}}} (-1)^i \sum_{\substack{j \in \mathbb{N} \\ j \mid n - i^2 \\ (n - i^2)/j \text{ odd}}} j$$

$$= 2 \sum_{\substack{i \in \mathbb{N} \\ 1 \le i < \sqrt{n}}} (-1)^i \sigma^*(n - i^2).$$

Adding these two evaluations, we obtain the formula of Theorem 4.2. □

Finally, adding the identities in Theorems 4.1 and 4.2, we obtain the following result.

Theorem 4.3. *Let n be a positive integer. Then*

$$\sum_{(i,j,k) \in A(n)} (-1)^i (i + j) = \sigma^*(n) + 2 \sum_{\substack{i \in \mathbb{N} \\ 1 \le i < \sqrt{n}}} (-1)^i \sigma^*(n - i^2).$$

In Chapter 6 we use Theorem 4.3 to give a recurrence relation for $\sigma^*(n)$, see Theorem 6.1.

Exercises 4

1. Let $n \in \mathbb{N}$. If $(i, j, k) \in A(n)$ prove that $i + j \equiv n \pmod 2$.

2. Let $n \in \mathbb{N}$. Prove that

$$\text{card } A(n) \geq 2\left[\sqrt{n}\right] - 1.$$

3. Let $n \in \mathbb{N}$. Prove that

$$\sum_{(i,j,k) \in A(n)} (-1)^i k = \sigma(n) - 2\sigma(n/2)$$

$$+2 \sum_{\substack{i \in \mathbb{N} \\ 1 \leq i < \sqrt{n}}} (-1)^i \left(\sigma(n - i^2) - 2\sigma((n - i^2)/2)\right).$$

4. Let $n \in \mathbb{N}$. Let $f : \mathbb{Z} \to \mathbb{C}$ be such that $f(-x) = -f(x)$ for all $x \in \mathbb{Z}$. Prove that

$$\sum_{(i,j,k) \in A(n)} f(i) = 0.$$

Deduce Theorem 4.1.

5. Let $n \in \mathbb{N}$. Prove that

$$\sum_{(i,j,k) \in A(n)} (-1)^i i^3 = 0.$$

6. Let $n \in \mathbb{N}$. Prove that

$$\sum_{(i,j,k) \in A(n)} (-1)^i i^2 j = 2 \sum_{\substack{i \in \mathbb{N} \\ 1 \leq i < \sqrt{n}}} (-1)^i i^2 \sigma^*(n - i^2).$$

7. Let $n \in \mathbb{N}$. Prove that

$$\sum_{(i,j,k) \in A(n)} (-1)^i i j^2 = 0.$$

8. Let $n \in \mathbb{N}$. Prove that

$$\sum_{(i,j,k) \in A(n)} (-1)^i j^3 = \sigma_3^*(n) + 2 \sum_{\substack{i \in \mathbb{N} \\ 1 \leq i < \sqrt{n}}} (-1)^i \sigma_3^*(n - i^2).$$

9. Let $n \in \mathbb{N}$. Deduce from Problems 5–8

$$\sum_{(i, j, k) \in A(n)} (-1)^i (i + j)^3 = \sigma_3^*(n) + 2 \sum_{\substack{i \in \mathbb{N} \\ 1 \leq i < \sqrt{n}}} (-1)^i \sigma_3^*(n - i^2)$$

$$+ 6 \sum_{\substack{i \in \mathbb{N} \\ 1 \leq i < \sqrt{n}}} (-1)^i i^2 \sigma^*(n - i^2).$$

10. Let $n \in \mathbb{N}$. Let $k \in \mathbb{N}$ be odd. Determine a formula analogous to that of Problem 9 for the sum

$$\sum_{(i, j, k) \in A(n)} (-1)^i (i + j)^k.$$

Notes on Chapter 4

The assertions of Theorems 4.1, 4.2 and 4.3 are given in the book by Uspensky and Heaslet [254, pp. 446–447] in their chapter on Liouville's methods.

5

An Identity of Liouville

We recall that a function $f : \mathbb{Z} \to \mathbb{C}$ is called an odd function if

$$f(-x) = -f(x) \tag{5.1}$$

for all integers x, and an even function if

$$f(-x) = f(x) \tag{5.2}$$

for all integers x. Taking $x = 0$ in (5.1) we see that $f(0) = 0$ for an odd function f. The function $f(x) = x^k$ ($k \in \mathbb{N}$) is an odd function when k is odd and an even function when k is even. From (3.15) we see that F_k is an even function for all $k \in \mathbb{N}$.

Let n be a positive integer. In this chapter we show that the sum

$$G(n) := \sum_{(i, j, k) \in A(n)} (-1)^i F(i + j), \tag{5.3}$$

where $A(n)$ is the set defined in (4.2) and $F : \mathbb{Z} \to \mathbb{C}$ is an odd function, has a very simple evaluation. Taking $n = 1, 2, 3, 4$ and 5 in (5.3), we find that

$$G(1) = F(1),$$
$$G(2) = -F(0) + F(2) - F(2) = 0,$$
$$G(3) = -F(1) + F(1) + F(3) - F(3) = 0,$$
$$G(4) = -F(0) - F(2) + F(4) - F(2) - F(4) = -2F(2),$$
$$G(5) = F(-1) - F(3) + F(1) + F(5) - F(5) + F(3) = 0,$$

as $F(-1) = -F(1)$.

This small amount of numerical evidence suggests that $G(n)$ is zero if n is not a perfect square and is $\pm\sqrt{n}\,F(\sqrt{n})$ if n is a perfect square. We prove this claim, which was first stated by Liouville. In proving this claim, we make use of the following simple principle. If S is a finite set, $\lambda : S \longrightarrow S$ is a bijection, and $f : S \longrightarrow \mathbb{C}$, then

$$\sum_{s \in S} f(s) = \sum_{s \in S} f(\lambda(s)).$$

The mapping $\lambda^{-1} : S \longrightarrow S$ is also a bijection, so

$$\sum_{s \in S} f(s) = \sum_{s \in S} f(\lambda^{-1}(s)).$$

More generally if S and S' are finite sets, $\lambda : S \longrightarrow S'$ is a bijection, and $f : S \longrightarrow \mathbb{C}$, then

$$\sum_{s \in S} f(s) = \sum_{s' \in S'} f(\lambda^{-1}(s')).$$

Theorem 5.1. *Let $F : \mathbb{Z} \to \mathbb{C}$ be an odd function. Let $n \in \mathbb{N}$. Then*

$$\sum_{(i,j,k) \in A(n)} (-1)^i F(i + j) = \begin{cases} (-1)^{\sqrt{n}+1} \sqrt{n} F(\sqrt{n}), & \text{if } n \in \mathbb{N}^2, \\ 0, & \text{if } n \notin \mathbb{N}^2. \end{cases}$$

Proof. We define three subsets U, V and W of $A(n)$ by

$$U = \{(i, j, k) \in A(n) \mid 2i + j - k > 0\},$$
$$V = \{(i, j, k) \in A(n) \mid 2i + j - k = 0\},$$
$$W = \{(i, j, k) \in A(n) \mid 2i + j - k < 0\}.$$

Clearly

$$A(n) = U \cup V \cup W$$

and

$$U \cap V = U \cap W = V \cap W = \emptyset$$

so that $A(n)$ is partitioned by U, V and W. Hence

$$\sum_{(i,j,k) \in A(n)} (-1)^i F(i + j) = \sum_{(i,j,k) \in U} (-1)^i F(i + j)$$
$$+ \sum_{(i,j,k) \in V} (-1)^i F(i + j) + \sum_{(i,j,k) \in W} (-1)^i F(i + j).$$

First we show that

$$\sum_{(i,j,k) \in U} (-1)^i F(i + j) = 0. \tag{5.4}$$

The mapping $\Theta : U \to U$ given by

$$\Theta((i, j, k)) = (-i + k, 2i + j - k, k)$$

is an involution, and thus a bijection on U, which preserves the quantity $i + j$. Hence we have

$$\sum_{(i,j,k) \in U} (-1)^i F(i + j) = \sum_{(i,j,k) \in U} (-1)^{-i+k} F(i + j).$$

As k is odd for $(i, j, k) \in U$, we have

$$\sum_{(i,j,k) \in U} (-1)^i F(i + j) = - \sum_{(i,j,k) \in U} (-1)^i F(i + j),$$

which proves (5.4).

Next we show that

$$\sum_{(i, j, k) \in V} (-1)^i F(i + j) = \begin{cases} (-1)^{m+1} m F(m), & \text{if } n = m^2, \ m \in \mathbb{N}, \\ 0, & \text{if } n \text{ is not a square.} \end{cases} \quad (5.5)$$

Clearly for $(i, j, k) \in V$ we see that j and k are both odd and $i = (-j + k)/2$. Hence

$$\sum_{(i, j, k) \in V} (-1)^i F(i + j) = \sum_{\substack{j, k \in \mathbb{N} \\ j, k \text{ odd} \\ ((j+k)/2)^2 = n}} (-1)^{(-j+k)/2} F((j + k)/2).$$

The right hand sum is 0 if n is not a perfect square. If n is a perfect square, say $n = m^2$, $m \in \mathbb{N}$, the right hand side is

$$\sum_{\substack{j, k \in \mathbb{N} \\ j, k \text{ odd} \\ j + k = 2m}} (-1)^{m+1} F(m) = (-1)^{m+1} m F(m),$$

completing the proof of (5.5).

Finally we prove that

$$\sum_{(i, j, k) \in W} (-1)^i F(i + j) = 0. \quad (5.6)$$

We define the sets P, Q, R and S as follows:

$P := \{(i, j, k) \in \mathbb{Z} \times \mathbb{N} \times \mathbb{N} \mid i^2 + jk = n, \ j \text{ odd}, k \text{ odd}, 2i + j - k < 0\},$

$Q := \{(i, j, k) \in \mathbb{Z} \times \mathbb{N} \times \mathbb{N} \mid i^2 + jk = n, \ j \text{ even}, k \text{ odd}, 2i + j - k < 0\},$

$R := \{(i, j, k) \in \mathbb{Z} \times \mathbb{N} \times \mathbb{N} \mid i^2 + jk = n, \ j \text{ odd}, k \text{ odd}, 2i + j - k > 0\},$

$S := \{(i, j, k) \in \mathbb{Z} \times \mathbb{N} \times \mathbb{N} \mid i^2 + jk = n, \ j \text{ even}, k \text{ odd}, 2i + j - k > 0\}.$

The sets P and Q partition W and the sets R and S partition U.

The mapping $\alpha : Q \to Q$ given by

$$\alpha((i, j, k)) = (-i - j, j, -2i - j + k)$$

is an involution, and thus a bijection. Hence

$$\sum_{(i, j, k) \in Q} (-1)^i F(i + j) = \sum_{(i, j, k) \in Q} (-1)^{-i-j} F(-i).$$

For $(i, j, k) \in Q$ we have j even so that $(-1)^{-i-j} = (-1)^i$, and as F is an odd function, we have $F(-i) = -F(i)$. Thus

$$\sum_{(i, j, k) \in Q} (-1)^i F(i + j) = - \sum_{(i, j, k) \in Q} (-1)^i F(i). \tag{5.7}$$

The mapping $\beta : P \to S$ given by

$$\beta((i, j, k)) = (i + j, -2i - j + k, j)$$

is a bijection. The inverse mapping $\beta^{-1} : S \to P$ is

$$\beta^{-1}((i, j, k)) = (i - k, k, 2i + j - k).$$

Hence

$$\sum_{(i,j,k)\in P} (-1)^i F(i + j) = \sum_{(i,j,k)\in S} (-1)^{i-k} F(i) = - \sum_{(i,j,k)\in S} (-1)^i F(i). \tag{5.8}$$

Adding (5.7) and (5.8), we obtain

$$\sum_{(i,j,k)\in W} (-1)^i F(i + j) = \sum_{(i,j,k)\in P\cup Q} (-1)^i F(i + j)$$

$$= \sum_{(i,j,k)\in P} (-1)^i F(i + j) + \sum_{(i,j,k)\in Q} (-1)^i F(i + j)$$

$$= - \sum_{(i,j,k)\in S} (-1)^i F(i) - \sum_{(i, j, k) \in Q} (-1)^i F(i)$$

$$= - \sum_{(i,j,k)\in Q\cup S} (-1)^i F(i).$$

Now

$$Q \cup S = \{(i, j, k) \in \mathbb{Z} \times \mathbb{N} \times \mathbb{N} \mid i^2 + jk = n, \ j \text{ even}, \ k \text{ odd}, \ 2i + j - k \neq 0\}.$$

As $2i + j - k \neq 0$ for j even and k odd, we deduce that

$$Q \cup S = \{(i, j, k) \in \mathbb{Z} \times \mathbb{N} \times \mathbb{N} \mid i^2 + jk = n, \ j \text{ even}, \ k \text{ odd}\}.$$

Since $(i, j, k) \in Q \cup S$ if and only if $(-i, j, k) \in Q \cup S$ and $(-1)^i F(i)$ is an odd function, we have

$$\sum_{(i,j,k) \in Q \cup S} (-1)^i F(i) = 0.$$

This completes the proof of (5.6).

Adding (5.4), (5.5) and (5.6), we obtain the theorem.   □

In the next chapter we use Theorem 5.1 in conjunction with Theorem 4.3 to obtain a recurrence relation for $\sigma^*(n)$, see Theorem 6.1.

Exercises 5

1. Prove that

$$F(x) = \begin{cases} 1, & \text{if } x \in \mathbb{N}, \\ 0, & \text{if } x = 0, \\ -1, & \text{if } x \in -\mathbb{N}, \end{cases}$$

is an odd function.

2. Prove that the mapping $\Theta : U \longrightarrow U$ defined in the proof of Theorem 5.1 by

$$\Theta((i, j, k)) = (-i + k, 2i + j - k, k)$$

is an involution.

3. Prove that the mapping $\alpha : Q \to Q$ defined in the proof of Theorem 5.1 by

$$\alpha((i, j, k)) = (-i - j, j, -2i - j + k)$$

is an involution.

4. Prove that the mapping $\beta : P \to S$ defined in the proof of Theorem 5.1 by

$$\beta((i, j, k)) = (i + j, -2i - j + k, j)$$

is a bijection.

5. Let $n \in \mathbb{N}$. Prove that

$$\sum_{\substack{(i, j, k) \in A(n) \\ i > j}} (-1)^i - \sum_{\substack{(i, j, k) \in A(n) \\ i < j}} (-1)^i = \begin{cases} (-1)^m m, & \text{if } n = m^2, m \in \mathbb{N}, \\ 0, & \text{otherwise.} \end{cases}$$

[Hint: Apply Theorem 5.1 to the function F defined in Problem 1.]

Notes on Chapter 5

Theorem 5.1 is due to Liouville [170, 7th article, p. 3], [170, 8th article, p. 80]. The proof of Theorem 5.1 given here is an expanded version of that presented in Williams [261]. Uspensky and Heaslet [254, p. 444] derive Theorem 5.1 as a special case of a more general identity.

6

A Recurrence Relation for $\sigma^*(n)$

Let n be a positive integer. In this chapter we use Theorems 4.3 and 5.1 to obtain a recurrence relation for the arithmetic function $\sigma^*(n)$. We choose $F(x) = x$ in Theorem 5.1. With this choice the left hand side of Theorem 5.1 becomes by Theorem 4.3

$$\sum_{(i,j,k) \in A(n)} (-1)^i F(i+j) = \sum_{(i,j,k) \in A(n)} (-1)^i (i+j)$$

$$= \sigma^*(n) + 2 \sum_{\substack{i \in \mathbb{N} \\ 1 \leq i < \sqrt{n}}} (-1)^i \sigma^*(n - i^2),$$

and the right hand side of Theorem 5.1 is $(-1)^{n+1} n s(n)$, where

$$s(n) = \begin{cases} 1, & \text{if } n \text{ is a perfect square, that is } n = m^2 \text{ for some } m \in \mathbb{N}, \\ 0, & \text{if } n \text{ is not a perfect square.} \end{cases}$$

Theorem 5.1 now gives the following recurrence relation for $\sigma^*(n)$.

Theorem 6.1. *Let n be a positive integer. Then*

$$\sigma^*(n) = 2 \sum_{\substack{i \in \mathbb{N} \\ 1 \leq i < \sqrt{n}}} (-1)^{i+1} \sigma^*(n - i^2) + (-1)^{n+1} n s(n).$$

Taking $n = 1, 2, 3, 4, 5$ in Theorem 6.1, we obtain recursively the first five values of $\sigma^*(n)$. We have

$$\sigma^*(1) = (-1)^{1+1} 1 = 1,$$
$$\sigma^*(2) = 2\sigma^*(1) = 2,$$
$$\sigma^*(3) = 2\sigma^*(2) = 4,$$
$$\sigma^*(4) = 2\sigma^*(3) - 4 = 4,$$
$$\sigma^*(5) = 2\sigma^*(4) - 2\sigma^*(1) = 8 - 2 = 6.$$

These values agree with those given in Chapter 3.

Appealing to Theorems 3.3 and 6.1, we obtain the following recurrence relation for $\sigma(n)$.

Theorem 6.2. *Let n be a positive integer. Then*

$$\sigma(n) = \sigma(n/2) + 2 \sum_{\substack{i \in \mathbb{N} \\ 1 \le i < \sqrt{n}}} (-1)^{i+1} \sigma(n - i^2)$$

$$- 2 \sum_{\substack{i \in \mathbb{N} \\ 1 \le i < \sqrt{n} \\ i \equiv n \, (\mathrm{mod} \, 2)}} (-1)^{i+1} \sigma((n - i^2)/2) + (-1)^{n+1} n s(n).$$

From Theorem 6.1 we can obtain a recurrence relation for $\sigma^*(n - i^2)$ and then use this recurrence relation in Theorem 6.1 to obtain a new recurrence relation for $\sigma^*(n)$ (Theorem 6.3). From Theorem 6.3 we obtain a congruence modulo 8 for $\sigma^*(n)$ (Theorem 6.4) and this congruence is used in Chapter 7 to determine those primes p which can be expressed in the form $a^2 + 2b^2$ for some integers a and b (Theorem 7.3). The statement of Theorem 6.3 is simplified by introducing the function $r(n)$ defined for $n \in \mathbb{N}$ by

$$r(n) := \sum_{\substack{(i, j) \in \mathbb{N}^2 \\ i^2 + j^2 = n}} 1.$$

It is easy to show that

$$r(n) = 0, \text{ if } n \equiv 3 \, (\mathrm{mod} \, 4) \tag{6.1}$$

and

$$r(n) \equiv 0 \, (\mathrm{mod} \, 2), \text{ if } n \neq 2k^2 \, (k \in \mathbb{N}). \tag{6.2}$$

Theorem 6.3. *Let* $n \in \mathbb{N}$. *Then*

$$\sigma^*(n) = 4 \sum_{\substack{(i, j) \in \mathbb{N}^2 \\ i^2 + j^2 < n}} (-1)^{i+j} \sigma^*(n - i^2 - j^2) + (-1)^n n(r(n) - s(n)).$$

Proof. Let $i \in \mathbb{N}$ satisfy $i < \sqrt{n}$ so that $n - i^2 \in \mathbb{N}$. Applying Theorem 6.1 to $n - i^2$, we obtain

$$\sigma^*(n - i^2) = 2 \sum_{\substack{j \in \mathbb{N} \\ 1 \le j < \sqrt{n - i^2}}} (-1)^{j+1} \sigma^*(n - i^2 - j^2)$$

$$+ (-1)^{n-i^2+1}(n - i^2) s(n - i^2).$$

Hence, by Theorem 6.1, we deduce

$$\sigma^*(n) = 4 \sum_{\substack{(i,j) \in \mathbb{N}^2 \\ i^2 + j^2 < n}} (-1)^{i+j} \sigma^*(n - i^2 - j^2)$$

$$+ 2(-1)^n \sum_{\substack{i \in \mathbb{N} \\ 1 \le i < \sqrt{n} \\ n - i^2 \in \mathbb{N}^2}} (n - i^2) + (-1)^{n+1} n s(n).$$

Finally, we have

$$\sum_{\substack{i \in \mathbb{N} \\ 1 \le i < \sqrt{n} \\ n - i^2 \in \mathbb{N}^2}} (n - i^2) = \sum_{\substack{(i,j) \in \mathbb{N}^2 \\ i^2 + j^2 = n}} j^2 = \sum_{\substack{(i,j) \in \mathbb{N}^2 \\ i^2 + j^2 = n}} i^2$$

$$= \frac{1}{2} \sum_{\substack{(i,j) \in \mathbb{N}^2 \\ i^2 + j^2 = n}} (i^2 + j^2) = \frac{n}{2} \sum_{\substack{(i,j) \in \mathbb{N}^2 \\ i^2 + j^2 = n}} 1 = \frac{n}{2} r(n),$$

and the theorem follows. $\square$

We close this chapter by observing that in the sum in Theorem 6.3 over all positive integers i and j such that $i^2 + j^2 < n$, to each summand for which i and j are unequal there corresponds an equal summand obtained by interchanging i and j. Thus

$$\sum_{\substack{(i,j) \in \mathbb{N}^2 \\ i^2 + j^2 < n \\ i \ne j}} (-1)^{i+j} \sigma^*(n - i^2 - j^2) \equiv 0 \,(\mathrm{mod}\ 2).$$

Hence

$$\sum_{\substack{(i,j) \in \mathbb{N}^2 \\ i^2 + j^2 < n}} (-1)^{i+j} \sigma^*(n - i^2 - j^2) \equiv \sum_{\substack{(i,j) \in \mathbb{N}^2 \\ i^2 + j^2 < n \\ i = j}} (-1)^{i+j} \sigma^*(n - i^2 - j^2) \,(\mathrm{mod}\ 2)$$

so that

$$\sum_{\substack{(i,j) \in \mathbb{N}^2 \\ i^2 + j^2 < n}} (-1)^{i+j} \sigma^*(n - i^2 - j^2) \equiv \sum_{\substack{i \in \mathbb{N} \\ 1 \le i < \sqrt{n/2}}} \sigma^*(n - 2i^2) \,(\mathrm{mod}\ 2).$$

Putting this congruence into Theorem 6.3, we obtain the following result.

Theorem 6.4. *Let* $n \in \mathbb{N}$. *Then*

$$\sigma^*(n) \equiv 4 \sum_{\substack{i \in \mathbb{N} \\ 1 \le i < \sqrt{n/2}}} \sigma^*(n - 2i^2) + (-1)^n n(r(n) - s(n)) \pmod{8}.$$

Exercises 6

1. Write a computer program based on Theorem 6.1 to calculate $\sigma^*(n)$ for all positive integers n up to 100.
2. Let $n \in \mathbb{N}$. Deduce from Theorem 6.1 that

$$\sigma^*(n) \equiv \begin{cases} 0 \pmod{2}, & \text{if } n \text{ is even}, \\ 0 \pmod{2}, & \text{if } n \text{ is odd and } n \notin \mathbb{N}^2, \\ 1 \pmod{2}, & \text{if } n \text{ is odd and } n \in \mathbb{N}^2. \end{cases}$$

   This gives another proof of Theorem 3.7.
3. Write a computer program based on Theorem 6.2 to calculate $\sigma(n)$ for all positive integers n up to 100.
4. Deduce from Theorem 6.2 the necessary and sufficient condition for $\sigma(n)$ to be odd given in Theorem 3.6.
5. Prove (6.1).
6. Prove (6.2).
7. Deduce from Problem 10 of Exercises 4 and Theorem 5.1 the following recurrence relation for $\sigma_3^*(n)$:

$$\sigma_3^*(n) = 2 \sum_{\substack{i \in \mathbb{N} \\ 1 \le i < \sqrt{n}}} (-1)^{i+1} \sigma_3^*(n - i^2)$$

$$+ 6 \sum_{\substack{i \in \mathbb{N} \\ 1 \le i < \sqrt{n}}} (-1)^{i+1} i^2 \sigma^*(n - i^2) + (-1)^{n+1} n^2 s(n).$$

8. Formulate and prove a recurrence relation analogous to Problem 7 for $\sigma_k^*(n)$, $k \in \mathbb{N}$, k odd.
9. Let $n \in \mathbb{N}$ be odd. Set

$$\rho(n) := \sum_{\substack{d \in \mathbb{N} \\ d \mid n}} (-1)^{(d-1)/2}.$$

Prove the recurrence relation for $\rho(n)$

$$\rho(n) = 2 \sum_{\substack{i \in \mathbb{N} \\ i < \sqrt{n}/2}} (-1)^{i+1} \rho(n - 4i^2) + \begin{cases} (-1)^{(n-1)/2}\sqrt{n}, & \text{if } n \in \mathbb{N}^2, \\ 0, & \text{if } n \notin \mathbb{N}^2. \end{cases}$$

10. Let $n \in \mathbb{N}$ be odd. Prove that

$$r(n) \equiv s(n) - n\sigma(n) \pmod{4}.$$

Notes on Chapter 6

The origins of Theorems 6.1 and 6.2 occur in the work of Liouville [170, 7th article, p. 7]. Theorem 6.2 gives a recurrence relation for $\sigma(n)$ involving squares. Leonhard Euler (1707–1783) gave the analogous one involving pentagonal numbers. Numbers of the form

$$\frac{1}{2}k(3k - 1), \quad k \in \mathbb{Z},$$

are called generalized pentagonal numbers. Arranged in increasing order, they are

$$0, 1, 2, 5, 7, 12, 15, 22, 26, 35, 40, 51, 57, \ldots.$$

Let $n \in \mathbb{N}$. Euler proved analytically the recurrence relation

$$\sigma(n) = \sum_{\substack{k \in \mathbb{Z}\setminus\{0\} \\ n - \frac{1}{2}k(3k-1) \in \mathbb{N}}} (-1)^{k-1} \sigma\left(n - \frac{1}{2}k(3k-1)\right)$$

$$+ \begin{cases} (-1)^{k-1}n, & \text{if } n = \frac{1}{2}k(3k-1) \text{ for some } k \in \mathbb{Z}, \\ 0, & \text{otherwise.} \end{cases}$$

The book of Uspensky and Heaslet [254, pp. 437–441] contains a proof of Euler's pentagonal recurrence relation using Liouville's methods.

The assertion of Problem 9 was stated by Liouville in [170, 8th article, p. 76].

7

The Girard-Fermat Theorem

In this chapter we use the recurrence relation for $\sigma^*(n)$ proved in Chapter 6 (Theorem 6.1) to prove the Girard-Fermat theorem, which asserts that every prime $p \equiv 1 \pmod 4$ is the sum of two integral squares. This theorem was discovered numerically by Girard in 1632 and proved by Fermat by his method of descent in 1654.

Theorem 7.1. *Let p be a prime satisfying $p \equiv 1 \pmod 4$. Then there exist integers a and b such that $p = a^2 + b^2$.*

Proof. Let p be a prime with $p \equiv 1 \pmod 4$. Then

$$\sigma^*(p) = \sum_{\substack{d \in \mathbb{N} \\ d \mid p \\ p/d \text{ odd}}} d = 1 + p \equiv 2 \pmod 4.$$

Hence, by Theorem 6.1, we obtain as $s(p) = 0$

$$\sum_{\substack{i \in \mathbb{N} \\ 1 \le i < \sqrt{p}}} (-1)^{i+1} \sigma^*(p - i^2) \equiv 1 \pmod 2.$$

As $(-1)^{i+1} \equiv 1 \pmod 2$ we deduce that

$$\sum_{\substack{i \in \mathbb{N} \\ 1 \le i < \sqrt{p}}} \sigma^*(p - i^2) \equiv 1 \pmod 2.$$

Hence there is an integer i between 1 and $\sqrt{p}$, say $i = b$, such that

$$\sigma^*(p - b^2) \equiv 1 \pmod 2.$$

By Theorem 3.7 $\sigma^*(k)$ is odd if and only if k is the square of an odd integer. Hence there is an odd integer a such that $p - b^2 = a^2$, that is,

$$p = a^2 + b^2, \quad a \text{ odd}, \quad b \text{ even}.$$

This completes the proof.   □

Before continuing we give another proof of the Girard-Fermat theorem, which is also very much in the spirit of Liouville. This proof was given by Heath-Brown in 1984.

Second proof of Theorem 7.1. We consider the set

$$S := \{(a, b, c) \in \mathbb{N} \times \mathbb{N} \times \mathbb{Z} \mid p = 4ab + c^2\}.$$

As p is a prime with $p \equiv 1 \pmod 4$ there exists $k \in \mathbb{N}$ such that $p = 4k + 1$. The set S is nonempty as $(k, 1, 1) \in S$. If $(a, b, c) \in S$ it is easy to check that $c \neq 0$ and $a - b + c \neq 0$ as p is a prime. Thus we have

$$S = S_+ \cup S_-, \quad S_+ \cap S_- = \emptyset,$$

where

$$S_+ := \{(a, b, c) \in S \mid c > 0\}, \quad S_- := \{(a, b, c) \in S \mid c < 0\},$$

and

$$S = S_1 \cup S_2, \quad S_1 \cap S_2 = \emptyset,$$

where

$$S_1 := \{(a, b, c) \in S \mid a - b + c > 0\}, \quad S_2 := \{(a, b, c) \in S \mid a - b + c < 0\}.$$

The mapping which sends $(a, b, c) \in S_+$ to $(a, b, -c) \in S_-$ is a bijection from S_+ to S_-. Hence card $S_+ = $ card S_-. But card $S_+ +$ card $S_- = $ card S, so

$$\text{card } S_+ = \frac{1}{2} \text{ card } S.$$

The mapping which sends $(a, b, c) \in S_1$ to $(b, a, -c) \in S_2$ is a bijection from S_1 to S_2. Hence card $S_1 = $ card S_2. But card $S_1 + $ card $S_2 = $ card S, so

$$\text{card } S_1 = \frac{1}{2} \text{ card } S.$$

Thus

$$\text{card } S_+ = \text{ card } S_1.$$

The mapping $\sigma : S_1 \longrightarrow S_1$ given by

$$\sigma((a, b, c)) = (a - b + c, b, 2b - c)$$

is a bijection. Moreover the only element of S_1 invariant under σ is $(k, 1, 1)$. Hence S_1 contains an odd number of elements as all of its elements α except $(k, 1, 1)$ can be grouped in pairs $\{\alpha, \sigma(\alpha)\}$. Thus S_+ contains an odd number of elements. Hence S_+ contains an element (a, b, c) with $a = b$ otherwise each element $(a, b, c) \in S_+$ could be paired with the distinct element $(b, a, c) \in S_+$ and card S_+ would be even. The triple (a, b, c) with $a = b$ gives $p = (2a)^2 + c^2$ as required. $\qquad\square$

Our next theorem determines all the pairs (a_1, b_1) of integers such that $p = a_1^2 + b_1^2$ for a prime p with $p \equiv 1 \pmod 4$ in terms of one pair (a, b) with $p = a^2 + b^2$.

Theorem 7.2. *Let p be a prime satisfying $p \equiv 1 \pmod 4$. By the Girard-Fermat theorem there are integers a and b such that $p = a^2 + b^2$. Then*

$$\{(a_1, b_1) \in \mathbb{Z}^2 \mid p = a_1^2 + b_1^2\}$$
$$= \{(a, b), (a, -b), (-a, b), (-a, -b), (b, a), (b, -a), (-b, a), (-b, -a)\}.$$

Proof. We begin by noting that as $p = a^2 + b^2 \equiv 1 \pmod 4$ is a prime, we have $a \neq 0$, $b \neq 0$ and $a \neq \pm b$. Using $p = a^2 + b^2 = a_1^2 + b_1^2$ we deduce

$$(aa_1 + bb_1)(aa_1 - bb_1) = a^2 a_1^2 - b^2 b_1^2$$
$$= a^2(p - b_1^2) - (p - a^2)b_1^2$$
$$= (a^2 - b_1^2)p$$
$$\equiv 0 \pmod p$$

so that (as p is a prime)

$$p \mid aa_1 + \epsilon bb_1$$

for some $\epsilon \in \{-1, 1\}$. Now

$$p^2 = (a^2 + b^2)(a_1^2 + b_1^2) = (aa_1 + \epsilon bb_1)^2 + (ab_1 - \epsilon ba_1)^2$$

so

$$p \mid ab_1 - \epsilon ba_1.$$

Thus

$$m := \frac{aa_1 + \epsilon bb_1}{p}, \quad n := \frac{ab_1 - \epsilon ba_1}{p}$$

are integers such that $1 = m^2 + n^2$. Hence $(m, n) = (\pm 1, 0)$ or $(0, \pm 1)$. The first possibility gives $a_1 = \pm a$, $b_1 = \pm \epsilon b$ and the second gives $a_1 = \mp \epsilon b$, $b_1 = \pm a$. As $a \neq 0$, $b \neq 0$ and $a \neq \pm b$, the eight pairs (a, b), $(a, -b)$,

$(-a, b), \ldots, (-b, -a)$ are all distinct. This completes the proof of the theorem. □

If m is an integer then we have $m^2 \equiv 0$ or $1 \pmod 4$. Thus, for integers m and n we have $m^2 + n^2 \equiv 0, 1$ or $2 \pmod 4$. Hence if p is a prime with $p \equiv 3 \pmod 4$ then $p \neq a^2 + b^2$ for any integers a and b. Since $2 = 1^2 + 1^2$ we have for a prime p

$$p = a^2 + b^2 \text{ for some integers } a \text{ and } b \iff p = 2 \text{ or } p \equiv 1 \pmod 4. \quad (7.1)$$

Now let p be a prime with $p \equiv 1 \pmod 4$. By the Girard-Fermat theorem there are integers a and b such that $p = a^2 + b^2$. Clearly one of a and b is odd and one is even. Interchanging a and b, if necessary, we may suppose that a is odd and b is even. Thus $a \equiv \pm 1 \pmod 4$. Further, replacing a by $-a$, if necessary, we may suppose that $a \equiv 1 \pmod 4$. Then, by Theorem 7.2, we see that a is uniquely determined by p, and the question arises "How is a given in terms of p?" Gauss showed that a is determined by a congruence modulo p and an inequality, namely,

$$a \equiv \frac{1}{2} \binom{(p-1)/2}{(p-1)/4} \pmod p, \quad |a| < p/2. \quad (7.2)$$

Jacobsthal showed that a is given by a sum of values of Legendre symbols, namely,

$$a = -\frac{1}{2} \sum_{m=1}^{p-1} \left(\frac{m^3 + m}{p} \right). \quad (7.3)$$

In our next theorem we use Theorem 6.4 to show that if p is a prime with $p \equiv 1$ or $3 \pmod 8$ then p can be expressed in the form $c^2 + 2d^2$ for some integers c and d.

Theorem 7.3. *Let p be a prime satisfying $p \equiv 1 \pmod 8$ or $p \equiv 3 \pmod 8$. Then there exist integers c and d such that $p = c^2 + 2d^2$.*

Proof. If $p \equiv 1 \pmod 8$ we have, by Theorem 7.2,

$$r(p) = \sum_{\substack{(i,j) \in \mathbb{N}^2 \\ i^2 + j^2 = p}} 1 = 2.$$

If $p \equiv 3 \pmod 8$ we have $r(p) = 0$ by (6.1). Thus, by Theorem 6.4, we deduce

$$\sigma^*(p) \equiv 4 \sum_{\substack{i \in \mathbb{N} \\ 1 \leq i < \sqrt{p/2}}} \sigma^*(p - 2i^2) - 2p \equiv 4 \sum_{\substack{i \in \mathbb{N} \\ 1 \leq i < \sqrt{p/2}}} \sigma^*(p - 2i^2) - 2 \pmod 8,$$

if $p \equiv 1 \pmod 8$, and

$$\sigma^*(p) \equiv 4 \sum_{\substack{i \in \mathbb{N} \\ 1 \leq i < \sqrt{p/2}}} \sigma^*(p - 2i^2) \pmod 8,$$

if $p \equiv 3 \pmod 8$. As $\sigma^*(p) = 1 + p$ we deduce in both cases that

$$\sum_{\substack{i \in \mathbb{N} \\ 1 \leq i < \sqrt{p/2}}} \sigma^*(p - 2i^2) \equiv 1 \pmod 2.$$

Hence there exists an integer i between 1 and $\sqrt{p/2}$, say $i = d$, such that

$$\sigma^*(p - 2d^2) \equiv 1 \pmod 2.$$

By Theorem 3.7 this implies that $p - 2d^2$ is the square of an odd integer. Hence there is an odd integer c such that $p - 2d^2 = c^2$, that is

$$p = c^2 + 2d^2, \quad c \text{ odd}.$$

This completes the proof of the theorem.   □

If m is an integer then $m^2 \equiv 0, 1$ or $4 \pmod 8$. Thus, if m and n are integers, we have $m^2 + 2n^2 \equiv 0, 1, 2, 3, 4$ or $6 \pmod 8$. Hence, if p is a prime with $p \equiv 5$ or $7 \pmod 8$ then $p \neq m^2 + 2n^2$ for any integers m and n. Since $2 = 0^2 + 2 \cdot 1^2$ we have for a prime p

$$p = c^2 + 2d^2 \text{ for some integers } c \text{ and } d \iff p = 2 \text{ or } p \equiv 1, 3 \pmod 8.$$
$$\tag{7.4}$$

Now let p be a prime with $p \equiv 1$ or $3 \pmod 8$. By Theorem 7.3 there are integers c and d such that $p = c^2 + 2d^2$. Clearly c is odd. Replacing c by $-c$ if necessary, we may suppose that $c \equiv 1 \pmod 4$. Then, by Problem 2 of the exercises for this chapter, c is uniquely determined by p. When $p \equiv 1 \pmod 8$ Jacobi and Stern have shown that c is given by

$$c \equiv \frac{1}{2}(-1)^{(p-1)/8} \binom{(p-1)/2}{(p-1)/8} \pmod p, \quad |c| < p/2. \tag{7.5}$$

When p is a prime with $p \equiv 3 \pmod 8$ the analogous determination is due to Eisenstein, namely,

$$c \equiv \frac{1}{2}(-1)^{(p+5)/8} \binom{(p-1)/2}{(p-3)/8} \pmod p, \quad |c| < p/2. \tag{7.6}$$

Exercises 7

1. Let p be a prime with $p \equiv 1$ or $3 \pmod 8$. By Theorem 7.3 there are integers c and d such that $p = c^2 + 2d^2$. Prove that

$$d \equiv \frac{1}{2}(p - 1) \pmod 2.$$

2. Let p be a prime with $p \equiv 1$ or $3 \pmod 8$. By Theorem 7.3 there are integers c and d such that $p = c^2 + 2d^2$. Prove that

$$\left\{ (c_1, d_1) \in \mathbb{Z}^2 \mid p = c_1^2 + 2d_1^2 \right\} = \left\{ (c, d), (c, -d), (-c, d), (-c, -d) \right\}.$$

3. In the second proof of Theorem 7.1, prove that $c \neq 0$ and $a - b + c \neq 0$.
4. In the second proof of Theorem 7.1, prove that the mapping from S_+ to S_- which sends (a, b, c) to $(a, b, -c)$ is a bijection.
5. In the second proof of Theorem 7.1, prove that the mapping from S_1 to S_2 which sends (a, b, c) to $(b, a, -c)$ is a bijection.
6. In the second proof of Theorem 7.1, prove that the mapping from S_1 to S_1 which sends (a, b, c) to $(a - b + c, b, 2b - c)$ is a bijection.
7. Prove that the mapping given in Problem 5 is an involution.
8. Let $n \in \mathbb{N}$. Prove that

$$\sigma^*(n) = 8 \sum_{\substack{(i, j, k) \in \mathbb{N}^3 \\ i^2 + j^2 + k^2 < n}} (-1)^{i+j+k+1} \sigma^*(n - i^2 - j^2 - k^2)$$

$$+ 4(-1)^{n+1} \sum_{\substack{(i, j) \in \mathbb{N}^2 \\ i^2 + j^2 < n}} (n - i^2 - j^2) s(n - i^2 - j^2)$$

$$+ (-1)^n n(r(n) - s(n)).$$

9. Let $n \in \mathbb{N}$. Define $r'(n)$ by

$$r'(n) = \sum_{\substack{(i, j, k) \in \mathbb{N}^3 \\ n = i^2 + j^2 + k^2}} i.$$

Prove that

$$\sum_{\substack{(i, j) \in \mathbb{N}^2 \\ i^2 + j^2 < n}} (n - i^2 - j^2) s(n - i^2 - j^2) = \frac{n}{3} r'(n).$$

10. Let $n \in \mathbb{N}$. Prove that

$$\sum_{\substack{(i, j, k) \in \mathbb{N}^3 \\ i^2 + j^2 + k^2 < n}} (-1)^{i+j+k+1} \sigma^*(n - i^2 - j^2 - k^2)$$

$$\equiv \sum_{\substack{i \in \mathbb{N} \\ 1 \le i < \sqrt{n/3}}} (-1)^{i+1} \sigma^*(n - 3i^2) \pmod{2}.$$

11. Deduce from Problems 8, 9, 10 that for $n \in \mathbb{N}$

$$\sigma^*(n) \equiv 8 \sum_{\substack{i \in \mathbb{N} \\ 1 \le i < \sqrt{n/3}}} \sigma^*(n - 3i^2)$$

$$+ \frac{4}{3}(-1)^{n+1} n r'(n) + (-1)^n n(r(n) - s(n)) \pmod{16}.$$

12. Let p be a prime with $p \equiv 1 \pmod 4$. By taking $n = p$ in the congruence of Problem 11, deduce that

$$r'(p) \equiv (p + 3)/4 \pmod 2.$$

13. Let p be a prime with $p \equiv 1 \pmod 8$. Deduce from Problem 12 that there exist positive integers x, y and z such that

$$p = x^2 + y^2 + z^2.$$

Notes on Chapter 7

Heath-Brown's proof of the Girard-Fermat theorem was given in 1984 in [129]. Varouchas [255] and Williams [260] have given presentations of Heath-Brown's proof. Zagier [273] has given a one-sentence proof. An algebraic proof of the Girard-Fermat theorem can be found for example in the book by Alaca and Williams [28, Theorem 2.5.1, p. 48].

Gauss' congruence (7.2) for a can be found in [107, Vol. 2, pp. 90–91]. Proofs of (7.2) are given in Barnes [37, pp. 3–6], Berndt, Evans and Williams [47, Theorem 6.4.2, p. 200], Cauchy [61, pp. 390–437], Jacobsthal [149, p. 15] and [150, p. 241]. Chowla, Dwork and Evans [73] (see also Chowla [72, Vol. III, pp. 1404–1410]) have determined $\binom{(p-1)/2}{(p-1)/4}$ modulo p^2 in terms of a. Recently Cosgrave and Dilcher [83] have determined $\binom{(p-1)/2}{(p-1)/4}$ modulo p^3.

Jacobsthal's formula (7.3) for a was given in Jacobsthal [149, p. 13], [150, p. 240]. Another proof is given in Berndt, Evans and Williams [47, Theorem 6.2.1, p. 190].

Other determinations of a and b in $p = a^2 + b^2$, p (prime) $\equiv 1$ (mod 4), have been given by Barnes [36], Brillhart [52], Hermite [130], [131, Vol. 1, p. 264], Legendre [164], Serret [240] and Smith [243], [245, Vol. 1, pp. 33–34]. Davenport [85] and Olds [214] have given very readable accounts of Legendre's continued fraction method of constructing a and b. The methods of Serret and Hermite are described in Lehmer [166]. Smith [243], [245, Vol. 1, pp. 33–34] found a by means of the continued fraction expansion of p/c for a certain integer c satisfying $1 \leq c < p/2$. Brillhart's algorithm is a beautiful modification of the methods of Serret and Hermite. Extensions of Brillhart's algorithm have been given in Hardy, Muskat and Williams [127], [128], Muskat [208], Williams [262].

Williams [264] has given a necessary and sufficient condition for $|a|$ to be larger than $|b|$ in $p = a^2 + b^2$, a odd, b even, when the norm of the fundamental unit of the real quadratic field $\mathbb{Q}(\sqrt{2p})$ is -1.

Jackson [145] has given a short proof of Theorem 7.3 in the case of primes $p \equiv 3 \pmod 8$.

The congruence (7.5) was proved by Jacobi [147], [148, Vol. VI, pp. 254–274] and Stern [249], see for example [47, Theorem 9.2.8, p. 272]. The congruence (7.6) was proved by Eisenstein [100], [101, Vol. II, pp. 506–535], see for example the book of Berndt, Evans and Williams [47, Theorem 12.9.7, p. 417]. Many congruences of the type given in (7.5) and (7.6) were given by Hudson and Williams [142]. For references to other results of this kind see Berndt, Evans and Williams [47].

8

A Second Identity of Liouville

We see from Theorem 7.2 and (7.1) that if p is a prime then the number of pairs (a, b) of integers such that $p = a^2 + b^2$ is

$$\begin{cases} 4, & \text{if } p = 2, \\ 8, & \text{if } p \equiv 1 \pmod 4, \\ 0, & \text{if } p \equiv 3 \pmod 4. \end{cases}$$

More generally Jacobi has given a formula for the number $r_2(n)$ of pairs (a, b) of integers such that $n = a^2 + b^2$ for an arbitrary positive integer n. We prove Jacobi's formula in Chapter 9 using Liouville's ideas, see Theorem 9.3. We require the following arithmetic identity of Liouville.

Theorem 8.1. *Let* $f : \mathbb{Z} \times \mathbb{Z} \to \mathbb{C}$ *be such that*

$$f(x, -y) = f(-x, y) = -f(x, y) \tag{8.1}$$

for all $x, y \in \mathbb{Z}$. *Then, for* $n \in \mathbb{N}$, *we have*

$$\sum_{(i, j, k) \in A(n)} (-1)^{(k-1)/2} f(-2i + k, i + j)$$

$$= \begin{cases} (-1)^m \sum_{r=1}^{m} (-1)^r f(2r - 1, m), & \text{if } n = m^2 \text{ for some } m \in \mathbb{N}, \\ 0, & \text{otherwise,} \end{cases}$$

where the set $A(n)$ *is defined in* (4.2).

Proof. Let $n \in \mathbb{N}$. We define the subsets U, V and W of $A(n)$ as in the proof of Theorem 5.1. Recall that the sets U, V and W partition the set $A(n)$.

First we show that

$$\sum_{(i, j, k) \in U} (-1)^{(k-1)/2} f(-2i + k, i + j) = 0. \tag{8.2}$$

67

Recall that the mapping $\Theta : U \to U$ defined in the proof of Theorem 5.1 by

$$\Theta((i, j, k)) = (-i + k, 2i + j - k, k)$$

is a bijection on U. Hence

$$\sum_{(i, j, k) \in U} (-1)^{(k-1)/2} f(-2i + k, i + j)$$

$$= \sum_{(i, j, k) \in U} (-1)^{(k-1)/2} f(2i - k, i + j)$$

$$= - \sum_{(i, j, k) \in U} (-1)^{(k-1)/2} f(-2i + k, i + j),$$

by (8.1), which proves (8.2).

Next we show that

$$\sum_{(i, j, k) \in V} (-1)^{(k-1)/2} f(-2i + k, i + j)$$

$$= \begin{cases} (-1)^m \displaystyle\sum_{r=1}^{m} (-1)^r f(2r - 1, m), & \text{if } n = m^2 \text{ for some } m \in \mathbb{N}, \\ 0, & \text{otherwise.} \end{cases}$$

$$(8.3)$$

For $(i, j, k) \in V$ we noted in the proof of Theorem 5.1 that j and k are both odd and $i = (-j + k)/2$. Hence

$$\sum_{(i, j, k) \in V} (-1)^{(k-1)/2} f(-2i + k, i + j)$$

$$= \sum_{\substack{j, k \in \mathbb{N} \\ j, k \text{ odd} \\ ((j+k)/2)^2 = n}} (-1)^{(k-1)/2} f(j, (j + k)/2).$$

This sum is 0 if n is not a perfect square. If n is a perfect square, say $n = m^2$ for some $m \in \mathbb{N}$, the sum is

$$\sum_{\substack{j, k \in \mathbb{N} \\ j, k \text{ odd} \\ (j+k)/2 = m}} (-1)^{(k-1)/2} f(j, m) = \sum_{r=1}^{m} (-1)^{m-r} f(2r - 1, m),$$

which gives (8.3).

Finally we show that

$$\sum_{(i,j,k) \in W} (-1)^{(k-1)/2} f(-2i + k, i + j) = 0. \tag{8.4}$$

We make use of the sets P, Q, R and S defined in the proof of Theorem 5.1. Recall that the involution $\alpha : Q \to Q$ is given by

$$\alpha((i, j, k)) = (-i - j, j, -2i - j + k)$$

and the bijection $\beta : P \to S$ by

$$\beta((i, j, k)) = (i + j, -2i - j + k, j).$$

Using α we obtain

$$\sum_{(i,j,k) \in Q} (-1)^{(k-1)/2} f(-2i + k, i + j)$$

$$= \sum_{(i,j,k) \in Q} (-1)^{-i+(k-j-1)/2} f(j + k, -i).$$

For $(i, j, k) \in Q$ we have j even. Hence, by Problem 1 of Exercises 4, we see that $i \equiv n \pmod 2$. Thus $(-1)^{-i} = (-1)^n$. By (8.1) we have $f(j + k, -i) = -f(j + k, i)$. Therefore

$$\sum_{(i,j,k) \in Q} (-1)^{(k-1)/2} f(-2i + k, i + j)$$

$$= -(-1)^n \sum_{(i,j,k) \in Q} (-1)^{(k-j-1)/2} f(j + k, i)$$

$$= (-1)^n \sum_{(i,j,k) \in Q} (-1)^{(k-j+1)/2} f(j + k, i)$$

$$= (-1)^n \sum_{(i,j,k) \in Q} (-1)^{(j-k-1)/2} f(j + k, i).$$

For $(i, j, k) \in S$, j is even, so by Problem 1 of Exercises 4, we have $(-1)^i = (-1)^n$. As

$$\beta^{-1}((i, j, k)) = (i - k, k, 2i + j - k),$$

we obtain

$$\sum_{(i,j,k) \in P} (-1)^{(k-1)/2} f(-2i + k, i + j) = (-1)^n \sum_{(i,j,k) \in S} (-1)^{(j-k-1)/2} f(j + k, i).$$

Since P and Q partition W, we have

$$\sum_{(i,j,k) \in W} (-1)^{(k-1)/2} f(-2i+k, i+j)$$

$$= \sum_{(i,j,k) \in P} (-1)^{(k-1)/2} f(-2i+k, i+j)$$

$$+ \sum_{(i,j,k) \in Q} (-1)^{(k-1)/2} f(-2i+k, i+j)$$

$$= (-1)^n \sum_{(i,j,k) \in S} (-1)^{(j-k-1)/2} f(j+k, i)$$

$$+ (-1)^n \sum_{(i,j,k) \in Q} (-1)^{(j-k-1)/2} f(j+k, i)$$

$$= (-1)^n \sum_{(i,j,k) \in Q \cup S} (-1)^{(j-k-1)/2} f(j+k, i)$$

$$= (-1)^n \sum_{\substack{(i,j,k) \in \mathbb{Z} \times \mathbb{N} \times \mathbb{N} \\ i^2 + jk = n \\ j \text{ even}, k \text{ odd} \\ 2i+j-k \neq 0}} (-1)^{(j-k-1)/2} f(j+k, i)$$

$$= (-1)^n \sum_{\substack{(i,j,k) \in \mathbb{Z} \times \mathbb{N} \times \mathbb{N} \\ i^2 + jk = n \\ j \text{ even}, k \text{ odd}}} (-1)^{(j-k-1)/2} f(j+k, i).$$

Replacing i by $-i$, we obtain by (8.1)

$$\sum_{\substack{(i,j,k) \in \mathbb{Z} \times \mathbb{N} \times \mathbb{N} \\ i^2 + jk = n \\ j \text{ even}, k \text{ odd}}} (-1)^{(j-k-1)/2} f(j+k, i)$$

$$= \sum_{\substack{(i,j,k) \in \mathbb{Z} \times \mathbb{N} \times \mathbb{N} \\ i^2 + jk = n \\ j \text{ even}, k \text{ odd}}} (-1)^{(j-k-1)/2} f(j+k, -i)$$

$$= - \sum_{\substack{(i,j,k) \in \mathbb{Z} \times \mathbb{N} \times \mathbb{N} \\ i^2 + jk = n \\ j \text{ even}, k \text{ odd}}} (-1)^{(j-k-1)/2} f(j+k, i).$$

Thus

$$\sum_{\substack{(i, j, k) \in \mathbb{Z} \times \mathbb{N} \times \mathbb{N} \\ i^2 + jk = n \\ j \text{ even}, k \text{ odd}}} (-1)^{(j-k-1)/2} f(j + k, i) = 0.$$

This completes the proof of (8.4).

Adding (8.2), (8.3) and (8.4), we obtain the assertion of the theorem. □

If k_1 and k_2 are positive odd integers then the function

$$f(x, y) = x^{k_1} y^{k_2}, \quad x, y \in \mathbb{Z},$$

satisfies the condition (8.1). Applying Theorem 8.1 to this function, we obtain the following result.

Theorem 8.2. *Let k_1 and k_2 be positive odd integers. Then, for all $n \in \mathbb{N}$, we have*

$$\sum_{(i, j, k) \in A(n)} (-1)^{(k-1)/2} (-2i + k)^{k_1} (i + j)^{k_2}$$

$$= \begin{cases} (-1)^m m^{k_2} \displaystyle\sum_{r=1}^{m} (-1)^r (2r - 1)^{k_1}, & \text{if } n = m^2 \text{ for some } m \in \mathbb{N}, \\ 0, & \text{otherwise.} \end{cases}$$

In Chapter 9 we apply Theorem 8.2 with $(k_1, k_2) = (1, 1)$ to obtain the number of representations of a positive integer n as the sum of two squares, see Theorem 9.3, and with $(k_1, k_2) = (1, 3)$ and $(3, 1)$ to obtain the number of representations of n as the sum of six squares, see Theorem 9.6. In a similar manner a formula for the number of representations of n as a sum of ten squares can be obtained from Theorem 8.2 by taking $(k_1, k_2) = (1, 5), (3, 3)$ and $(5, 1)$. However in order to determine the number of representations of n as the sum of four squares, we need the following analogue of Theorem 8.2. It is convenient to define for $j, k \in \mathbb{N}$

$$s(j, k) := \begin{cases} 1, & \text{if } j \equiv k \equiv 0 \pmod 2, \\ -1, & \text{otherwise.} \end{cases}$$

Theorem 8.3. *Let k_1 be a positive odd integer and k_2 a positive even integer. Then, for all $n \in \mathbb{N}$, we have*

$$\sum_{\substack{(i,j,k) \in \mathbb{Z} \times \mathbb{N} \times \mathbb{N} \\ i^2 + jk = n}} s(j,k) \left(2(-2i+k)^{k_1}(i+j)^{k_2} - (j+k)^{k_1} i^{k_2} \right)$$

$$= \begin{cases} 2m^{k_2} \displaystyle\sum_{r=1}^{2m-1} (-1)^r r^{k_1} - (-1)^m 2^{k_1} m^{k_1} \displaystyle\sum_{r=-m+1}^{m-1} (-1)^r r^{k_2}, & \text{if } n = m^2, m \in \mathbb{N}, \\[4mm] 0, & \text{otherwise.} \end{cases}$$

Theorem 8.3 can be proved in a similar manner to Theorem 8.2. A proof of Theorem 8.3 is outlined in the problems at the end of the chapter. We deduce Jacobi's four squares theorem from Theorem 8.3 by taking $(k_1, k_2) = (1, 2)$, see Theorem 9.5. We also derive this theorem in another way in Theorem 11.1. Jacobi's formula for the number of representations of n as the sum of eight squares can also be deduced from Theorem 8.3. However we do not do this but rather give a shorter proof, see Theorem 19.1.

When applying Theorems 8.2 and 8.3 the following two results are often useful in simplifying the terms that occur.

Theorem 8.4. *Let $n \in \mathbb{N}$. Let $g : \mathbb{Z}^3 \to \mathbb{C}$ be a function such that*

$$g(-i, j, k) = -g(i, j, k), \quad i, j, k \in \mathbb{Z}.$$

Then

$$\sum_{(i,j,k) \in A(n)} g(i, j, k) = 0.$$

Proof. As

$$(i, j, k) \in A(n) \iff (-i, j, k) \in A(n),$$

and $g(-i, j, k) = -g(i, j, k)$, we have

$$\sum_{(i,j,k) \in A(n)} g(i, j, k) = \sum_{(-i,j,k) \in A(n)} g(-i, j, k) = - \sum_{(i,j,k) \in A(n)} g(i, j, k)$$

so that

$$\sum_{(i,j,k) \in A(n)} g(i, j, k) = 0$$

as asserted. $\qquad\qquad\qquad\qquad\qquad\qquad\qquad\qquad\qquad\qquad\qquad\qquad\qquad\square$

Theorem 8.5. *Let $n \in \mathbb{N}$. Let $h : \mathbb{Z}^3 \to \mathbb{C}$ be a function such that*

$$h(i, k, j) = -h(i, j, k).$$

Then

$$\sum_{\substack{(i, j, k) \in \mathbb{Z} \times \mathbb{N} \times \mathbb{N} \\ i^2 + jk = n}} h(i, j, k) = 0.$$

Proof. As (i, j, k) is a solution of $i^2 + jk = n$ if and only if (i, k, j) is also a solution, we have

$$\sum_{\substack{(i, j, k) \in \mathbb{Z} \times \mathbb{N} \times \mathbb{N} \\ i^2 + jk = n}} h(i, j, k) = \sum_{\substack{(i, k, j) \in \mathbb{Z} \times \mathbb{N} \times \mathbb{N} \\ i^2 + kj = n}} h(i, k, j)$$

$$= - \sum_{\substack{(i, j, k) \in \mathbb{Z} \times \mathbb{N} \times \mathbb{N} \\ i^2 + jk = n}} h(i, j, k)$$

so

$$\sum_{\substack{(i, j, k) \in \mathbb{Z} \times \mathbb{N} \times \mathbb{N} \\ i^2 + jk = n}} h(i, j, k) = 0$$

as asserted. $\square$

Exercises 8

1. Let $f : \mathbb{Z}^2 \longrightarrow \mathbb{C}$ satisfy (8.1). Prove that

$$f(-x, -y) = f(x, y)$$

for all $(x, y) \in \mathbb{Z} \times \mathbb{Z}$.

2. Let $n \in \mathbb{N}$. Let $f : \mathbb{Z}^3 \longrightarrow \mathbb{C}$ be such that

$$f(-x, y, z) = -f(x, y, z) \tag{8.5}$$

and

$$f(x, -y, -z) = f(x, y, z) \tag{8.6}$$

for all $(x, y, z) \in \mathbb{Z}^3$. Use the ideas of the proof of Theorem 8.1 to prove that

$$\sum_{\substack{(i, j, k) \in \mathbb{Z} \times \mathbb{N} \times \mathbb{N} \\ n = i^2 + jk}} (2f(-2i + k, i + j, 2i + 2j - k) - f(j + k, i, j - k))$$

$$= \begin{cases} \displaystyle\sum_{r=1}^{2m-1} (2f(r, m, r) - f(2m, r - m, 2r - 2m)), & \text{if } n = m^2, m \in \mathbb{N}, \\ 0, & \text{otherwise.} \end{cases}$$

3. Let $n \in \mathbb{N}$. Let $f : \mathbb{Z}^3 \longrightarrow \mathbb{C}$ satisfy conditions (8.5) and (8.6) for all $(x, y, z) \in \mathbb{Z}^3$. Apply the identity of Problem 1 to the function $(-1)^x f(x, y, z)$ to obtain

$$\sum_{\substack{(i, j, k) \in \mathbb{Z} \times \mathbb{N} \times \mathbb{N} \\ n = i^2 + jk}} (2(-1)^k f(-2i + k, i + j, 2i + 2j - k)$$

$$-(-1)^{j+k} f(j + k, i, j - k))$$

$$= \begin{cases} \displaystyle\sum_{r=1}^{2m-1} (2(-1)^r f(r, m, r) - f(2m, r - m, 2r - 2m)), & \text{if } n = m^2, m \in \mathbb{N}, \\ 0, & \text{otherwise.} \end{cases}$$

4. Let $n \in \mathbb{N}$. Let $f : \mathbb{Z}^3 \longrightarrow \mathbb{C}$ satisfy conditions (8.5) and (8.6) for all $(x, y, z) \in \mathbb{Z}^3$. Deduce from Problems 1 and 2 that

$$2 \sum_{\substack{(i, j, k) \in \mathbb{Z} \times \mathbb{N} \times \mathbb{N} \\ n = i^2 + jk \\ k \equiv 0 \,(\mathrm{mod}\ 2)}} f(-2i + k, i + j, 2i + 2j - k)$$

$$- \sum_{\substack{(i, j, k) \in \mathbb{Z} \times \mathbb{N} \times \mathbb{N} \\ n = i^2 + jk \\ j \equiv k \,(\mathrm{mod}\ 2)}} f(j + k, i, j - k)$$

$$= \begin{cases} \displaystyle 2\sum_{r=1}^{m-1} f(2r, m, 2r) - \sum_{r=1}^{2m-1} f(2m, r - m, 2r - 2m), & \text{if } n = m^2, m \in \mathbb{N}, \\ 0, & \text{otherwise.} \end{cases}$$

5. Let $n \in \mathbb{N}$. Let $f : \mathbb{Z}^3 \longrightarrow \mathbb{C}$ satisfy conditions (8.5) and (8.6) for all $(x, y, z) \in \mathbb{Z}^3$. Deduce from Problems 1 and 2 that

$$2 \sum_{\substack{(i, j, k) \in \mathbb{Z} \times \mathbb{N} \times \mathbb{N} \\ n = i^2 + jk \\ k \equiv 1 \,(\mathrm{mod}\, 2)}} f(-2i + k, i + j, 2i + 2j - k)$$

$$- \sum_{\substack{(i, j, k) \in \mathbb{Z} \times \mathbb{N} \times \mathbb{N} \\ n = i^2 + jk \\ j + k \equiv 1 \,(\mathrm{mod}\, 2)}} f(j + k, i, j - k) = 2 \sum_{\substack{r = 1 \\ r \equiv 1 \,(\mathrm{mod}\, 2)}}^{2m - 1} f(r, m, r).$$

6. Let $n \in \mathbb{N}$. Let $f : \mathbb{Z}^3 \longrightarrow \mathbb{C}$ satisfy

$$f(-x, y) = -f(x, -y) = -f(x, y)$$

for all $(x, y, z) \in \mathbb{Z}^3$. Deduce from Problem 4 that

$$\sum_{(i, j, k) \in A(n)} f(-2i + k, i + j) = \sum_{r=1}^{m} f(2r - 1, m) + \sum_{\substack{(i, j, k) \in \mathbb{Z} \times \mathbb{N} \times \mathbb{N} \\ n = i^2 + jk \\ j \equiv 0 \,(\mathrm{mod}\, 2), \ k \equiv 1 \,(\mathrm{mod}\, 2)}} f(j + k, i).$$

7. Let $n \in \mathbb{N}$. Let $f : \mathbb{Z}^3 \longrightarrow \mathbb{C}$ satisfy conditions (8.5) and (8.6) for all $(x, y, z) \in \mathbb{Z}^3$. Apply the identity of Problem 3 to the function

$$\begin{cases} (-1)^{(x+z)/2} f(x, y, z), & \text{if } x \equiv z \equiv 0 \,(\mathrm{mod}\, 2), \\ \\ 0, & \text{if } x \equiv 1 \,(\mathrm{mod}\, 2) \text{ or } z \equiv 1 \,(\mathrm{mod}\, 2). \end{cases}$$

to deduce that

$$2 \sum_{\substack{(i, j, k) \in \mathbb{Z} \times \mathbb{N} \times \mathbb{N} \\ n = i^2 + jk \\ k \equiv 0 \,(\mathrm{mod}\, 2)}} (-1)^j f(-2i + k, i + j, 2i + 2j - k)$$

$$- \sum_{\substack{(i, j, k) \in \mathbb{Z} \times \mathbb{N} \times \mathbb{N} \\ n = i^2 + jk \\ j \equiv k \,(\mathrm{mod}\, 2)}} (-1)^j f(j + k, i, j - k)$$

$$= \begin{cases} 2 \sum_{r=1}^{m-1} f(2r, m, 2r) - \sum_{r=1}^{2m-1} (-1)^r f(2m, r - m, 2r - 2m), & \text{if } n = m^2, m \in \mathbb{N}, \\ \\ 0, & \text{otherwise.} \end{cases}$$

8. Let $n \in \mathbb{N}$. Let $f : \mathbb{Z}^3 \longrightarrow \mathbb{C}$ satisfy conditions (8.5) and (8.6) for all $(x, y, z) \in \mathbb{Z}^3$. Deduce from Problems 4 and 6 that

$$\sum_{\substack{(i, j, k) \in \mathbb{Z} \times \mathbb{N} \times \mathbb{N} \\ n = i^2 + jk}} s(j, k)(2f(-2i + k, i + j, 2i + 2j - k) - f(j + k, i, j - k))$$

$$= \begin{cases} \displaystyle\sum_{r=1}^{2m-1} (-1)^r (2f(r, m, r) - f(2m, r - m, 2r - 2m)), & \text{if } n = m^2, m \in \mathbb{N}, \\ 0, & \text{otherwise.} \end{cases}$$

9. Let k_1 be a positive odd integer and k_2 a positive even integer. By taking $f(x, y, z) = x^{k_1} y^{k_2}$ in the identity of Problem 7 deduce Theorem 8.3.

Notes on Chapter 8

Theorem 8.1 is due to Liouville [170, 8th article, p. 74]. Similar formulae are given by Liouville in [170, 9th, 10th and 11th articles]. Nathanson [209, Theorem 13.7, p. 420] deduces the identity of Theorem 8.1 from a more general identity [209, Theorem 13.1, p. 402] of Liouville. This more general identity was given by Liouville in [170, 12th article]. In their books Uspensky and Heaslet [254] and Venkov [256] have given treatments of these identities of Liouville.

9

Sums of Two, Four and Six Squares

For $k \in \mathbb{N}$ and $n \in \mathbb{N}_0$ we denote the number of representations of n as the sum of k squares by $r_k(n)$, that is

$$r_k(n) := \operatorname{card}\left\{(x_1, \ldots, x_k) \in \mathbb{Z}^k \mid n = x_1^2 + \cdots + x_k^2\right\}.$$

As there is only one way of expressing 0 as a sum of k squares, namely, $0 = 0^2 + \cdots + 0^2$, we have $r_k(0) = 1$ for all $k \in \mathbb{N}$. We begin by giving a recursion formula for $r_k(n)$.

Theorem 9.1. *Let* $n, k \in \mathbb{N}$. *Then*

$$\sum_{\substack{i \in \mathbb{Z} \\ |i| \le \sqrt{n}}} (n - (k+1)i^2)r_k(n - i^2) = 0.$$

Proof. We have

$$r_{k+1}(n) = \sum_{\substack{(x_1, \ldots, x_{k+1}) \in \mathbb{Z}^{k+1} \\ n = x_1^2 + \cdots + x_{k+1}^2}} 1$$

$$= \sum_{\substack{x_{k+1} \in \mathbb{Z} \\ |x_{k+1}| \le \sqrt{n}}} \sum_{\substack{(x_1, \ldots, x_k) \in \mathbb{Z}^k \\ n - x_{k+1}^2 = x_1^2 + \cdots + x_k^2}} 1,$$

that is

$$r_{k+1}(n) = \sum_{\substack{i \in \mathbb{Z} \\ |i| \le \sqrt{n}}} r_k(n - i^2).$$

Also

$$nr_{k+1}(n) = \sum_{\substack{(x_1,\ldots,x_{k+1}) \in \mathbb{Z}^{k+1} \\ n = x_1^2 + \cdots + x_{k+1}^2}} n$$

$$= \sum_{\substack{(x_1,\ldots,x_{k+1}) \in \mathbb{Z}^{k+1} \\ n = x_1^2 + \cdots + x_{k+1}^2}} (x_1^2 + \cdots + x_{k+1}^2)$$

$$= (k+1) \sum_{\substack{(x_1,\ldots,x_{k+1}) \in \mathbb{Z}^{k+1} \\ n = x_1^2 + \cdots + x_{k+1}^2}} x_{k+1}^2$$

$$= (k+1) \sum_{\substack{x_{k+1} \in \mathbb{Z} \\ |x_{k+1}| \le \sqrt{n}}} x_{k+1}^2 \sum_{\substack{(x_1,\ldots,x_k) \in \mathbb{Z}^k \\ n - x_{k+1}^2 = x_1^2 + \cdots + x_k^2}} 1,$$

that is

$$nr_{k+1}(n) = (k+1) \sum_{\substack{i \in \mathbb{Z} \\ |i| \le \sqrt{n}}} i^2 r_k(n - i^2).$$

Hence

$$n \sum_{\substack{i \in \mathbb{Z} \\ |i| \le \sqrt{n}}} r_k(n - i^2) = (k+1) \sum_{\substack{i \in \mathbb{Z} \\ |i| \le \sqrt{n}}} i^2 r_k(n - i^2)$$

from which the asserted recursion formula follows. □

Our next theorem shows that if we can find a function $D : \mathbb{N}_0 \to \mathbb{Z}$ with $D(0) = 1$ satisfying the same recursion as $r_k(n)$ then $r_k(n) = D(n)$ for all $n \in \mathbb{N}_0$.

Theorem 9.2. *Let $k \in \mathbb{N}$. Let $D : \mathbb{N}_0 \to \mathbb{Z}$ be such that*

$$\sum_{\substack{i \in \mathbb{Z} \\ |i| \le \sqrt{n}}} (n - (k+1)i^2) D(n - i^2) = 0, \quad n \in \mathbb{N}, \tag{9.1}$$

and

$$D(0) = 1.$$

Then

$$D(n) = r_k(n), \quad n \in \mathbb{N}_0.$$

Proof. Set

$$E(n) := D(n) - r_k(n), \quad n \in \mathbb{N}_0. \tag{9.2}$$

Clearly

$$E(0) = D(0) - r_k(0) = 1 - 1 = 0.$$

From Theorem 9.1, (9.1) and (9.2), we deduce

$$\sum_{\substack{i \in \mathbb{Z} \\ |i| \le \sqrt{n}}} (n - (k+1)i^2)E(n - i^2) = 0, \quad n \in \mathbb{N}.$$

Thus

$$E(n) = -\frac{1}{n} \sum_{\substack{i \in \mathbb{Z} \\ 0 < |i| \le \sqrt{n}}} (n - (k+1)i^2)E(n - i^2)$$

from which it follows by a simple induction argument on n that

$$E(n) = 0, \quad n \in \mathbb{N}_0,$$

proving the required assertion. $\qquad\qquad\square$

We now use Theorems 8.2 and 9.2 to obtain Jacobi's formula for the number of representations of a positive integer n as the sum of two squares, that is for the quantity

$$r_2(n) := \operatorname{card} \left\{ (x, y) \in \mathbb{Z} \times \mathbb{Z} \mid n = x^2 + y^2 \right\}.$$

Jacobi's formula is conveniently stated in terms of the Legendre-Jacobi-Kronecker symbol for discriminant -4, which is defined for $d \in \mathbb{N}$ by

$$\left(\frac{-4}{d}\right) = \begin{cases} +1, & \text{if } d \equiv 1 \pmod{4}, \\ -1, & \text{if } d \equiv 3 \pmod{4}, \\ 0, & \text{if } d \equiv 0 \pmod{2}. \end{cases}$$

Theorem 9.3. *Let $n \in \mathbb{N}$. Then*

$$r_2(n) = 4 \sum_{\substack{d \in \mathbb{N} \\ d \mid n}} \left(\frac{-4}{d}\right).$$

Proof. Choosing $(k_1, k_2) = (1, 1)$ in Theorem 8.2, we obtain

$$\sum_{(i, j, k) \in A(n)} (-1)^{(k-1)/2}(-2i + k)(i + j)$$

$$= \begin{cases} (-1)^m m \sum_{r=1}^{m} (-1)^r (2r - 1), & \text{if } n = m^2, \ m \in \mathbb{N}, \\ 0, & \text{otherwise.} \end{cases}$$

Now for every positive integer ℓ we have

$$\sum_{r=1}^{\ell} (-1)^r (2r - 1) = \sum_{r=1}^{\ell} ((-1)^r r - (-1)^{r-1}(r - 1)) = (-1)^\ell \ell.$$

Hence

$$\sum_{(i, j, k) \in A(n)} (-1)^{(k-1)/2}(-2i + k)(i + j) = ns(n), \quad n \in \mathbb{N}.$$

Expanding $(-2i + k)(i + j)$ and using $jk = n - i^2$, we deduce

$$\sum_{(i, j, k) \in A(n)} (-1)^{(k-1)/2}(-2ij + ik + n - 3i^2) = ns(n). \qquad (9.3)$$

By Theorem 8.4 we have

$$\sum_{(i, j, k) \in A(n)} (-1)^{(k-1)/2}(2ij - ik) = 0. \qquad (9.4)$$

Adding (9.3) and (9.4), we deduce

$$\sum_{(i, j, k) \in A(n)} (-1)^{(k-1)/2}(n - 3i^2) = ns(n). \qquad (9.5)$$

Thus

$$\sum_{\substack{i \in \mathbb{Z} \\ |i| < \sqrt{n}}} (n - 3i^2) \sum_{\substack{k \in \mathbb{N} \\ k \mid n - i^2 \\ k \equiv 1 \,(\text{mod } 2)}} (-1)^{(k-1)/2} = ns(n).$$

Let

$$D(s) := 4 \sum_{\substack{d \in \mathbb{N} \\ d \mid s \\ d \equiv 1 \,(\mathrm{mod}\ 2)}} (-1)^{(d-1)/2} = 4 \sum_{\substack{d \in \mathbb{N} \\ d \mid s}} \left(\frac{-4}{d} \right), \quad s \in \mathbb{N},$$

and $D(0) := 1$. Hence for all $n \in \mathbb{N}$ we have

$$\sum_{\substack{i \in \mathbb{Z} \\ |i| < \sqrt{n}}} (n - 3i^2) D(n - i^2) = 4ns(n).$$

If n is a perfect square then

$$\sum_{\substack{i \in \mathbb{Z} \\ |i| \le \sqrt{n}}} (n - 3i^2) D(n - i^2) = \sum_{\substack{i \in \mathbb{Z} \\ |i| < \sqrt{n}}} (n - 3i^2) D(n - i^2) + 2(-2n) D(0)$$

$$= 4n - 4n$$

$$= 0.$$

If n is not a perfect square then

$$\sum_{\substack{i \in \mathbb{Z} \\ |i| \le \sqrt{n}}} (n - 3i^2) D(n - i^2) = \sum_{\substack{i \in \mathbb{Z} \\ |i| < \sqrt{n}}} (n - 3i^2) D(n - i^2) = 0.$$

Hence, for all positive integers n, we have

$$\sum_{\substack{i \in \mathbb{Z} \\ |i| \le \sqrt{n}}} (n - 3i^2) D(n - i^2) = 0.$$

Thus, by Theorem 9.2 with $k = 2$, we deduce

$$r_2(n) = D(n), \quad n \in \mathbb{N}_0,$$

completing the proof of the theorem. $\qquad\qquad\square$

Example 9.1. We calculate the number of ways of expressing 306 as the sum of two squares. By Theorem 9.3 we have

$$
r_2(306) = 4 \sum_{\substack{d \in \mathbb{N} \\ d \mid 306}} \left(\frac{-4}{d}\right)
$$

$$
= 4 \sum_{\substack{d \in \mathbb{N} \\ d \mid 306 \\ d \text{ odd}}} \left(\frac{-4}{d}\right)
$$

$$
= 4 \sum_{\substack{d \in \mathbb{N} \\ d \mid 153}} \left(\frac{-4}{d}\right)
$$

$$
= 4 \left(\left(\frac{-4}{1}\right) + \left(\frac{-4}{3}\right) + \left(\frac{-4}{9}\right) + \left(\frac{-4}{17}\right) + \left(\frac{-4}{51}\right) + \left(\frac{-4}{153}\right) \right)
$$

$$
= 4(1 - 1 + 1 + 1 - 1 + 1)
$$

$$
= 8.
$$

The eight representations are given by

$$
306 = (\pm 9)^2 + (\pm 15)^2 = (\pm 15)^2 + (\pm 9)^2.
$$

From Theorem 9.3 we see that

$$
r_2(n) = 4(d_{1,4}(n) - d_{3,4}(n)).
$$

Appealing to (3.7) (with $m = 4$), we have

$$
d_{0,4}(n) = d(n/4).
$$

From (3.8) (with $m = 4$), we deduce

$$
d_{0,4}(n) + d_{1,4}(n) + d_{2,4}(n) + d_{3,4}(n) = d(n)
$$

and from (3.9) (with $e = k = 2$ and $m = 4$) and (3.10)

$$
d_{2,4}(n) = d_{1,2}(n/2) = d(n/2) - d(n/4).
$$

Solving these four equations for $d_{0,4}(n)$, $d_{1,4}(n)$, $d_{2,4}(n)$ and $d_{3,4}(n)$, we obtain the following theorem.

Theorem 9.4. *For $n \in \mathbb{N}$*

$$d_{0,4}(n) = d(n/4),$$

$$d_{1,4}(n) = \frac{1}{2}d(n) - \frac{1}{2}d(n/2) + \frac{1}{8}r_2(n),$$

$$d_{2,4}(n) = d(n/2) - d(n/4),$$

$$d_{3,4}(n) = \frac{1}{2}d(n) - \frac{1}{2}d(n/2) - \frac{1}{8}r_2(n).$$

Example 9.2. We give a simple upper bound for $r_2(n)$. As $d_{3,4}(n) \geq 0$ we see from Theorem 9.4 that

$$r_2(n) \leq 4d(n) - 4d(n/2)$$

for all $n \in \mathbb{N}$. Moreover the upper bound is achieved if and only if the only possible primes dividing n are 2 and primes congruent to 1 modulo 4.

Next we deduce from Theorems 8.3, 8.4, 8.5 and 9.2 Jacobi's formula for the number of representations of $n \in \mathbb{N}$ as the sum of four squares, that is for the quantity

$$r_4(n) = \operatorname{card}\{(x, y, z, t) \in \mathbb{Z}^4 \mid n = x^2 + y^2 + z^2 + t^2\}.$$

Theorem 9.5. *For $n \in \mathbb{N}$*

$$r_4(n) = 8\sigma(n) - 32\sigma(n/4).$$

Proof. First we note that for $m \in \mathbb{N}$ we have

$$\sum_{r=1}^{2m-1}(-1)^r r = \sum_{\substack{r=1 \\ r \equiv 0 \,(\mathrm{mod}\, 2)}}^{2m-1}(-1)^r r + \sum_{\substack{r=1 \\ r \equiv 1 \,(\mathrm{mod}\, 2)}}^{2m-1}(-1)^r r$$

$$= \sum_{s=1}^{m-1} 2s - \sum_{s=1}^{m}(2s - 1)$$

$$= m(m - 1) - m^2$$

$$= -m,$$

and

$$\sum_{r=-m+1}^{m-1}(-1)^r r^2 = 2\sum_{r=1}^{m-1}(-1)^r r^2$$

$$= \sum_{r=1}^{m-1}\left((-1)^r r(r + 1) - (-1)^{r-1}(r - 1)r\right)$$

$$= (-1)^{m-1}(m - 1)m,$$

so that

$$2m^2 \sum_{r=1}^{2m-1} (-1)^r r - (-1)^m 2m \sum_{r=-m+1}^{m-1} (-1)^r r^2 = -2m^3 + 2m^2(m-1) = -2m^2.$$

Now we take $(k_1, k_2) = (1, 2)$ in Theorem 8.3 to obtain

$$\sum_{\substack{(i,j,k) \in \mathbb{Z} \times \mathbb{N} \times \mathbb{N} \\ i^2 + jk = n}} s(j,k)(2(-2i+k)(i+j)^2 - (j+k)i^2)$$

$$= \begin{cases} 2m^2 \displaystyle\sum_{r=1}^{2m-1} (-1)^r r - (-1)^m 2m \displaystyle\sum_{r=-m+1}^{m-1} (-1)^r r^2, & \text{if } n = m^2, \ m \in \mathbb{N}, \\ 0, & \text{otherwise,} \end{cases}$$

$$= \begin{cases} -2n, & \text{if } n = m^2, \ m \in \mathbb{N}, \\ 0, & \text{otherwise.} \end{cases}$$

The left hand side is

$$\sum_{\substack{(i,j,k) \in \mathbb{Z} \times \mathbb{N} \times \mathbb{N} \\ i^2 + jk = n}} s(j,k)(-4i^3 + i^2 k - 9i^2 j + 4ijk - 4ij^2 + 2j^2 k)$$

$$= \sum_{\substack{(i,j,k) \in \mathbb{Z} \times \mathbb{N} \times \mathbb{N} \\ i^2 + jk = n}} s(j,k)(i^2 k - 9i^2 j + 2j^2 k) \quad \text{(by Theorem 8.4)}$$

$$= \sum_{\substack{(i,j,k) \in \mathbb{Z} \times \mathbb{N} \times \mathbb{N} \\ i^2 + jk = n}} s(j,k)(i^2 k - 9i^2 j + 2j(n - i^2))$$

$$= \sum_{\substack{(i,j,k) \in \mathbb{Z} \times \mathbb{N} \times \mathbb{N} \\ i^2 + jk = n}} s(j,k)(2(n - 5i^2)j - i^2(j-k))$$

$$= 2 \sum_{\substack{(i,j,k) \in \mathbb{Z} \times \mathbb{N} \times \mathbb{N} \\ i^2 + jk = n}} s(j,k)(n - 5i^2)j \quad \text{(by Theorem 8.5)}$$

$$= 2 \sum_{\substack{i \in \mathbb{Z} \\ |i| < \sqrt{n}}} (n - 5i^2) \sum_{\substack{(j,k) \in \mathbb{N}^2 \\ jk = n - i^2}} s(j,k)j.$$

Thus

$$\sum_{\substack{i \in \mathbb{Z} \\ |i| < \sqrt{n}}} (n - 5i^2) \sum_{\substack{(j,k) \in \mathbb{N}^2 \\ jk = n - i^2}} s(j,k)j = -ns(n),$$

for all $n \in \mathbb{N}$. Now define $D : \mathbb{N}_0 \to \mathbb{Z}$ by

$$D(0) := 1$$

and

$$D(n) := -8 \sum_{\substack{(j, k) \in \mathbb{N}^2 \\ jk = n}} s(j, k)j, \quad n \in \mathbb{N},$$

so that

$$\sum_{\substack{i \in \mathbb{Z} \\ |i| < \sqrt{n}}} (n - 5i^2)D(n - i^2) = 8ns(n), \quad n \in \mathbb{N}.$$

If n is a square, say $n = m^2$, $m \in \mathbb{N}$, then

$$\sum_{\substack{i \in \mathbb{Z} \\ |i| \leq \sqrt{n}}} (n - 5i^2)D(n - i^2) = \sum_{\substack{i \in \mathbb{Z} \\ |i| < \sqrt{n}}} (n - 5i^2)D(n - i^2) - 8nD(0)$$

$$= 8n - 8n = 0.$$

If n is not a square then

$$\sum_{\substack{i \in \mathbb{Z} \\ |i| \leq \sqrt{n}}} (n - 5i^2)D(n - i^2) = \sum_{\substack{i \in \mathbb{Z} \\ |i| < \sqrt{n}}} (n - 5i^2)D(n - i^2) = 0.$$

Thus

$$\sum_{\substack{i \in \mathbb{Z} \\ |i| \leq \sqrt{n}}} (n - 5i^2)D(n - i^2) = 0$$

for all $n \in \mathbb{N}$. By Theorem 9.2 with $k = 4$, we deduce that

$$r_4(n) = D(n), \quad n \in \mathbb{N}_0.$$

Finally, for $n \in \mathbb{N}$, we have

$$\sum_{\substack{(j,k) \in \mathbb{N}^2 \\ jk = n}} s(j,k)j = 2 \sum_{\substack{(j,k) \in \mathbb{N}^2 \\ jk = n \\ j \equiv k \equiv 0 \,(\mathrm{mod}\,2)}} j - \sum_{\substack{(j,k) \in \mathbb{N}^2 \\ jk = n}} j$$

$$= 4 \sum_{\substack{(j,k) \in \mathbb{N}^2 \\ jk = n/4}} j - \sum_{\substack{(j,k) \in \mathbb{N}^2 \\ jk = n}} j$$

$$= 4 \sum_{\substack{j \in \mathbb{N} \\ j \mid n/4}} j - \sum_{\substack{j \in \mathbb{N} \\ j \mid n}} j$$

$$= 4\sigma(n/4) - \sigma(n)$$

so that

$$r_4(n) = -8(4\sigma(n/4) - \sigma(n)) = 8\sigma(n) - 32\sigma(n/4),$$

which is Jacobi's formula for $r_4(n)$.   □

Another proof of Jacobi's four squares theorem is given in Chapter 11.

Example 9.3. By Theorem 9.5 the integer 8 has

$$r_4(8) = 8\sigma(8) - 32\sigma(2) = 8 \cdot 15 - 32 \cdot 3 = 120 - 96 = 24$$

representations as the sum of four squares.

Our next objective is to use Theorems 8.2, 8.4 and 9.2 to prove Jacobi's formula for

$$r_6(n) = \mathrm{card}\{(x_1, x_2, x_3, x_4, x_5, x_6) \in \mathbb{Z}^6 \mid n = x_1^2 + x_2^2 + x_3^2 + x_4^2 + x_5^2 + x_6^2\}.$$

Theorem 9.6. *Let $n \in \mathbb{N}$. Then*

$$r_6(n) = 16 \sum_{\substack{d \in \mathbb{N} \\ d \mid n}} \left(\frac{-4}{n/d}\right) d^2 - 4 \sum_{\substack{d \in \mathbb{N} \\ d \mid n}} \left(\frac{-4}{d}\right) d^2.$$

Proof. Let $u(r) := 4r^3 - 3r$ so that $u(0) = 0$. We note that $u(r) + u(r - 1) = (2r - 1)^3$. Hence for $m \in \mathbb{N}$ we have

$$\sum_{r=1}^{m} (-1)^r (2r - 1)^3 = \sum_{r=1}^{m} ((-1)^r u(r) - (-1)^{r-1} u(r - 1))$$

$$= (-1)^m u(m) = (-1)^m (4m^3 - 3m).$$

Now take $(k_1, k_2) = (3, 1)$ in Theorem 8.2. We obtain

$$\sum_{(i,j,k)\in A(n)} (-1)^{(k-1)/2}(-2i+k)^3(i+j)$$

$$= \begin{cases} (-1)^m m \sum_{r=1}^{m}(-1)^r(2r-1)^3, & \text{if } n = m^2, m \in \mathbb{N}, \\ 0, & \text{otherwise}, \end{cases}$$

$$= (4n^2 - 3n)s(n).$$

Expanding $(-2i+k)^3(i+j)$ we obtain

$$-8i^4 + 12i^3k - 6i^2k^2 + ik^3 - 8i^3j + 12i^2jk - 6ijk^2 + jk^3.$$

By Theorem 8.4 we see that

$$\sum_{(i,j,k)\in A(n)} (-1)^{(k-1)/2}(12i^3k + ik^3 - 8i^3j - 6ijk) = 0.$$

Thus

$$\sum_{(i,j,k)\in A(n)} (-1)^{(k-1)/2}(-8i^4 - 6i^2k^2 + 12i^2jk + jk^3) = (4n^2 - 3n)s(n).$$

$$(9.6)$$

Similarly by taking $(k_1, k_2) = (1, 3)$ in Theorem 8.2 we deduce

$$\sum_{(i,j,k)\in A(n)} (-1)^{(k-1)/2}(-2i^4 - 6i^2j^2 + 3i^2jk + j^3k) = n^2s(n). \qquad (9.7)$$

Multiplying (9.7) by 4, and then subtracting (9.6) from it, we obtain

$$\sum_{(i,j,k)\in A(n)} (-1)^{(k-1)/2}(-24i^2j^2 + 6i^2k^2 - jk^3 + 4j^3k) = 3ns(n).$$

For $(i, j, k) \in A(n)$ we have

$$-24i^2j^2 + 6i^2k^2 - jk^3 + 4j^3k = (jk - 6i^2)(4j^2 - k^2) = (n - 7i^2)(4j^2 - k^2).$$

Thus

$$\sum_{\substack{i \in \mathbb{Z} \\ |i| < \sqrt{n}}} (n - 7i^2) \sum_{\substack{j, k \in \mathbb{N} \\ jk = n - i^2 \\ k \equiv 1 \pmod 2}} (-1)^{(k-1)/2}(4j^2 - k^2) = 3ns(n).$$

Let

$$D(n) := 4 \sum_{\substack{(j, k) \in \mathbb{N}^2 \\ jk = n \\ k \equiv 1 \pmod 2}} (-1)^{(k-1)/2}(4j^2 - k^2), \quad n \in \mathbb{N},$$

and

$$D(0) := 1.$$

Hence

$$\sum_{\substack{i \in \mathbb{Z} \\ |i| < \sqrt{n}}} (n - 7i^2)D(n - i^2) = 12ns(n), \quad n \in \mathbb{N}.$$

If n is a perfect square then

$$\sum_{\substack{i \in \mathbb{Z} \\ |i| \le \sqrt{n}}} (n - 7i^2)D(n - i^2) = \sum_{\substack{i \in \mathbb{Z} \\ |i| < \sqrt{n}}} (n - 7i^2)D(n - i^2) - 12nD(0)$$

$$= 12n - 12n = 0.$$

If n is not a perfect square then

$$\sum_{\substack{i \in \mathbb{Z} \\ |i| \le \sqrt{n}}} (n - 7i^2)D(n - i^2) = \sum_{\substack{i \in \mathbb{Z} \\ |i| < \sqrt{n}}} (n - 7i^2)D(n - i^2) = 0.$$

Thus for all $n \in \mathbb{N}$ we have

$$\sum_{\substack{i \in \mathbb{Z} \\ |i| \le \sqrt{n}}} (n - 7i^2)D(n - i^2) = 0.$$

Thus, by Theorem 9.2 with $k = 6$, we have

$$r_6(n) = D(n), \quad n \in \mathbb{N}_0.$$

Finally we observe for $n \in \mathbb{N}$ that

$$D(n) = 4 \sum_{\substack{(j, k) \in \mathbb{N}^2 \\ jk = n}} \left(\frac{-4}{k}\right)(4j^2 - k^2)$$

$$= 16 \sum_{\substack{(j, k) \in \mathbb{N}^2 \\ jk = n}} \left(\frac{-4}{k}\right)j^2 - 4 \sum_{\substack{(j, k) \in \mathbb{N}^2 \\ jk = n}} \left(\frac{-4}{k}\right)k^2$$

$$= 16 \sum_{\substack{j \in \mathbb{N} \\ j \mid n}} \left(\frac{-4}{n/j}\right)j^2 - 4 \sum_{\substack{k \in \mathbb{N} \\ k \mid n}} \left(\frac{-4}{k}\right)k^2$$

$$= 16 \sum_{\substack{d \in \mathbb{N} \\ d \mid n}} \left(\frac{-4}{n/d}\right)d^2 - 4 \sum_{\substack{d \in \mathbb{N} \\ d \mid n}} \left(\frac{-4}{d}\right)d^2.$$

This gives the asserted result. $\qquad \square$

Example 9.4. The number of representations of 6 as a sum of six squares is

$$r_6(6) = 16\left(\left(\frac{-4}{6}\right)1^2 + \left(\frac{-4}{3}\right)2^2 + \left(\frac{-4}{2}\right)3^2 + \left(\frac{-4}{1}\right)6^2\right)$$

$$\qquad - 4\left(\left(\frac{-4}{1}\right)1^2 + \left(\frac{-4}{2}\right)2^2 + \left(\frac{-4}{3}\right)3^2 + \left(\frac{-4}{6}\right)6^2\right)$$

$$= 16(-4 + 36) - 4(1 - 9)$$

$$= 544.$$

This is easily checked as

$$6 = 1^2 + 1^2 + 2^2 + 0^2 + 0^2 + 0^2$$

gives rise to

$$\frac{6!}{2!3!}2^3 = 480$$

representations and

$$6 = 1^2 + 1^2 + 1^2 + 1^2 + 1^2 + 1^2$$

gives rise to $2^6 = 64$ representations, for a total of $480 + 64 = 544$.

Let $k, n \in \mathbb{N}$. If $x_1, \ldots, x_k$ are integers such that

$$n = x_1^2 + \cdots + x_k^2, \quad \gcd(x_1, \ldots, x_k) = 1,$$

then $(x_1, \ldots, x_k) \in \mathbb{Z}^k$ is called a primitive representation of n as the sum of k squares. The number of primitive representations of n as the sum of k squares is denoted by $p_k(n)$, that is

$$p_k(n) := \text{card}\left\{(x_1, \ldots, x_k) \in \mathbb{Z}^k \mid n = x_1^2 + \cdots + x_k^2, \, \gcd(x_1, \ldots, x_k) = 1\right\}.$$

Theorem 9.7. *Let* $k, n \in \mathbb{N}$. *Then*

$$r_k(n) = \sum_{\substack{d \in \mathbb{N} \\ d^2 \mid n}} p_k(n/d^2).$$

Proof. If $n = x_1^2 + \cdots + x_k^2$ and $d \mid x_i$ for $i = 1, 2, \ldots, k$ then $d^2 \mid n$ and $x_i = dy_i$, $y_i \in \mathbb{Z}$ $(i = 1, 2, \ldots, k)$ with $n/d^2 = y_1^2 + \cdots + y_k^2$. Hence

$$r_k(n) = \sum_{\substack{(x_1, \ldots, x_k) \in \mathbb{Z}^k \\ n = x_1^2 + \cdots + x_k^2}} 1$$

$$= \sum_{\substack{d \in \mathbb{N} \\ d^2 \mid n}} \sum_{\substack{(x_1, \ldots, x_k) \in \mathbb{Z}^k \\ n = x_1^2 + \cdots + x_k^2 \\ \gcd(x_1, \ldots, x_k) = d}} 1$$

$$= \sum_{\substack{d \in \mathbb{N} \\ d^2 \mid n}} \sum_{\substack{(y_1, \ldots, y_k) \in \mathbb{Z}^k \\ n/d^2 = y_1^2 + \cdots + y_k^2 \\ \gcd(y_1, \ldots, y_k) = 1}} 1$$

$$= \sum_{\substack{d \in \mathbb{N} \\ d^2 \mid n}} p_k(n/d^2),$$

as asserted.   □

In order to determine $p_k(n)$ from $r_k(n)$ we are going to invert the formula of Theorem 9.7. To do this we make use of the Möbius function. We recall that the Möbius function $\mu(n)$ is defined for $n \in \mathbb{N}$ by

$$\mu(n) = \begin{cases} 1, & \text{if } n = 1, \\ (-1)^k, & \text{if } n = p_1 p_2 \cdots p_k, \text{ where } p_1, \ldots, p_k \text{ are distinct primes,} \\ 0, & \text{otherwise.} \end{cases}$$

It is easy to check that $\mu(n)$ is a multiplicative function of n.

Let $m \in \mathbb{N}$. The positive integer n is said to be m-free if there does not exist a prime p with $p^m \mid n$. The only positive integer which is 1-free is 1. The 2-free integers are precisely the squarefree integers.

Theorem 9.8. *Let $m, n \in \mathbb{N}$. Then*

$$\sum_{\substack{d \in \mathbb{N} \\ d^m \mid n}} \mu(d) = \begin{cases} 1, & \text{if } n \text{ is } m\text{-free,} \\ 0, & \text{otherwise.} \end{cases}$$

Proof. Set

$$G(n) := \sum_{\substack{d \in \mathbb{N} \\ d^m \mid n}} \mu(d), \quad n \in \mathbb{N}.$$

Clearly $G(1) = 1$ so the theorem is true for $n = 1$. Thus we may suppose that $n > 1$. Then there exists at least one prime p dividing n. Let $a(p) \in \mathbb{N}$ be such that $p^{a(p)} \| n$. We have

$$G(p^a) = \sum_{\substack{d \in \mathbb{N} \\ d^m \mid p^{a(p)}}} \mu(d) = \sum_{\substack{x \in \mathbb{N}_0 \\ mx \le a(p)}} \mu(p^x) = \sum_{\substack{x \in \mathbb{N}_0 \\ x \le a(p)/m \\ x = 0, 1}} (-1)^x$$

$$= \begin{cases} 1 + (-1) = 0, & \text{if } a(p) \ge m, \\ 1, & \text{if } a(p) < m. \end{cases}$$

As μ is multiplicative it is easy to show that G is also multiplicative. Hence

$$G(n) = \prod_{\substack{p \text{ (prime)} \mid n \\ p^{a(p)} \| n}} G(p^{a(p)}) = \prod_{\substack{p \text{ (prime)} \mid n \\ p^{a(p)} \| n \\ a(p) \ge m}} 0 \prod_{\substack{p \text{ (prime)} \mid n \\ p^{a(p)} \| n \\ a(p) < m}} 1.$$

Thus $G(n) = 1$ if n is m-free and $G(n) = 0$ otherwise. $\qquad\square$

Taking $m = 1$ and $m = 2$ in Theorem 9.8 we obtain

$$\sum_{\substack{d \in \mathbb{N} \\ d \mid n}} \mu(d) = \begin{cases} 1, & \text{if } n = 1, \\ 0, & \text{if } n > 1, \end{cases} \tag{9.8}$$

and

$$\sum_{\substack{d \in \mathbb{N} \\ d^2 \mid n}} \mu(d) = \begin{cases} 1, & \text{if } n \text{ is squarefree}, \\ 0, & \text{otherwise}. \end{cases} \tag{9.9}$$

Property (9.8) is a fundamental property of the Möbius function and we use it to prove the following inversion formula, which will enable us to invert the formula of Theorem 9.7.

Theorem 9.9. *Let $m \in \mathbb{N}$. Let f be an arithmetic function. Define the arithmetic function F by*

$$F(n) = \sum_{\substack{d \in \mathbb{N} \\ d^m \mid n}} f(n/d^m), \quad n \in \mathbb{N}.$$

Then

$$f(n) = \sum_{\substack{d \in \mathbb{N} \\ d^m \mid n}} \mu(d) F(n/d^m), \quad n \in \mathbb{N}.$$

Proof. We have

$$
\sum_{\substack{d \in \mathbb{N} \\ d^m \mid n}} \mu(d) F(n/d^m) = \sum_{\substack{d \in \mathbb{N} \\ d^m \mid n}} \mu(d) \sum_{\substack{e \in \mathbb{N} \\ e^m \mid n/d^m}} f(n/(d^m e^m))
$$

$$
= \sum_{\substack{(d, e) \in \mathbb{N}^2 \\ d^m e^m \mid n}} \mu(d) f(n/(d^m e^m))
$$

$$
= \sum_{\substack{k \in \mathbb{N} \\ k^m \mid n}} f(n/k^m) \sum_{\substack{(d, e) \in \mathbb{N}^2 \\ de = k}} \mu(d)
$$

$$
= \sum_{\substack{k \in \mathbb{N} \\ k^m \mid n}} f(n/k^m) \sum_{\substack{d \in \mathbb{N} \\ d \mid k}} \mu(d)
$$

$$
= \sum_{\substack{k \in \mathbb{N} \\ k^m \mid n \\ k = 1}} f(n/k^m) \quad \text{(by (9.8))}
$$

$$
= f(n),
$$

which is the assertion of the theorem. $\qquad\square$

We now use Theorem 9.9 to determine p_k in terms of r_k.

Theorem 9.10. *Let* $k, n \in \mathbb{N}$. *Then*

$$
p_k(n) = \sum_{\substack{d \in \mathbb{N} \\ d^2 \mid n}} \mu(d) r_k(n/d^2).
$$

Proof. We take $m = 2$ and $f(n) = p_k(n)$ in Theorem 9.9. By Theorem 9.7 we have $F(n) = r_k(n)$. The required result now follows by Theorem 9.9. $\qquad\square$

Next we use (9.9) and Theorem 9.10 to determine $p_2(n)$ ($n \in \mathbb{N}$).

Theorem 9.11. *Let* $n \in \mathbb{N}$. *Let* t *denote the number of distinct primes* $p \equiv 1 \pmod 4$ *dividing* n. *Then*

$$
p_2(n) = \begin{cases} 0, & \text{if } 4 \mid n \text{ or there exists a prime } q \equiv 3 \pmod 4 \text{ dividing } n, \\ 2^{t+2}, & \text{otherwise.} \end{cases}
$$

Proof. If $4 \mid n$ and $n = x^2 + y^2$ then $2 \mid x$ and $2 \mid y$ so $p_2(n) = 0$.

If q is a prime with $q \equiv 3 \pmod 4$ dividing $n = x^2 + y^2$ then $q \mid x$ and $q \mid y$ so $p_2(n) = 0$.

Hence we may assume that $n = 2^a p_1{}^{a_1} \cdots p_t{}^{a_t}$, where $a = 0, 1$; $t \in \mathbb{N}_0$; $p_1, \ldots, p_t$ are distinct primes $\equiv 1 \pmod 4$, and $a_1, \ldots, a_t \in \mathbb{N}$. Appealing to

Theorems 9.3 and 9.10, we obtain

$$p_2(n) = \sum_{\substack{d \in \mathbb{N} \\ d^2 \mid n}} \mu(d) r_2(n/d^2) = 4 \sum_{\substack{d \in \mathbb{N} \\ d^2 \mid n}} \mu(d) \sum_{\substack{e \in \mathbb{N} \\ e \mid n/d^2}} \left(\frac{-4}{e} \right)$$

$$= 4 \sum_{\substack{(d, e) \in \mathbb{N}^2 \\ d^2 e \mid n}} \mu(d) \left(\frac{-4}{e} \right) = 4 \sum_{\substack{k \in \mathbb{N} \\ k \mid n}} \sum_{\substack{d \in \mathbb{N} \\ d^2 \mid k}} \mu(d) \left(\frac{-4}{k/d^2} \right).$$

As $4 \nmid n$ we have $4 \nmid k$ so d is odd. Thus $\left(\dfrac{-4}{d^2} \right) = 1$. Hence by (9.9) we obtain

$$p_2(n) = 4 \sum_{\substack{k \in \mathbb{N} \\ k \mid n}} \left(\frac{-4}{k} \right) \sum_{\substack{d \in \mathbb{N} \\ d^2 \mid k}} \mu(d) = 4 \sum_{\substack{k \in \mathbb{N} \\ k \mid n \\ k \text{ squarefree}}} \left(\frac{-4}{k} \right) = 4 \sum_{\substack{k \in \mathbb{N} \\ k \mid n \\ k \text{ squarefree} \\ k \text{ odd}}} \left(\frac{-4}{k} \right).$$

As $\left(\dfrac{-4}{p} \right) = 1$ for every odd prime p dividing n, we have

$$\sum_{\substack{k \in \mathbb{N} \\ k \mid n \\ k \text{ squarefree} \\ k \text{ odd}}} \left(\frac{-4}{k} \right) = \sum_{\substack{k \in \mathbb{N} \\ k \mid n \\ k \text{ squarefree} \\ \text{odd}}} 1 = 2^t.$$

Thus $p_2(n) = 4 \cdot 2^t = 2^{t+2}$, as asserted. $\qquad\square$

Example 9.5. We choose $n = 850 = 2 \cdot 5^2 \cdot 17$ so that n is not divisible by 4 or a prime $\equiv 3 \pmod 4$ and $t = 2$. Thus, by Theorem 9.11, the number of primitive representations of 850 as the sum of two squares is $2^{t+2} = 2^4 = 16$. Eight of these representations arise from $850 = 3^2 + 29^2$ by change of order and sign, and eight from $850 = 11^2 + 27^2$, for a total of $8 + 8 = 16$. By Theorem 9.3 the total number of representations of 850 as a sum of two squares is

$$4 \sum_{\substack{d \in \mathbb{N} \\ d \mid 850}} \left(\frac{-4}{d} \right) = 4 \sum_{\substack{d \in \mathbb{N} \\ d \mid 425}} \left(\frac{-4}{d} \right) = 4 \sum_{\substack{d \in \mathbb{N} \\ d \mid 425}} 1 = 4 \times 6 = 24.$$

The eight nonprimitive representations arise from $850 = 15^2 + 25^2$.

Our final result of this chapter is a proof of Aubry's formula for $p_4(n)$.

Theorem 9.12. *Let $n \in \mathbb{N}$. Set $n = 2^\alpha N$, where $\alpha \in \mathbb{N}_0$, $N \in \mathbb{N}$ and $N \equiv 1 \pmod 2$. Define*

$$
k(\alpha) = \begin{cases}
1, & \text{if } \alpha = 0, \\
3, & \text{if } \alpha = 1, \\
2, & \text{if } \alpha = 2, \\
0, & \text{if } \alpha \geq 3.
\end{cases}
$$

Then

$$
p_4(n) = 8k(\alpha) \prod_{\substack{p(\text{prime}) \mid N \\ p^\beta \parallel N}} p^{\beta-1}(p+1).
$$

Proof. From Theorem 9.5 we have

$$
r_4(m) = 8 \sum_{\substack{d \in \mathbb{N} \\ d \mid m \\ 4 \nmid d}} d, \quad m \in \mathbb{N}.
$$

Hence, by Theorem 9.10 (with $k = 4$), we obtain

$$
p_4(n) = \sum_{\substack{d \in \mathbb{N} \\ d^2 \mid n}} \mu(d) r_4(n/d^2)
$$

$$
= 8 \sum_{\substack{d \in \mathbb{N} \\ d^2 \mid n}} \mu(d) \sum_{\substack{e \in \mathbb{N} \\ e \mid n/d^2 \\ 4 \nmid e}} e
$$

$$
= 8 \sum_{\substack{(d,e) \in \mathbb{N}^2 \\ d^2 e \mid n \\ 4 \nmid e}} \mu(d) e
$$

$$
= 8 \sum_{\substack{k \in \mathbb{N} \\ k \mid n}} \sum_{\substack{(d,e) \in \mathbb{N}^2 \\ d^2 e = k \\ 4 \nmid e}} \mu(d) e
$$

$$
= 8 \sum_{\substack{k \in \mathbb{N} \\ k \mid n}} k \sum_{\substack{d \in \mathbb{N} \\ d^2 \mid k \\ 4 \nmid k/d^2}} \frac{\mu(d)}{d^2},
$$

that is

$$p_4(n) = 8 \sum_{\substack{k \in \mathbb{N} \\ k \mid n}} k H(k), \tag{9.10}$$

where

$$H(k) := \sum_{\substack{d \in \mathbb{N} \\ d^2 \mid k \\ 4 \nmid k/d^2}} \frac{\mu(d)}{d^2}, \quad k \in \mathbb{N}.$$

Clearly $H(1) = 1$. Let $k_1, k_2 \in \mathbb{N}$ be such that $\gcd(k_1, k_2) = 1$. Then

$$H(k_1 k_2) = \sum_{\substack{d \in \mathbb{N} \\ d^2 \mid k_1 k_2 \\ 4 \nmid k_1 k_2/d^2}} \frac{\mu(d)}{d^2}$$

$$= \sum_{\substack{(d_1, d_2) \in \mathbb{N}^2 \\ d_1^2 \mid k_1, \, d_2^2 \mid k_2 \\ 4 \nmid k_1/d^2, \, 4 \nmid k_2/d^2}} \frac{\mu(d_1 d_2)}{d_1^2 d_2^2}$$

$$= \sum_{\substack{d_1 \in \mathbb{N} \\ d_1^2 \mid k_1 \\ 4 \nmid k_1/d_1^2}} \frac{\mu(d_1)}{d_1^2} \sum_{\substack{d_2 \in \mathbb{N} \\ d_2^2 \mid k_2 \\ 4 \nmid k_2/d_2^2}} \frac{\mu(d_2)}{d_2^2}$$

$$= H(k_1) H(k_2).$$

Thus H is a multiplicative arithmetic function. First we determine $H(p^a)$ for an odd prime p and $a \in \mathbb{N}_0$. We have

$$H(p^a) = \sum_{\substack{d \in \mathbb{N} \\ d^2 \mid p^a}} \frac{\mu(d)}{d^2} = \begin{cases} 1, & \text{if } a = 0, 1, \\ 1 - \dfrac{1}{p^2}, & \text{if } a \geq 2. \end{cases}$$

Next we determine $H(2^a)$ for $a \in \mathbb{N}_0$. We have

$$H(2^a) = \sum_{\substack{d \in \mathbb{N} \\ d^2 \mid 2^a \\ 4 \nmid 2^a/d^2}} \frac{\mu(d)}{d^2} = \sum_{\substack{b = 0 \\ (a-1)/2 \leq b \leq a/2}}^{1} \frac{(-1)^b}{2^{2b}} = m(a),$$

where

$$m(a) := \begin{cases} 1, & \text{if } a = 0, 1, \\ -\frac{1}{4}, & \text{if } a = 2, 3, \\ 0, & \text{if } a \geq 4. \end{cases}$$

Combining these results, we obtain as H is multiplicative

$$H(k) = m(a) \prod_{\substack{p \text{ (prime)} \neq 2 \\ p^2 \mid k}} \left(1 - \frac{1}{p^2}\right), \quad k \in \mathbb{N}, \quad 2^a \parallel k.$$

Putting this formula into (9.10), and splitting up the resulting sum depending on the power of 2 dividing k, we obtain

$$p_4(n) = 8 \sum_{\substack{k \mid n \\ 2 \nmid k}} k \prod_{p^2 \mid k} \left(1 - \frac{1}{p^2}\right) + 8 \sum_{\substack{k \mid n \\ 2 \parallel k}} k \prod_{p^2 \mid k} \left(1 - \frac{1}{p^2}\right)$$

$$- 2 \sum_{\substack{k \mid n \\ 2^2 \parallel k}} k \prod_{\substack{p^2 \mid k \\ p \neq 2}} \left(1 - \frac{1}{p^2}\right) - 2 \sum_{\substack{k \mid n \\ 2^3 \parallel k}} k \prod_{\substack{p^2 \mid k \\ p \neq 2}} \left(1 - \frac{1}{p^2}\right).$$

Set $n = 2^\alpha N$, where $\alpha \in \mathbb{N}_0$ and $N \in \mathbb{N}$ is odd, and let

$$J(N) := \sum_{\substack{k \in \mathbb{N} \\ k \mid N}} k \prod_{p^2 \mid k} \left(1 - \frac{1}{p^2}\right).$$

Clearly $J(1) = 1$. Then

$$p_4(n) = (8 + 16F_2(n) - 8F_4(n) - 16F_8(n))J(N) = 8k(\alpha)J(N). \quad (9.11)$$

It is easy to check that $J(N)$ is a multiplicative function of N. We determine $J(p^\beta)$ for an odd prime p and $\beta \in \mathbb{N}$. We have

$$J(p^\beta) = \sum_{\substack{k \in \mathbb{N} \\ k \mid p^\beta}} k \prod_{p^2 \mid k} \left(1 - \frac{1}{p^2}\right)$$

$$= 1 + p + \sum_{j=2}^{\beta} p^\beta \left(1 - \frac{1}{p^2}\right)$$

$$= p^{\beta-1}(p + 1).$$

Hence

$$J(N) = \prod_{\substack{p \text{ (prime)} \mid N \\ p^\beta \| N}} p^{\beta-1}(p+1). \tag{9.12}$$

The asserted result now follows from (9.11) and (9.12). ☐

Example 9.6. We choose $n = 2500 = 2^2 \cdot 5^4$. By Theorem 9.12 the number of primitive representations of 2500 as the sum of four squares is

$$p_4(2500) = 8 \cdot 2 \cdot 5^{4-1}(5+1) = 12000.$$

Exercises 9

1. Prove that when $n = p$ (prime) Theorem 9.3 agrees with the result stated at the beginning of Chapter 8.
2. Deduce from Theorem 9.3 that $r_2(2n) = r_2(n)$ for all $n \in \mathbb{N}$.
3. Prove that $\frac{1}{4}r_2(n)$ is a multiplicative function of $n \in \mathbb{N}$.
4. (i) Prove that $r_2(p^a) = 4(a+1)$ if p is a prime with $p \equiv 1 \pmod 4$ and $a \in \mathbb{N}_0$.
   (ii) Prove that $r_2(p^a) = 2 + 2(-1)^a$ if p is a prime with $p \equiv 3 \pmod 4$ and $a \in \mathbb{N}_0$.
   (iii) Prove that $r_2(2^a) = 4$ if $a \in \mathbb{N}_0$.
5. Let $a \in \mathbb{N}_0$. Determine all integers x and y such that $2^a = x^2 + y^2$.
6. Deduce from Problems 3 and 4 a formula for $r_2(n)$ ($n \in \mathbb{N}$) in terms of the prime power decomposition of n.
7. Let $n \in \mathbb{N}$. If $n = x^2 + y^2$ for some integers x and y, and p is a prime dividing n with $p \equiv 3 \pmod 4$, prove that p divides both x and y.
8. Deduce from Theorem 9.3 and Problem 9 of Exercises 6 that if n ($\in \mathbb{N}$) is odd then

$$r_2(n) = 2 \sum_{i=1}^{[\sqrt{n}/2]} r_2(n - 4i^2) + 4(-1)^{(n-1)/2}\sqrt{n}s(n).$$

9. How many representations does a prime have as a sum of four squares?
10. Prove that $\frac{1}{8}r_4(n)$ is a multiplicative function of n ($n \in \mathbb{N}$).
11. Is $\frac{1}{12}r_6(n)$ a multiplicative function of n ($n \in \mathbb{N}$)?
12. For $q \in \mathbb{Q}$ with $|q| < 1$ define

$$\varphi(q) := \sum_{n=-\infty}^{\infty} q^{n^2}.$$

We note that

$$\varphi(q) = \theta_3(0, q),$$

where the theta functions θ_1, θ_2, θ_3 and θ_4 were defined in Chapter 2. Let k be a positive integer with $k \geq 2$. Prove that

$$\varphi^k(q) = \sum_{n=0}^{\infty} r_k(n)q^n.$$

13. Use Problem 12 with $k = 2$ and Theorem 9.3 to show that

$$\varphi^2(q) = 1 + 4 \sum_{n=1}^{\infty} \left(\frac{-4}{n}\right) \frac{q^n}{1 - q^n}.$$

(A series of the form $\displaystyle\sum_{n=1}^{\infty} a_n \frac{q^n}{1 - q^n}$, where the a_n are independent of q, is is called a Lambert series.)

14. Use Problem 12 with $k = 4$ and Theorem 9.5 to show that

$$\varphi^4(q) = 1 + 8 \sum_{\substack{n=1 \\ 4 \nmid n}}^{\infty} \frac{nq^n}{1 - q^n}.$$

15. Use Problem 12 with $k = 6$ and Theorem 9.6 to prove that

$$\varphi^6(q) = 1 + 16 \sum_{n=1}^{\infty} \frac{n^2 q^n}{1 + q^{2n}} - 4 \sum_{\substack{n=1 \\ 2 \nmid n}}^{\infty} \frac{(-1)^{(n-1)/2} n^2 q^n}{1 - q^n}.$$

16. Let $n \in \mathbb{N}$ be such that $n \equiv 0 \pmod 8$. From Theorem 9.12 we see that $p_4(n) = 0$. Prove this result from first principles.

17. Use Theorems 9.6 and 9.10 to determine a formula for $p_6(n)$ valid for all positive integers n.

Notes on Chapter 9

Theorem 9.1 can be found in the books of Venkov [256, p. 205] and Nathanson [209, p. 424]. Theorem 9.2 is given in [209, p. 425]. Theorem 9.3 is implicit in the work of Jacobi [146, §§. 40–42], [148, Vol. I, pp. 159–170]. The elementary arithmetic proof of Theorem 9.3 that we give is based on the treatment given in Nathanson's book [209, pp. 428–429]. Hirschhorn [134] has deduced Theorem 9.3 from Jacobi's triple product identity. Theorem 9.5 is due to Jacobi and is given implicitly in [146, §§40–42], [148, Vol. I, pp. 159–170]. The proof given here is based on that given in [209, pp. 428–429]. A very simple arithmetic proof of Theorem 9.5 has been given by Spearman and Williams [246]. An arithmetic proof is given in Landau's book [162, pp. 146–150]. Many proofs of Theorem 9.5 are known; see for example Andrews, Ekhad and Zeilberger [31], Bhargava and Adiga [49], Carlitz [58], [59], Cooper and Lam [82], Hirschhorn [134], and Venkov [256, pp. 209–210]. Theorem 9.6 is also due to Jacobi [146, §§40–42], [148, Vol. I, pp. 159–170]. Other proofs have been given by Alaca, Alaca and Williams [16], Carlitz [58], Chan [65], Cooper and Lam [82], McAfee and Williams [198], and Venkov [256, pp. 206–207]. For the determination of $r_8(n)$ using Liouville's methods, see Nathanson [209, pp. 441–445] and Venkov [256, pp. 210–211] and for the determination of $r_{10}(n)$, see Nathanson [209, pp. 446–452] and Venkov [256, pp. 207–209].

Formula (9.8) is well known and can be found for example in the book of Niven, Zuckerman and Montgomery [211, p. 193]. Theorem 9.10 is a slight generalization of the Möbius inversion formula, see for example [211, p. 194]. A proof of Theorem 9.11, which shows that $p_2(n)/4$ is the number of solutions of the congruence $x^2 \equiv -1 \pmod{n}$ is given in Landau's book [162, p. 136]. Theorem 9.12 is due to Aubry [33], see also Dickson's *History* [93, Vol. II, pp. 302–303].

10

A Third Identity of Liouville

Let n be a positive integer and let $f : \mathbb{Z} \to \mathbb{C}$ be an even function. In 1858 Liouville stated the surprising result that the sum

$$\sum_{\substack{(a, b, x, y) \in \mathbb{N}^4 \\ ax + by = n}} (f(a - b) - f(a + b))$$

can be evaluated in terms of sums over the positive integers d dividing n.

Theorem 10.1. *Let n be a positive integer and let $f : \mathbb{Z} \to \mathbb{C}$ be an even function. Then*

$$\sum_{\substack{(a, b, x, y) \in \mathbb{N}^4 \\ ax + by = n}} (f(a - b) - f(a + b))$$

$$= f(0)(\sigma(n) - d(n)) + \sum_{\substack{d \in \mathbb{N} \\ d \mid n}} \left(1 + \frac{2n}{d} - d\right) f(d) - 2 \sum_{\substack{d \in \mathbb{N} \\ d \mid n}} \left(\sum_{v=1}^{d} f(v)\right).$$

Before proving Theorem 10.1 we make a few remarks about the set of solutions of $ax + by = n$ ($n \in \mathbb{N}$) in positive integers a, b, x and y, that is the set

$$B(n) := \{(a, b, x, y) \in \mathbb{N}^4 \mid ax + by = n\}. \tag{10.1}$$

We have

$$B(1) = \emptyset,$$
$$B(2) = \{(1, 1, 1, 1)\},$$
$$B(3) = \{(1, 1, 1, 2), (1, 1, 2, 1), (1, 2, 1, 1), (2, 1, 1, 1)\},$$
$$B(4) = \{(1, 1, 1, 3), (1, 1, 3, 1), (1, 3, 1, 1), (3, 1, 1, 1),$$
$$(1, 1, 2, 2), (1, 2, 2, 1), (2, 1, 1, 2), (2, 2, 1, 1)\}.$$

We observe from (10.1) that if $(a, b, x, y) \in B(n)$ then (a, y, x, b), (b, a, y, x), (b, x, y, a), (x, b, a, y), (x, y, a, b), (y, a, b, x) and (y, x, b, a) also belong to $B(n)$. It is clear that card $B(n)$ increases rapidly with n. A theorem of Ingham enables us to give some idea of the rate of growth of card $B(n)$. Ingham showed in 1927 that

$$\sum_{r=1}^{n-1} d(r)d(n-r) = \frac{6}{\pi^2} \sigma(n)(\log n)^2 \left\{ 1 + O\left(\frac{\log \log n}{\log n} \right) \right\},$$

as $n \to \infty$. Hence

$$\text{card } B(n) = \sum_{\substack{(a, b, x, y) \in \mathbb{N}^4 \\ ax + by = n}} 1$$

$$= \sum_{\substack{(r, s) \in \mathbb{N}^2 \\ r + s = n}} \sum_{\substack{(a, x) \in \mathbb{N}^2 \\ ax = r}} \sum_{\substack{(b, y) \in \mathbb{N}^2 \\ by = s}} 1$$

$$= \sum_{r=1}^{n-1} \left(\sum_{\substack{a \in \mathbb{N} \\ a \mid r}} 1 \right) \left(\sum_{\substack{b \in \mathbb{N} \\ b \mid n-r}} 1 \right)$$

$$= \sum_{r=1}^{n-1} d(r)d(n-r)$$

$$= \frac{6}{\pi^2} \sigma(n)(\log n)^2 \left\{ 1 + O\left(\frac{\log \log n}{\log n} \right) \right\},$$

as $n \to \infty$. Now for all $n \in \mathbb{N}$ we have

$$\sigma(n) = \sum_{\substack{d \in \mathbb{N} \\ d \mid n}} d \geq n$$

and

$$\sigma(n) = \sum_{\substack{d \in \mathbb{N} \\ d \mid n}} \frac{n}{d} = n \sum_{\substack{d \in \mathbb{N} \\ d \mid n}} \frac{1}{d} \le n \sum_{d=1}^{n} \frac{1}{d} \le n(\log n + 1)$$

so that for large n there are positive constants C_1 and C_2 such that

$$C_1 n(\log n)^2 \le \text{card } B(n) \le C_2 n(\log n)^3.$$

The arithmetic property of the set $B(n)$ that we use to prove Theorem 10.1 is given in Theorem 10.2.

Theorem 10.2. *Let n and k be positive integers. Then*

$$2 \sum_{\substack{(a,b,x,y) \in B(n) \\ a-b=k}} 1 - \sum_{\substack{(a,b,x,y) \in B(n) \\ a+b=k}} 1 = \left(1 - k + \frac{2n}{k}\right) F_k(n) - 2 \sum_{\substack{v \in \mathbb{N} \\ v \mid n \\ v \ge k}} 1.$$

Proof. For $k \in \mathbb{N}$ and $n \in \mathbb{N}$ we define the following six sums:

$$A_1 := \sum_{\substack{(b,x,y) \in \mathbb{N}^3 \\ kx+by=n \\ x<y}} 1, \quad A_2 := \sum_{\substack{(b,x,y) \in \mathbb{N}^3 \\ kx+by=n \\ x=y}} 1, \quad A_3 := \sum_{\substack{(b,x,y) \in \mathbb{N}^3 \\ kx+by=n \\ x>y}} 1,$$

$$B_1 := \sum_{\substack{(b,x,y) \in \mathbb{N}^3 \\ kx+by=n \\ b<k}} 1, \quad B_2 := \sum_{\substack{(b,x,y) \in \mathbb{N}^3 \\ kx+by=n \\ b=k}} 1, \quad B_3 := \sum_{\substack{(b,x,y) \in \mathbb{N}^3 \\ kx+by=n \\ b>k}} 1.$$

Clearly

$$A_1 + A_2 + A_3 = \sum_{\substack{(b,x,y) \in \mathbb{N}^3 \\ kx+by=n}} 1$$

and

$$B_1 + B_2 + B_3 = \sum_{\substack{(b,x,y) \in \mathbb{N}^3 \\ kx+by=n}} 1$$

so that

$$A_1 + A_2 + A_3 = B_1 + B_2 + B_3. \tag{10.2}$$

Next we show that the sums A_3 and B_3 are equal. We have

$$A_3 = \sum_{\substack{(b, x, y) \in \mathbb{N}^3 \\ kx + by = n \\ x > y}} 1 = \sum_{\substack{(b, x', y) \in \mathbb{N}^3 \\ k(x' + y) + by = n}} 1$$

$$= \sum_{\substack{(b, x, y) \in \mathbb{N}^3 \\ kx + (k + b)y = n}} 1 = \sum_{\substack{(b', x, y) \in \mathbb{N}^3 \\ kx + b'y = n \\ b' > k}} 1$$

so that

$$A_3 = B_3. \tag{10.3}$$

From (10.2) and (10.3) we deduce that

$$A_1 + A_2 = B_1 + B_2. \tag{10.4}$$

The sum A_2 is easy to evaluate. We have

$$A_2 = \sum_{\substack{(b, x) \in \mathbb{N}^2 \\ (k + b)x = n}} 1 = \sum_{\substack{b \in \mathbb{N} \\ k + b \mid n}} 1 = \sum_{\substack{v \in \mathbb{N} \\ v \mid n \\ v > k}} 1,$$

that is

$$A_2 = \sum_{\substack{v \in \mathbb{N} \\ v \mid n \\ v \geq k}} 1 - F_k(n). \tag{10.5}$$

The sum B_2 is also easy to evaluate. We have

$$B_2 = \sum_{\substack{(x, y) \in \mathbb{N}^2 \\ k(x + y) = n}} 1.$$

If $k \nmid n$ then $B_2 = 0$. If $k \mid n$ then

$$B_2 = \sum_{\substack{(x, y) \in \mathbb{N}^2 \\ x + y = n/k}} 1 = \frac{n}{k} - 1.$$

Thus

$$B_2 = \left(\frac{n}{k} - 1\right) F_k(n) = \frac{n}{k} F_k(n) - F_k(n). \tag{10.6}$$

Putting the values of A_2 and B_2 from (10.5) and (10.6) respectively into (10.4), we obtain

$$A_1 - B_1 = \frac{n}{k} F_k(n) - \sum_{\substack{v \in \mathbb{N} \\ v \mid n \\ v \geq k}} 1. \tag{10.7}$$

We are now ready to relate the sums $\displaystyle\sum_{\substack{(a,b,x,y) \in B(n) \\ a-b=k}} 1$ and $\displaystyle\sum_{\substack{(a,b,x,y) \in B(n) \\ a+b=k}} 1$ to the

sums A_1 and B_1. First we have

$$\sum_{\substack{(a,b,x,y) \in B(n) \\ a-b=k}} 1 = \sum_{\substack{(a,b,x,y) \in \mathbb{N}^4 \\ ax+by=n \\ a-b=k}} 1$$

$$= \sum_{\substack{(b,x,y) \in \mathbb{N}^3 \\ (b+k)x+by=n}} 1$$

$$= \sum_{\substack{(b,x,y) \in \mathbb{N}^3 \\ kx+b(x+y)=n}} 1$$

$$= \sum_{\substack{(b,x,y') \in \mathbb{N}^3 \\ kx+by'=n \\ x<y'}} 1$$

so that

$$\sum_{\substack{(a,b,x,y) \in B(n) \\ a-b=k}} 1 = A_1. \tag{10.8}$$

Secondly we have

$$\sum_{\substack{(a,b,x,y) \in B(n) \\ a+b=k}} 1 = \sum_{\substack{(a,b,x,y) \in \mathbb{N}^4 \\ ax+by=n \\ a+b=k}} 1$$

$$= \sum_{\substack{(a,b,x,y) \in \mathbb{N}^4 \\ ax+by=n \\ a+b=k \\ x<y}} 1 + \sum_{\substack{(a,b,x,y) \in \mathbb{N}^4 \\ ax+by=n \\ a+b=k \\ x>y}} 1 + \sum_{\substack{(a,b,x,y) \in \mathbb{N}^4 \\ ax+by=n \\ a+b=k \\ x=y}} 1$$

$$= 2 \sum_{\substack{(a,b,x,y) \in \mathbb{N}^4 \\ ax+by=n \\ a+b=k \\ x<y}} 1 + \sum_{\substack{(a,b,x) \in \mathbb{N}^3 \\ kx=n \\ a+b=k}} 1$$

$$= 2 \sum_{\substack{(a,b,x,y') \in \mathbb{N}^4 \\ ax+b(x+y')=n \\ a+b=k}} 1 + \sum_{\substack{(a,b) \in \mathbb{N}^2 \\ a+b=k}} F_k(n)$$

$$= 2 \sum_{\substack{(b,x,y) \in \mathbb{N}^3 \\ kx+by=n \\ b<k}} 1 + (k-1)F_k(n),$$

that is

$$\sum_{\substack{(a,b,x,y) \in B(n) \\ a+b=k}} 1 = 2B_1 + (k-1)F_k(n). \tag{10.9}$$

From (10.8), (10.9) and (10.7), we deduce

$$2 \sum_{\substack{(a,b,x,y) \in B(n) \\ a-b=k}} 1 - \sum_{\substack{(a,b,x,y) \in B(n) \\ a+b=k}} 1 = 2A_1 - 2B_1 - (k-1)F_k(n)$$

$$= \frac{2n}{k} F_k(n) - 2 \sum_{\substack{v \in \mathbb{N} \\ v \mid n \\ v \geq k}} 1 - (k-1)F_k(n)$$

$$= \left(1 - k + \frac{2n}{k}\right) F_k(n) - 2 \sum_{\substack{v \in \mathbb{N} \\ v \mid n \\ v \geq k}} 1,$$

which is the assertion of Theorem 10.2. □

We are now ready to use Theorem 10.2 to prove Theorem 10.1.

Proof of Theorem 10.1. The proof is achieved by collecting together all the terms on the left hand side of the theorem having a common value k for the argument of f for each $k \in \mathbb{Z}$. We obtain

$$\sum_{\substack{(a,b,x,y) \in \mathbb{N}^4 \\ ax+by=n}} (f(a-b) - f(a+b))$$

$$= \sum_{\substack{(a,b,x,y) \in \mathbb{N}^4 \\ ax+by=n}} f(a-b) - \sum_{\substack{(a,b,x,y) \in \mathbb{N}^4 \\ ax+by=n}} f(a+b)$$

$$= \sum_{k \in \mathbb{Z}} f(k) \sum_{\substack{(a,b,x,y) \in \mathbb{N}^4 \\ ax+by=n \\ a-b=k}} 1 - \sum_{k \in \mathbb{N}} f(k) \sum_{\substack{(a,b,x,y) \in \mathbb{N}^4 \\ ax+by=n \\ a+b=k}} 1$$

$$= f(0) \sum_{\substack{(a,b,x,y) \in \mathbb{N}^4 \\ ax+by=n \\ a=b}} 1 + \sum_{k \in \mathbb{N}} f(k) \sum_{\substack{(a,b,x,y) \in \mathbb{N}^4 \\ ax+by=n \\ a-b=k}} 1$$

$$+ \sum_{k \in \mathbb{N}} f(-k) \sum_{\substack{(a,b,x,y) \in \mathbb{N}^4 \\ ax+by=n \\ a-b=-k}} 1 - \sum_{k \in \mathbb{N}} f(k) \sum_{\substack{(a,b,x,y) \in \mathbb{N}^4 \\ ax+by=n \\ a+b=k}} 1.$$

Now

$$\sum_{\substack{(a,b,x,y) \in \mathbb{N}^4 \\ ax+by=n \\ a=b}} 1 = \sum_{\substack{(a,x,y) \in \mathbb{N}^3 \\ a(x+y)=n}} 1 = \sum_{\substack{a \in \mathbb{N} \\ a \mid n}} \sum_{\substack{(x,y) \in \mathbb{N}^2 \\ x+y=n/a}} 1$$

$$= \sum_{\substack{a \in \mathbb{N} \\ a \mid n}} \left(\frac{n}{a} - 1 \right) = \sigma(n) - d(n).$$

Also

$$\sum_{k \in \mathbb{N}} f(-k) \sum_{\substack{(a,b,x,y) \in \mathbb{N}^4 \\ ax+by=n \\ a-b=-k}} 1 = \sum_{k \in \mathbb{N}} f(k) \sum_{\substack{(a,b,x,y) \in \mathbb{N}^4 \\ ax+by=n \\ b-a=k}} 1$$

$$= \sum_{k \in \mathbb{N}} f(k) \sum_{\substack{(a,b,x,y) \in \mathbb{N}^4 \\ ax+by=n \\ a-b=k}} 1,$$

as f is even. Thus, appealing to Theorem 10.2, we obtain

$$\sum_{\substack{(a,b,x,y)\,\in\,\mathbb{N}^4 \\ ax+by=n}} (f(a-b)-f(a+b))-f(0)(\sigma(n)-d(n))$$

$$= 2 \sum_{k\,\in\,\mathbb{N}} f(k) \sum_{\substack{(a,b,x,y)\,\in\,B(n) \\ a-b=k}} 1 - \sum_{k\,\in\,\mathbb{N}} f(k) \sum_{\substack{(a,b,x,y)\,\in\,B(n) \\ a+b=k}} 1$$

$$= \sum_{k\,\in\,\mathbb{N}} f(k)\left(2 \sum_{\substack{(a,b,x,y)\,\in\,B(n) \\ a-b=k}} 1 - \sum_{\substack{(a,b,x,y)\,\in\,B(n) \\ a+b=k}} 1 \right)$$

$$= \sum_{k\,\in\,\mathbb{N}} f(k)\left(\left(1-k+\frac{2n}{k}\right) F_k(n) - 2 \sum_{\substack{v\,\in\,\mathbb{N} \\ v\,|\,n \\ v\,\geq\,k}} 1 \right)$$

$$= \sum_{\substack{k\,\in\,\mathbb{N} \\ k\,|\,n}} \left(1+\frac{2n}{k}-k\right) f(k) - 2 \sum_{\substack{v\,\in\,\mathbb{N} \\ v\,|\,n}} \left(\sum_{k=1}^{v} f(k) \right),$$

which completes the proof of Theorem 10.1. $\qquad\square$

Theorem 10.1 is a special case of the following more general identity of Liouville, which can be proved in a similar manner, see Problems 4 and 5. Another proof is given Chapter 13.

Theorem 10.3. *Let n be a positive integer and let $f : \mathbb{Z}^2 \to \mathbb{C}$ be a function satisfying*

$$f(x,y)=f(-x,y)=f(x,-y), \quad x,y \in \mathbb{Z}.$$

Then

$$\sum_{\substack{(a,b,x,y)\,\in\,\mathbb{N}^4 \\ ax+by=n}} (f(a-b,x+y)-f(a+b,x-y))$$

$$= \sum_{\substack{d\,\in\,\mathbb{N} \\ d\,|\,n}} (d-1)(f(0,d)-f(d,0)) + 2 \sum_{\substack{d\,\in\,\mathbb{N} \\ d\,|\,n}} \sum_{\substack{e\,\in\,\mathbb{N} \\ e\,<\,n/d}} (f(d,e)-f(e,d)).$$

Our next theorem results from applying Theorem 10.1 to the function $f(x)=F_k(x)$.

Theorem 10.4. *Let $k, n \in \mathbb{N}$. Then*

$$\sum_{\substack{(a, b, x, y) \in \mathbb{N}^4 \\ ax + by = n}} (F_k(a - b) - F_k(a + b))$$

$$= \frac{(k - 2)}{k}(\sigma(n) - k\sigma(n/k)) - \sum_{\ell=1}^{k-1}\left(1 - \frac{2\ell}{k}\right) d_{\ell,k}(n).$$

Proof. Taking $f(x) = F_k(x)$ in Theorem 10.1, and appealing to Example 3.7 and Problems 6, 7 and 8 in Exercises 3, as well as to Theorem 3.8, we obtain the asserted formula. $\square$

We remark that

$$\sum_{\substack{(a, b, x, y) \in \mathbb{N}^4 \\ ax + by = n}} (F_k(a - b) - F_k(a + b)) = \sum_{\substack{(a, b, x, y) \in \mathbb{N}^4 \\ ax + by = n \\ a \equiv b \,(\mathrm{mod}\, k)}} 1 - \sum_{\substack{(a, b, x, y) \in \mathbb{N}^4 \\ ax + by = n \\ a \equiv -b \,(\mathrm{mod}\, k)}} 1.$$

Example 10.1. Taking $k = 3$ in Theorem 10.4 we obtain

$$\sum_{\substack{(a, b, x, y) \in \mathbb{N}^4 \\ ax + by = n}} (F_3(a - b) - F_3(a + b)) = \frac{1}{3}\sigma(n) - \sigma(n/3) - \frac{1}{3}(d_{1,3}(n) - d_{2,3}(n)).$$

Example 10.2. Taking $k = 4$ in Theorem 10.4 we obtain

$$\sum_{\substack{(a, b, x, y) \in \mathbb{N}^4 \\ ax + by = n}} (F_4(a - b) - F_4(a + b)) = \frac{1}{2}\sigma(n) - 2\sigma(n/4) - \frac{1}{2}(d_{1,4}(n) - d_{3,4}(n)).$$

Example 10.3. Taking $k = 7$ in Theorem 10.4 we obtain

$$\sum_{\substack{(a, b, x, y) \in \mathbb{N}^4 \\ ax + by = n}} (F_7(a - b) - F_7(a + b)) = \frac{5}{7}\sigma(n) - 5\sigma(n/7) - \frac{5}{7}d_{1,7}(n) - \frac{3}{7}d_{2,7}(n)$$

$$- \frac{1}{7}d_{3,7}(n) + \frac{1}{7}d_{4,7}(n) + \frac{3}{7}d_{5,7}(n) + \frac{5}{7}d_{6,7}(n).$$

As an application of Liouville's identity (Theorem 10.1) we prove Bouniakowsky's theorem. To do this we require the following special case of Theorem 10.1.

Theorem 10.5. *Let p be a prime. Then*

$$\sum_{m=1}^{p-1} \sigma(m)\sigma(p - m) = \frac{1}{12}(5p - 6)(p^2 - 1).$$

Proof. We take $n = p$ and $f(x) = x^2$ in Theorem 10.1. The left hand side is

$$\sum_{\substack{(a,b,x,y) \in \mathbb{N}^4 \\ ax+by=p}} ((a-b)^2 - (a+b)^2) = -4 \sum_{\substack{(a,b,x,y) \in \mathbb{N}^4 \\ ax+by=p}} ab$$

$$= -4 \sum_{m=1}^{p-1} \sum_{\substack{a \in \mathbb{N} \\ a \mid m}} a \sum_{\substack{b \in \mathbb{N} \\ b \mid p-m}} b$$

$$= -4 \sum_{m=1}^{p-1} \sigma(m)\sigma(p-m).$$

The right hand side is

$$\sum_{\substack{d \in \mathbb{N} \\ d \mid p}} \left(\left(1 + \frac{2p}{d} - d\right) d^2 - 2 \sum_{v=1}^{d} v^2 \right).$$

As the only divisors d of the prime p are $d = 1$ and p, and $\sum_{v=1}^{d} v^2 = \frac{1}{6}d(d+1)(2d+1)$, the right hand side is

$$\frac{1}{3}(-5p^3 + 6p^2 + 5p - 6) = -\frac{1}{3}(5p - 6)(p^2 - 1).$$

Equating the left and right hand sides, we obtain the required formula. $\square$

We now use Theorem 10.5 to prove Bouniakowsky's theorem.

Theorem 10.6. *Let q be a prime with $q \equiv 7 \pmod{16}$. Then there exist positive odd integers x, y, e and a prime $p \equiv 5 \pmod 8$ such that $p \nmid y$, $e \equiv 1 \pmod 4$ and $q = 2x^2 + p^e y^2$. Furthermore the number of such representations is odd.*

Proof. By Theorem 10.5 we have

$$6 \sum_{i=1}^{(q-1)/2} \sigma(i)\sigma(q-i) = 3 \sum_{i=1}^{q-1} \sigma(i)\sigma(q-i) = \frac{1}{4}(5q-6)(q^2-1) \equiv 4 \pmod 8$$

so that

$$\sum_{i=1}^{(q-1)/2} \sigma(i)\sigma(q-i) \equiv 2 \pmod 4. \tag{10.10}$$

Suppose that $\sigma(i)\sigma(q-i) \equiv 1 \pmod 2$ for some $i \in \{1, 2, \ldots, (q-1)/2\}$. Then $\sigma(i)$ and $\sigma(q-i)$ are both odd so there exist positive integers x and y such that $i = x^2$ or $2x^2$ and $q - i = y^2$ or $2y^2$. Hence

$$q = x^2 + y^2, \; x^2 + 2y^2, \; 2x^2 + y^2 \text{ or } 2x^2 + 2y^2.$$

But none of these is possible as $q \equiv 7 \pmod 8$ and

$$x^2 + y^2 \equiv 0, 1 \text{ or } 2 \pmod 4,$$
$$x^2 + 2y^2, \ 2x^2 + y^2 \equiv 0, 1, 2, 3, 4 \text{ or } 6 \pmod 8,$$
$$2x^2 + 2y^2 \equiv 0 \pmod 2.$$

Hence $\sigma(i)\sigma(q-i) \equiv 0 \pmod 2$ for all $i \in \{1, 2, \ldots, (q-1)/2\}$. If $\sigma(i)\sigma(q-i) \equiv 0 \pmod 4$ for all $i \in \{1, 2, \ldots, (q-1)/2\}$ then

$$\sum_{i=1}^{(q-1)/2} \sigma(i)\sigma(q-i) \equiv \sum_{i=1}^{(q-1)/2} 0 \equiv 0 \pmod 4,$$

which contradicts (10.10). Hence there exists $i \in \{1, 2, \ldots, (q-1)/2\}$ such that

$$\sigma(i)\sigma(q-i) \equiv 2 \pmod 4,$$

that is, there exists $i \in \{1, 2, \ldots, q-1\}$ such that

$$\sigma(i) \equiv 1 \pmod 2, \ \sigma(q-i) \equiv 2 \pmod 4.$$

As $\sigma(i) \equiv 1 \pmod 2$ there exists $x \in \mathbb{N}$ such that $i = x^2$ or $2x^2$. As $\sigma(q-i) \equiv 2 \pmod 4$ there exist $y \in \mathbb{N}, e \in \mathbb{N}$ with $e \equiv 1 \pmod 4$, and a prime $p \equiv 1 \pmod 4$ with $p \nmid y$, such that $q - i = p^e y^2$ or $2p^e y^2$. Hence

$$q = x^2 + p^e y^2, \ x^2 + 2p^e y^2, \ 2x^2 + p^e y^2 \text{ or } 2x^2 + 2p^e y^2.$$

Now

$$x^2 + p^e y^2 \equiv x^2 + y^2 \equiv 0, 1, 2 \pmod 4,$$
$$x^2 + 2p^e y^2 \equiv x^2 + 2y^2 \equiv 0, 1, 2, 3, 4, 6 \pmod 8,$$

and

$$2x^2 + 2p^e y^2 \equiv 0 \pmod 2.$$

As $q \equiv 7 \pmod 8$ we deduce that $q \neq x^2 + p^e y^2, \ x^2 + 2p^e y^2$ and $2x^2 + 2p^e y^2$. Hence $q = 2x^2 + p^e y^2$. Clearly y is odd so $7 \equiv q \equiv 2x^2 + p \pmod 8$. Hence x is odd and $p \equiv 5 \pmod 8$.

Let A_0 denote the number of $i \in \{1, 2, \ldots, (q-1)/2\}$ such that $\sigma(i)\sigma(q-i) \equiv 0 \pmod 4$ and A_2 the number of $i \in \{1, 2, \ldots, (q-1)/2\}$ such that $\sigma(i)\sigma(q-i) \equiv 2 \pmod 4$. Clearly $A_0 + A_2 = (q-1)/2$ so that $A_0 = (q-1)/2 - A_2$. By (10.10) we have

$$((q-1)/2 - A_2) \times 0 + A_2 \times 2 \equiv 2 \pmod 4$$

so that

$$A_2 \equiv 1 \ (\text{mod } 2).$$

As the number of representations of q in the stated form is equal to A_2, it is odd as asserted. ☐

Example 10.4. For the prime $q = 71 \equiv 7 \ (\text{mod } 16)$ there is a single representation of the type specified in Bouniakowsky's theorem, namely,

$$71 = 2 \cdot 3^2 + 53^1 \cdot 1^2.$$

Example 10.5. For the prime $q = 103 \equiv 7 \ (\text{mod } 16)$ there are three representations of the type specified in Bouniakowsky's theorem, namely,

$$103 = 2 \cdot 1^2 + 101^1 \cdot 1^2 = 2 \cdot 5^2 + 53^1 \cdot 1^2 = 2 \cdot 7^2 + 5^1 \cdot 1^2.$$

Exercises 10

1. What does Theorem 10.1 give with $f(a) = (-1)^a$, $a \in \mathbb{Z}$?
2. Apply Theorem 10.1 with $f(a) = |a|$ to obtain the identity

$$\sum_{\substack{(a, b, x, y) \in \mathbb{N}^4 \\ ax + by = n \\ a < b}} a = \frac{1}{2}\sigma_2(n) + \frac{1}{2}\sigma(n) - nd(n).$$

3. Deduce Theorem 10.1 from Theorem 10.3.
4. Let $k, \ell, n \in \mathbb{N}$. Define

$$G_{k,\ell}(n) := \begin{cases} 1, & \text{if } k \mid n \text{ and } \ell < n/k, \\ 0, & \text{otherwise.} \end{cases}$$

Prove that

$$\sum_{\substack{(a, b, x, y) \in \mathbb{N}^4 \\ ax + by = n \\ a - b = k \\ x + y = \ell}} 1 - \sum_{\substack{(a, b, x, y) \in \mathbb{N}^4 \\ ax + by = n \\ a + b = k \\ x - y = \ell}} 1 = G_{k,\ell}(n) - G_{\ell,k}(n).$$

5. Use the identity of Problem 4 to prove Theorem 10.3.
6. Prove Theorem 10.2 by summing the identity of Problem 4 over $\ell \in \mathbb{N}$.

7. Let $k, n \in \mathbb{N}$ with k even. Prove that

$$2 \sum_{\substack{(a,b,x,y) \in B(n) \\ a-b=k \\ a,b \text{ odd}}} 1 - \sum_{\substack{(a,b,x,y) \in B(n) \\ a+b=k \\ a,b \text{ odd}}} 1 = -\frac{k}{2} F_k(n) - 2 \sum_{\substack{v \in \mathbb{N} \\ v \mid n \\ v > k \\ v \text{ odd}}} 1.$$

8. Let $f : \mathbb{Z} \to \mathbb{C}$ be an even function. Let $n \in \mathbb{N}$. Deduce from Problem 7 that

$$\sum_{\substack{(a,b,x,y) \in \mathbb{N}^4 \\ ax+by=n \\ a,b \text{ odd}}} (f(a-b) - f(a+b))$$

$$= f(0)(\sigma^*(n) - d^*(n)) - \frac{1}{2} \sum_{\substack{d \in \mathbb{N} \\ d \mid n \\ d \text{ even}}} df(d) - 2 \sum_{\substack{d \in \mathbb{N} \\ d \mid \mathbb{N} \\ d \text{ odd}}} \sum_{\substack{k \in \mathbb{N} \\ 1 \le k < d \\ k \text{ even}}} f(k).$$

9. Let $k, n \in \mathbb{N}$. Prove that

$$2 \sum_{\substack{(a,b,x,y) \in B(n) \\ a-b=k \\ x,y \text{ odd}}} 1 - \sum_{\substack{(a,b,x,y) \in B(n) \\ a+b=k \\ x,y \text{ odd}}} 1 = \left(\frac{n}{k} - k\right)(F_k(n) - F_k(n/2)).$$

10. Let $f : \mathbb{Z} \to \mathbb{C}$ be an even function. Let $n \in \mathbb{N}$. Deduce from Problem 9 that

$$\sum_{\substack{(a,b,x,y) \in \mathbb{N}^4 \\ ax+by=n \\ x,y \text{ odd}}} (f(a-b) - f(a+b))$$

$$= f(0)\sigma(n/2) + \sum_{\substack{d \in \mathbb{N} \\ d \mid n}} \left(\frac{n}{d} - d\right) f(d) - \sum_{\substack{d \in \mathbb{N} \\ d \mid n/2}} \left(\frac{n}{d} - d\right) f(d).$$

11. Let k and n be even positive integers. Prove that

$$2 \sum_{\substack{(a,b,x,y) \in B(n) \\ a-b=k \\ a,b,x,y \text{ odd}}} 1 - \sum_{\substack{(a,b,x,y) \in B(n) \\ a+b=k \\ a,b,x,y \text{ odd}}} 1 = -\frac{k}{2}(F_k(n) - F_k(n/2)).$$

12. Let $f : \mathbb{Z} \to \mathbb{C}$ be an even function. Let $n \in \mathbb{N}$ be even. Deduce from Problem 11 that

$$\sum_{\substack{(a, b, x, y) \in \mathbb{N}^4 \\ ax + by = n \\ a, b, x, y \text{ odd}}} (f(a - b) - f(a + b))$$

$$= \frac{1}{2} f(0)\sigma^*(n) - \frac{1}{2} \sum_{\substack{d \in \mathbb{N} \\ d \mid n}} df(d) + \frac{1}{2} \sum_{\substack{d \in \mathbb{N} \\ d \mid n/2}} df(d).$$

13. Let $f : \mathbb{Z} \to \mathbb{C}$ be an even function. Let r be a positive integer and N a positive odd integer. Prove that the identity of Problem 12 can be expressed in the form

$$\sum_{\substack{(a, b, x, y) \in \mathbb{N}^4 \\ ax + by = 2^r N \\ a, b, x, y \text{ odd}}} (f(a - b) - f(a + b)) = 2^{r-1} \sum_{\substack{d \in \mathbb{N} \\ d \mid N}} d(f(0) - f(2^r d)).$$

14. Let r be a positive integer and N a positive odd integer. Let $t \in \mathbb{R}$. Deduce from Problem 13 that

$$\sum_{\substack{(a, b, x, y) \in \mathbb{N}^4 \\ ax + by = 2^r N \\ a, b, x, y \text{ odd}}} \sin at \sin bt = 2^{r-1} \sum_{\substack{d \in \mathbb{N} \\ d \mid N}} d \sin^2(2^{r-1} dt).$$

15. Let N be a positive odd integer. Deduce from Problem 14 that

$$\sum_{\substack{(a, b, x, y) \in \mathbb{N}^4 \\ ax + by = 2N \\ a, b, x, y \text{ odd}}} (-1)^{(a-1)/2 + (b-1)/2} = \sigma(N).$$

16. Let N be a positive odd integer. Let $f : \mathbb{Z}^2 \to \mathbb{C}$ be such that

$$f(x, -y) = f(-x, y) = f(x, y)$$

for all $(x, y) \in \mathbb{Z}^2$. Prove that

$$\sum_{\substack{(a, b, x, y) \in \mathbb{N}^4 \\ ax + by = 2N \\ a, b, x, y \text{ odd}}} (f(a - b, x + y) - f(a + b, x - y))$$

$$= \sum_{\substack{d \in \mathbb{N} \\ d \mid N}} d(f(0, 2d) - f(2d, 0)).$$

17. Let N be a positive odd integer. Let $f : \mathbb{Z} \to \mathbb{C}$ be an even function. Apply the result of Problem 16 to the function

$$f(x, y) = \begin{cases} (-1)^{y/2} f(x), & \text{if } y \equiv 0 \ (\mathrm{mod}\ 2), \\ 0, & \text{if } y \equiv 1 \ (\mathrm{mod}\ 2), \end{cases}$$

to deduce

$$\sum_{\substack{(a, b, x, y) \in \mathbb{N}^4 \\ ax + by = 2N \\ a, b, x, y \text{ odd}}} (-1)^{(x-1)/2 + (y-1)/2} (f(a - b) + f(a + b))$$

$$= \sum_{\substack{d \in \mathbb{N} \\ d \mid N}} d(f(0) + f(2d)).$$

Notes on Chapter 10

Ingham's asymptotic formula was proved in [144, p. 208]. A proof of

$$\sum_{r=1}^{n-1} d(r)d(n - r) \sim \frac{6}{\pi^2} \sigma(n) \log^2 n, \text{ as } n \to \infty,$$

is given in Nathanson's book [209, Theorem 7.11, p. 248]. Theorem 10.1 was first stated by Liouville [170, 4th article, p. 247], [170, 5th article, p. 275]. Proofs have been given by McAfee [197], Meissner [201], Pepin [220, p. 93] and Piuma [225]. The proof given here is an unpublished proof of Spearman and Williams. Theorem 10.3 was stated in Liouville [170, 5th article, p. 284]. A slightly different form of Theorem 10.4 is given in Williams [269, Theorem 3.2, p. 798]. A special case of Example 10.2 is used in Huard, Ou, Spearman and Williams [137, p. 260]. Example 10.3 is given in Williams [269, Theorem 4.1, p. 800]. Theorem 10.5 is given in [170, 4th article, p. 249]. Bouniakowsky's theorem (Theorem 10.6) is discussed by Liouville in [170, 4th article, pp. 249–250] and mentioned in Dickson's *History* [93, Vol. II, p. 331]. It is also given as an exercise in the book of Moreno and Wagstaff [207, p. 59]. The formula of Problem 8 is taken from Alaca, Alaca, McAfee and Williams [9, Theorem 5.7, p. 17]. The formula of Problem 10 was stated by Liouville [170, 1st article, p. 144], [170, 2nd article, p. 194]. Proofs have been given by McAfee [197], Pepin [219, p. 159] and Alaca, Alaca, McAfee and Williams [9]. The formula of Problem 12 was stated by Liouville [170, 2nd article, p. 194] in the form given in Problem 13. Proofs have been given by Alaca, Alaca, McAfee and Williams [9], Baskakov [39, p. 344], Bugaev [53, p. 9], Deltour [86, p. 123], Humbert [143], Mathews [195], McAfee [197], Pepin [220, p. 94] and Smith

[244, p. 346–348], [245, Vol. I, p. 348]. The identities of Problems 13–17 are due to Liouville [170, 1st and 2nd articles].

Multiplying the identity of Theorem 10.1 by q^n and summing over $n \in \mathbb{N}$, we obtain

$$\sum_{a,b=1}^{\infty} \frac{(f(a-b) - f(a+b))}{(1-q^a)(1-q^b)} q^{a+b}$$

$$= 2\sum_{s=1}^{\infty} \frac{f(s)q^s}{(1-q^s)^2} + \sum_{s=1}^{\infty} \left((s-1)(f(0) - f(s)) - 2\sum_{k=1}^{s} f(k)\right) \frac{q^s}{1-q^s},$$

where $f : \mathbb{Z} \to \mathbb{C}$ is an even function. Taking $f(x) = x^2$ we deduce

$$\left(\sum_{n=1}^{\infty} \frac{nq^n}{1-q^n}\right)^2 = \frac{1}{12}\sum_{n=1}^{\infty} (5n^3 + n)\frac{q^n}{1-q^n} - \frac{1}{2}\sum_{n=1}^{\infty} \frac{n^2q^n}{(1-q^n)^2}.$$

This formula appears in a slightly different form in Alaca, Alaca, McAfee and Williams [9, p. 57].

Multiplying the identity of Problem 13 by $q^{2^r N}$ and summing over $r, N \in \mathbb{N}$ with N odd, we obtain

$$\sum_{\substack{a,b=1 \\ a,b \text{ odd}}}^{\infty} \frac{(f(a-b) - f(a+b))}{(1-q^{2a})(1-q^{2b})} q^{a+b} = \sum_{s=1}^{\infty} s(f(0) - f(2s))\frac{q^{2s}}{1-q^{4s}},$$

where $f : \mathbb{Z} \to \mathbb{C}$ is an even function. This identity is due to Liouville [170, 1st article]. Taking $f(x) = x^2$ we deduce

$$\left(\sum_{\substack{n=1 \\ n \text{ odd}}}^{\infty} \frac{nq^n}{1-q^{2n}}\right)^2 = \sum_{n=1}^{\infty} \frac{n^3 q^{2n}}{1-q^{4n}},$$

which is Theorem 17.9 (a) of Alaca, Alaca, McAfee and Williams [9, p. 66].

The extension of Theorem 10.1 to three variables is due to McAfee and Williams [200, Theorem 1.2, p. 35]. The four variable extension is due to Alaca, Alaca, McAfee and Williams [9, Theorem 6.4, p. 20]. No extension to five or more variables is known.

11

Jacobi's Four Squares Formula

There is a simple convolution formula relating the number $r_4(n)$ of representations of a positive integer n as the sum of four squares to the number $r_2(n)$ of representations of n as the sum of two squares. We have

$$r_4(n) = \sum_{\substack{(x_1, x_2, x_3, x_4) \in \mathbb{Z}^4 \\ n = x_1^2 + x_2^2 + x_3^2 + x_4^2}} 1$$

$$= \sum_{k=0}^{n} \left(\sum_{\substack{(x_1, x_2) \in \mathbb{Z}^2 \\ x_1^2 + x_2^2 = k}} 1 \right) \left(\sum_{\substack{(x_3, x_4) \in \mathbb{Z}^2 \\ x_3^2 + x_4^2 = n - k}} 1 \right),$$

so that

$$r_4(n) = \sum_{k=0}^{n} r_2(k) r_2(n - k). \tag{11.1}$$

In this chapter we use Jacobi's formula for $r_2(n)$ (Theorem 9.3) in (11.1) and then use the identity of Liouville proved in Example 10.2 to give another proof of Jacobi's formula for $r_4(n)$, which was first proved in Theorem 9.5.

Theorem 11.1. *Let $n \in \mathbb{N}$. Then*

$$r_4(n) = 8\sigma(n) - 32\sigma(n/4).$$

Proof. Let $n \in \mathbb{N}$. Since $r_2(0) = 1$, we can rewrite (11.1) as

$$r_4(n) - 2r_2(n) = \sum_{k=1}^{n-1} r_2(k) r_2(n - k).$$

116

Recalling from Theorem 9.3 that

$$r_2(n) = 4 \sum_{\substack{d \in \mathbb{N} \\ d \mid n}} \left(\frac{-4}{d}\right) = 4(d_{1,4}(n) - d_{3,4}(n)), \qquad (11.2)$$

we obtain by Theorem 3.10

$$r_4(n) - 2r_2(n) = 16 \sum_{k=1}^{n-1} \left(\sum_{\substack{a \in \mathbb{N} \\ a \mid k}} \left(\frac{-4}{a}\right)\right)\left(\sum_{\substack{b \in \mathbb{N} \\ b \mid n-k}} \left(\frac{-4}{b}\right)\right)$$

$$= 16 \sum_{\substack{(a, b, x, y) \in \mathbb{N}^4 \\ ax + by = n}} \left(\frac{-4}{ab}\right)$$

$$= 16 \sum_{\substack{(a, b, x, y) \in \mathbb{N}^4 \\ ax + by = n}} (F_4(a - b) - F_4(a + b)).$$

Appealing to Example 10.2 and (11.2), we have

$$\sum_{\substack{(a, b, x, y) \in \mathbb{N}^4 \\ ax + by = n}} (F_4(a - b) - F_4(a + b)) = \frac{1}{2}\sigma(n) - 2\sigma(n/4) - \frac{1}{8}r_2(n).$$

Finally

$$r_4(n) - 2r_2(n) = 8\sigma(n) - 32\sigma(n/4) - 2r_2(n)$$

from which Jacobi's formula for $r_4(n)$ follows. □

Example 11.1. The number of representations of 20 as the sum of four squares is by Theorem 11.1

$$r_4(20) = 8\sigma(20) - 32\sigma(5) = 8 \times 42 - 32 \times 6 = 336 - 192 = 144.$$

These arise by permutation and change of sign as follows: 48 from $0^2 + 0^2 + 2^2 + 4^2$ and 96 from $1^2 + 1^2 + 3^2 + 3^2$ for a total of 144.

If $(x_1, x_2, x_3, x_4) \in \mathbb{Z}^4$ is a solution of $n = x_1^2 + x_2^2 + x_3^2 + x_4^2$ then each of x_1, x_2, x_3 and x_4 belongs to $\{0, \pm 1, \ldots, \pm[\sqrt{n}]\}$. Once x_1, x_2 and x_3 are specified, the equation $n = x_1^2 + x_2^2 + x_3^2 + x_4^2$ shows that there are at most two possibilities for x_4. Thus an upper bound for $r_4(n)$ is

$$2(2[\sqrt{n}] + 1)^3 \le 2(3\sqrt{n})^3 = 54n^{3/2}.$$

In our next theorem we improve this bound.

Theorem 11.2. *For all* $n \in \mathbb{N}$ *we have*

$$r_4(n) \leq 8n(\log n + 1).$$

Proof. By Theorem 11.1 we have

$$r_4(n) \leq 8\sigma(n).$$

The asserted estimate now follows by appealing to the upper bound for $\sigma(n)$ given in Chapter 10. □

Next we prove Lagrange's theorem.

Theorem 11.3. *Every positive integer is the sum of four squares.*

Proof. Let $n \in \mathbb{N}$. If $n \equiv 1 \pmod 2$ or $n \equiv 2 \pmod 4$ then $\sigma(n/4) = 0$ and $r_4(n) = 8\sigma(n) \geq 8$. Suppose now that $n \equiv 0 \pmod 4$ so that $n/4 \in \mathbb{N}$. Set $\ell = d(n/4)$. Let $d_1, \ldots, d_\ell$ be the distinct positive integers dividing $n/4$. Then $4d_1, \ldots, 4d_\ell$ are distinct positive divisors of n. As 1 is also a divisor of n, we have

$$\sigma(n) \geq 1 + 4d_1 + \cdots + 4d_\ell > 4(d_1 + \cdots + d_\ell) = 4\sigma(n/4)$$

so that

$$r_4(n) = 8(\sigma(n) - 4\sigma(n/4)) > 0.$$

Thus $r_4(n) > 0$ for all $n \in \mathbb{N}$. □

Example 11.2. Let p be a prime. Let $\ell = \ell(p)$ denote the least positive integer such that $p = x_1^2 + \cdots + x_\ell^2$ for positive integers $x_1, \ldots, x_\ell$. We show that

$$\ell(p) = \begin{cases} 2, & \text{if } p = 2 \text{ or } p \equiv 1 \pmod 4, \\ 3, & \text{if } p \equiv 3 \pmod 8, \\ 4, & \text{if } p \equiv 7 \pmod 8. \end{cases}$$

Since $2 = 1^2 + 1^2$ it is clear that $\ell(2) = 2$. If p is a prime with $p \equiv 1 \pmod 4$ then by Theorem 7.1 we have $p = x_1^2 + x_2^2$ for integers x_1 and x_2. Clearly neither x_1 nor x_2 is zero. This shows that $\ell(p) = 2$ for primes p with $p \equiv 1 \pmod 4$. If p is a prime with $p \equiv 3 \pmod 8$ then by Theorem 7.3 we have $p = x_1^2 + 2x_2^2$ for integers x_1 and x_2. Thus $p = x_1^2 + x_2^2 + x_3^2$ with $x_2 = x_3$. By (7.1) none of x_1, x_2, x_3 is zero. This proves that $\ell(p) = 3$ for primes p with $p \equiv 3 \pmod 8$. Finally let p be a prime with $p \equiv 7 \pmod 8$. By Theorem 11.3 there are integers x_1, x_2, x_3 and x_4 such that $p = x_1^2 + x_2^2 + x_3^2 + x_4^2$. Suppose one of x_1, x_2, x_3 and x_4 is zero. Without loss of generality we may suppose that

$x_4 = 0$. Then $p = x_1^2 + x_2^2 + x_3^2 \not\equiv 7 \pmod 8$, a contradiction. Thus $\ell(p) = 4$ for primes p with $p \equiv 7 \pmod 8$.

Our final theorem of this chapter gives two aspects of the distribution of the values of $\{r_4(n) \mid n \in \mathbb{N}\}$.

Theorem 11.4. (i) *There is an infinite sequence of positive integers n having a constant value of $r_4(n)$.*

(ii) *There is an infinite sequence of positive odd integers n such that $r_4(n) \to +\infty$ as $n \to +\infty$.*

Proof. (i) For $k \in \mathbb{N}$ with $k \geq 2$ we have

$$r_4(2^k) = 8\sigma(2^k) - 32\sigma(2^{k-2}) = 8(2^{k+1} - 1) - 32(2^{k-1} - 1) = 24.$$

Thus $\{n = 2^k \mid k \in \mathbb{N}, k \geq 2\}$ is an infinite sequence of positive integers having a constant value of 24 for $r_4(n)$.

(ii) Let p_k ($k \in \mathbb{N}$) denote the k-th prime. For $k \geq 2$ we have $p_k \geq p_2 = 3$ so that $\log p_k \geq \log 3 > 1$ and thus $[\log p_k] \geq 1$. Set

$$n_k = p_k^{[\log p_k]}, \quad k \geq 2.$$

Then

$$\log n_k = [\log p_k] \log p_k \leq \log^2 p_k$$

so that

$$\sqrt{\log n_k} \leq \log p_k.$$

Hence, by Bernoulli's inequality, we obtain

$$\sigma(n_k) = \sigma(p_k^{[\log p_k]}) = \frac{p_k^{[\log p_k]+1} - 1}{p_k - 1} > [\log p_k] + 1 > \log p_k \geq \sqrt{\log n_k}.$$

Thus

$$r_4(n_k) = 8\sigma(n_k) > 8\sqrt{\log n_k}.$$

Hence $r_4(n) > 8\sqrt{\log n}$ for an infinite sequence of positive odd integers n. This completes the proof. $\qquad\square$

Exercises 11

1. Let $n \in \mathbb{N}$. Define $\alpha \in \mathbb{N}_0$ and $N \in \mathbb{N}$ with $2 \nmid N$ by $n = 2^\alpha N$. Prove that

$$r_4(n) = \begin{cases} 8\sigma(N), & \text{if } \alpha = 0, \\ 24\sigma(N), & \text{if } \alpha \geq 1. \end{cases}$$

Deduce Theorem 11.3.

2. Prove that Theorem 11.1 can be written in the form

$$r_4(n) = 8 \sum_{\substack{d \in \mathbb{N} \\ d \mid n \\ 4 \nmid d}} d.$$

3. Let $n \in \mathbb{N}$. Let $e, k \in \mathbb{N}$ satisfy $2 \leq e \leq k - 2$. Prove that

$$r_k(n) = \sum_{\ell=0}^{n} r_e(\ell) r_{k-e}(n - \ell).$$

4. Let $k, n \in \mathbb{N}$. Prove that

$$r_k(n) \leq 2 \cdot 3^{k-1} n^{(k-1)/2}.$$

5. Use Lagrange's theorem to prove that every $n \in \mathbb{N}$ can be expressed in the form $x^2 + y^2 + 2z^2 + 2t^2$ for some integers x, y, z and t.

6. Let n be a positive odd integer. Use the ideas of the proof of Theorem 11.1 to prove

$$\text{card}\{(x_1, x_2, x_3, x_4) \in \mathbb{N}^4 \mid 4n = x_1^2 + x_2^2 + x_3^2 + x_4^2, \; x_1, x_2, x_3, x_4 \text{ odd}\}$$
$$= \sigma(n).$$

7. Let n be a positive odd integer. Prove from first principles that

$$r_4(2n) = 3r_4(n).$$

8. Let $n \in \mathbb{N}$. Prove from first principles that

$$r_4(4n) = r_4(2n).$$

9. Let n be a positive odd integer. Use the result of Problem 6 to prove

$$r_4(4n) = 16\sigma(n) + r_4(n).$$

10. Deduce from Problems 6–9 that for a positive odd integer n

$$r_4(n) = 8\sigma(n).$$

11. Deduce from Problems 6–10 that for a positive even integer n

$$r_4(n) = 24\sigma(n_1),$$

where n_1 is the largest odd divisor of n.

12. Deduce Theorem 11.1 from Problems 10 and 11.

13. Let $k \in \mathbb{N}_0$. Determine all $(x_1, x_2, x_3, x_4) \in \mathbb{Z}^4$ such that

$$2^k = x_1^2 + x_2^2 + x_3^2 + x_4^2.$$

14. Prove that there does not exist a positive constant c such that

$$r_4(n) \geq c\sigma(n)$$

for all $n \in \mathbb{N}$.

Notes on Chapter 11

Theorem 11.1 is implicit in the work of Jacobi [146, §§40–42], [148, Vol. I, pp. 159–170]. Many proofs of this theorem have appeared in the literature, see the notes for Chapter 9.

Theorem 11.3 was stated by Bachet in 1621. It was first proved in 1770 by Lagrange based on ideas of Euler. It is now known as Lagrange's (four squares) theorem. A proof using the ideas of Euler and Lagrange is presented in the book of Hardy and Wright [126, pp. 302–303]. Many proofs of Lagrange's theorem have been given, see for example Davenport [84], Dixon [96] and the book of Niven, Zuckerman and Montgomery [211, pp. 317–318].

Lagrange's theorem is a simple consequence of the three squares theorem due to Legendre (1798) and Gauss (1801), which asserts that a positive integer n is the sum of three squares if and only if it is not of the form $4^a(8b + 7)$ for any $a, b \in \mathbb{N}_0$. For if $n \neq 4^a(8b + 7)$ then n is the sum of three squares and thus the sum of four squares, one of them being 0^2. Otherwise, if $n = 4^a(8b + 7)$ then $n - 4^a = 4^a(8b + 6)$ is the sum of three squares and hence n is the sum of four squares, one of them being $(2^a)^2$. An elementary proof of the three squares theorem using the ideas of Liouville is presented in the book of Uspensky and Heaslet [254, pp. 465–474]. Ankeny [32] has given a short, elementary proof of the three squares theorem based on a method due to H. Davenport [84] in the geometry of numbers.

Gauss actually gave a formula for the number $p_3(n)$ of primitive representations of n as a sum of three squares. His theorem may be formulated as follows: Let d be a negative integer with $d \equiv 0$ or $1 \pmod{4}$. Let $h(d)$ denote the number of equivalence classes of positive-definite, primitive, integral, binary quadratic forms of discriminant d under the action of the modular group. Then for $n \in \mathbb{N}$

with $n \neq 1, 3$ we have

$$
p_3(n) = \begin{cases} 0, & \text{if } n \equiv 0, 4, 7 \ (\mathrm{mod}\ 8), \\ 12h(-4n), & \text{if } n \equiv 1, 2, 5, 6 \ (\mathrm{mod}\ 8), \\ 24h(-n), & \text{if } n \equiv 3 \ (\mathrm{mod}\ 8). \end{cases}
$$

We note that $p_3(1) = 6$ and $p_3(3) = 8$. For a discussion and proof of Gauss's formula, the reader should consult Grosswald's book [119, Chapter 4, pp. 38–65]. Using Dirichlet's class number formula for $h(d)$ in the form

$$
h(d) = \frac{w(d)\sqrt{|d|}}{2\pi} L(1; d),
$$

where

$$
L(1; d) := \sum_{m=1}^{\infty} \left(\frac{d}{m}\right) \frac{1}{m}
$$

and $w(d) := 6$, 4 or 2 according as $d = -3$, $d = -4$ or $d < -4$ respectively, we obtain the following formula for $p_3(n)$, namely,

$$
p_3(n) = \pi^{-1} c(n) \sqrt{n} L(1; -4n), \tag{11.3}
$$

where

$$
c(n) := \begin{cases} 0, & \text{if } n \equiv 0, 4, 7 \ (\mathrm{mod}\ 8), \\ 16, & \text{if } n \equiv 3 \ (\mathrm{mod}\ 8), \\ 24, & \text{if } n \equiv 1, 2, 5, 6 \ (\mathrm{mod}\ 8). \end{cases}
$$

Formula (11.3) is immediate except in the case $n \equiv 3 \ (\mathrm{mod}\ 8)$. In this case we note that

$$
\begin{aligned}
L(1; -4n) &= \sum_{m=1}^{\infty} \left(\frac{-4n}{m}\right) \frac{1}{m} \\
&= \sum_{\substack{m=1 \\ m \text{ odd}}}^{\infty} \left(\frac{-n}{m}\right) \frac{1}{m} \\
&= \sum_{m=1}^{\infty} \left(\frac{-n}{m}\right) \frac{1}{m} - \sum_{\substack{m=1 \\ m \text{ even}}}^{\infty} \left(\frac{-n}{m}\right) \frac{1}{m} \\
&= L(1; -n) - \sum_{m=1}^{\infty} \left(\frac{-n}{2m}\right) \frac{1}{2m} \\
&= L(1; -n) + \frac{1}{2} L(1; -n) \quad \left(\text{as } \left(\frac{-n}{2}\right) = \left(\frac{5}{2}\right) = -1\right) \\
&= \frac{3}{2} L(1; -n)
\end{aligned}
$$

so that

$$L(1; -n) = \frac{2}{3} L(1; -4n).$$

Formula (11.3) is also true for the exceptional values $n = 1$ and $n = 3$. For $n = 1$ we have

$$
\begin{aligned}
\pi^{-1} c(1) \sqrt{1} L(1; -4) &= \frac{24}{\pi} \sum_{m=1}^{\infty} \left(\frac{-4}{m} \right) \frac{1}{m} \\
&= \frac{24}{\pi} \left(1 - \frac{1}{3} + \frac{1}{5} - \frac{1}{7} + \cdots \right) \\
&= \frac{24}{\pi} \cdot \frac{\pi}{4} = 6 = p_3(1)
\end{aligned}
$$

and for $n = 3$

$$
\begin{aligned}
\pi^{-1} c(3) \sqrt{3} L(1; -12) &= \frac{16}{\pi} \sqrt{3} \sum_{m=1}^{\infty} \left(\frac{-12}{m} \right) \frac{1}{m} \\
&= \frac{16}{\pi} \sqrt{3} \left(1 - \frac{1}{5} + \frac{1}{7} - \frac{1}{11} + \cdots \right) \\
&= \frac{16}{\pi} \sqrt{3} \cdot \frac{\pi}{2\sqrt{3}} = 8 = p_3(3).
\end{aligned}
$$

From Theorem 9.7 with $k = 3$ the total number $r_3(n)$ of representations of a positive integer n as a sum of three squares is given by

$$r_3(n) = \sum_{\substack{d \in \mathbb{N} \\ d^2 \mid n}} p_3(n/d^2). \tag{11.4}$$

Bateman's formula [40] for $r_3(n)$ follows using (11.3) in (11.4). Let $n \in \mathbb{N}$. Define $a \in \mathbb{N}_0$ and $n_1 \in \mathbb{N}$ with $4 \nmid n_1$ by $n = 4^a n_1$. Then

$$r_3(n) = \frac{16}{\pi} \sqrt{n} q(n) P(n) L(1; -4n), \tag{11.5}$$

where

$$
q(n) := \begin{cases}
0, & \text{if } n_1 \equiv 7 \pmod 8, \\
2^{-a}, & \text{if } n_1 \equiv 3 \pmod 8, \\
3 \cdot 2^{-a-1}, & \text{if } n \equiv 1, 2, 5 \text{ or } 6 \pmod 8,
\end{cases}
$$

and

$$P(n) := \prod_{\substack{p \text{ (prime)} > 2 \\ p^{2b} \| n \text{ or } p^{2b+1} \| n}} \left(1 + \sum_{j=1}^{b-1} p^{-j} + p^{-b} \left(1 - \left(\frac{-n/p^{2b}}{p} \right) \frac{1}{p} \right)^{-1} \right).$$

For $n \in \mathbb{N}$ we have $q(4n) = q(n)/2$, $P(4n) = P(n)$ and $L(1; -16n) = L(1; -4n)$ so that

$$r_3(4n) = r_3(n), \quad n \in \mathbb{N},$$

a result stated by Hirschhorn and Sellers [135]. For $n \in \mathbb{N}$ with $n \equiv 1 \pmod 3$ we have $q(9n) = q(n)$, $P(9n) = \frac{5}{4}P(n)$ and $L(1; -36n) = \frac{4}{3}L(1; -4n)$ so (11.5) gives

$$r_3(9n) = 5r_3(n),$$

which is a particular case of a result given by Hirschhorn and Sellers. Their paper contains many more results of this kind.

Hirschhorn and Sellers [135] have given the generating function for various subsequences of $\{r_3(n) \mid n \in \mathbb{N}\}$, for example

$$\sum_{n=0}^{\infty} r_3(3n+2)q^n = 12 \prod_{n=1}^{\infty} \frac{(1-q^{2n})^4(1-q^{6n})^3}{(1-q^n)^2(1-q^{4n})^2}$$

and

$$\sum_{n=0}^{\infty} r_3(8n+5)q^n = 24 \prod_{n=1}^{\infty} \frac{(1-q^{4n})^4}{(1-q^n)^2}.$$

Bernoulli's inequality, which was used in the proof of Theorem 11.4 (ii), asserts that for $x \in \mathbb{R}$ with $x > 1$ and $p \in \mathbb{R}$ with $p > 1$ we have

$$\frac{x^p - 1}{x - 1} > p,$$

see for example Ferrar's book [105, Theorem 43, p. 252].

12

Besge's Formula

The objective of this chapter is to determine the value of the convolution sum

$$\sum_{m=1}^{n-1} \sigma(m)\sigma(n-m)$$

for an arbitrary positive integer n. (We evaluated this sum when n is a prime p in Theorem 10.5.) The value of this convolution sum first appeared in a letter from Besge to Liouville in 1862. Dickson in his *History of the Theory of Numbers* erroneously attributes it to Lebesgue. Rankin attributed it to Besgue. Lützen asserts that Besge/Besgue is a pseudonym for Liouville. The evaluation also appears in the work of Glaisher, Lahiri, Lehmer, Ramanujan and Skoruppa. We obtain Besge's formula as a simple application of Liouville's identity (Theorem 10.1).

Theorem 12.1. *Let $n \in \mathbb{N}$. Then*

$$\sum_{m=1}^{n-1} \sigma(m)\sigma(n-m) = \frac{5}{12}\sigma_3(n) + \left(\frac{1}{12} - \frac{1}{2}n\right)\sigma(n).$$

Proof. We take $f(x) = x^2$ in Theorem 10.1. The left hand side of Theorem 10.1 is

$$\sum_{\substack{(a,b,x,y)\in\mathbb{N}^4 \\ ax+by=n}} (f(a-b) - f(a+b)) = \sum_{\substack{(a,b,x,y)\in\mathbb{N}^4 \\ ax+by=n}} ((a-b)^2 - (a+b)^2)$$

$$= -4 \sum_{\substack{(a,b,x,y)\in\mathbb{N}^4 \\ ax+by=n}} ab$$

$$= -4 \sum_{m=1}^{n-1} \left(\sum_{\substack{a \in \mathbb{N} \\ a \mid m}} a \right) \left(\sum_{\substack{b \in \mathbb{N} \\ b \mid n-m}} b \right)$$

$$= -4 \sum_{m=1}^{n-1} \sigma(m) \sigma(n - m).$$

As $f(0) = 0$ the right hand side of Theorem 10.1 is

$$\sum_{\substack{d \in \mathbb{N} \\ d \mid n}} \left(1 + \frac{2n}{d} - d \right) f(d) - 2 \sum_{\substack{d \in \mathbb{N} \\ d \mid n}} \left(\sum_{v=1}^{d} f(v) \right)$$

$$= \sum_{\substack{d \in \mathbb{N} \\ d \mid n}} d^2 + 2n \sum_{\substack{d \in \mathbb{N} \\ d \mid n}} d - \sum_{\substack{d \in \mathbb{N} \\ d \mid n}} d^3 - 2 \sum_{\substack{d \in \mathbb{N} \\ d \mid n}} \left(\sum_{v=1}^{d} v^2 \right)$$

$$= \sigma_2(n) + 2n\sigma(n) - \sigma_3(n) - 2 \sum_{\substack{d \in \mathbb{N} \\ d \mid n}} \left(\frac{1}{3}d^3 + \frac{1}{2}d^2 + \frac{1}{6}d \right)$$

$$= \sigma_2(n) + 2n\sigma(n) - \sigma_3(n) - \frac{2}{3}\sigma_3(n) - \sigma_2(n) - \frac{1}{3}\sigma(n)$$

$$= -\frac{5}{3}\sigma_3(n) - \frac{1}{3}\sigma(n) + 2n\sigma(n).$$

Equating the left and right hand sides of Theorem 10.1, we obtain

$$-4 \sum_{m=1}^{n-1} \sigma(m) \sigma(n - m) = -\frac{5}{3}\sigma_3(n) - \frac{1}{3}\sigma(n) + 2n\sigma(n)$$

from which Besge's formula follows on dividing both sides by -4.   □

An immediate consequence of Theorem 12.1 is Glaisher's theorem, which gives the power series expansion in powers of q of the square of the series

$$\sigma(1)q + \sigma(2)q^2 + \sigma(3)q^3 + \cdots.$$

Theorem 12.2. *For $q \in \mathbb{C}$ with $|q| < 1$ we have*

$$\left(\sum_{n=1}^{\infty} \sigma(n)q^n \right)^2 = \sum_{n=1}^{\infty} \left(\frac{5}{12}\sigma_3(n) + \frac{1}{12}\sigma(n) - \frac{1}{2}n\sigma(n) \right) q^n.$$

Liouville generalized Besge's formula in the following way.

Theorem 12.3. *Let $k \in \mathbb{N}$ and $n \in \mathbb{N}$. Then*

$$\sum_{s=0}^{k-1} \binom{2k}{2s+1} \left(\sum_{m=1}^{n-1} \sigma_{2k-2s-1}(m)\sigma_{2s+1}(n-m) \right) = \frac{2k+3}{4k+2}\sigma_{2k+1}(n)$$

$$+ \left(\frac{k}{6} - n \right)\sigma_{2k-1}(n) + \frac{1}{2k+1} \sum_{j=2}^{k} \binom{2k+1}{2j} B_{2j}\sigma_{2k+1-2j}(n).$$

Proof. We take $f(x) = x^{2k}$ ($k \in \mathbb{N}$) in Theorem 10.1. For $n \in \mathbb{N}$ the left hand side of Theorem 10.1 becomes by the binomial theorem

$$\sum_{\substack{(a,b,x,y) \in \mathbb{N}^4 \\ ax+by=n}} \left((a-b)^{2k} - (a+b)^{2k} \right)$$

$$= \sum_{\substack{(a,b,x,y) \in \mathbb{N}^4 \\ ax+by=n}} \left(\sum_{r=0}^{2k} \binom{2k}{r}(-1)^r a^{2k-r}b^r - \sum_{r=0}^{2k} \binom{2k}{r} a^{2k-r}b^r \right)$$

$$= -2 \sum_{\substack{(a,b,x,y) \in \mathbb{N}^4 \\ ax+by=n}} \sum_{\substack{r=0 \\ r \text{ odd}}}^{2k} \binom{2k}{r} a^{2k-r}b^r$$

$$= -2 \sum_{s=0}^{k-1} \binom{2k}{2s+1} \sum_{\substack{(a,b,x,y) \in \mathbb{N}^4 \\ ax+by=n}} a^{2k-2s-1}b^{2s+1}$$

$$= -2 \sum_{s=0}^{k-1} \binom{2k}{2s+1} \sum_{m=1}^{n-1} \sigma_{2k-2s-1}(m)\sigma_{2s+1}(n-m).$$

By Problem 28 of Exercises 3, the right hand side of Theorem 10.1 is

$$\sum_{\substack{d \in \mathbb{N} \\ d \mid n}} \left(1 + \frac{2n}{d} - d \right)d^{2k} - 2\sum_{\substack{d \in \mathbb{N} \\ d \mid n}} \left(\sum_{v=1}^{d} v^{2k} \right)$$

$$= \sigma_{2k}(n) + 2n\sigma_{2k-1}(n) - \sigma_{2k+1}(n) - 2\sigma_{2k}(n) - 2\sum_{\substack{d \in \mathbb{N} \\ d \mid n}} \left(\sum_{v=1}^{d-1} v^{2k} \right)$$

$$= -\sigma_{2k+1}(n) - \sigma_{2k}(n) + 2n\sigma_{2k-1}(n) - \frac{2}{2k+1} \sum_{j=0}^{2k} \binom{2k+1}{j} B_j \sigma_{2k+1-j}(n).$$

Now, as $B_0 = 1$, $B_1 = -1/2$, $B_2 = 1/6$, $B_{2r+1} = 0$ $(r \in \mathbb{N})$, we have

$$\sum_{j=0}^{2k} \binom{2k+1}{j} B_j \sigma_{2k+1-j}(n) = \sigma_{2k+1}(n) - \frac{2k+1}{2}\sigma_{2k}(n) + \frac{k(2k+1)}{6}\sigma_{2k-1}(n)$$

$$+ \sum_{j=2}^{k} \binom{2k+1}{2j} B_{2j} \sigma_{2k+1-2j}(n).$$

Thus the right hand side is

$$-\frac{2k+3}{2k+1}\sigma_{2k+1}(n) - \left(\frac{k}{3} - 2n\right)\sigma_{2k-1}(n) - \frac{2}{2k+1}\sum_{j=2}^{k}\binom{2k+1}{2j} B_{2j}\sigma_{2k+1-2j}(n).$$

Equating the left and right hand sides, and then dividing both sides by -2, we obtain the assertion of the theorem. □

Example 12.1. Taking $k = 1$ in Theorem 12.3 we obtain

$$2\sum_{m=1}^{n-1}\sigma(m)\sigma(n-m) = \frac{5}{6}\sigma_3(n) + \left(\frac{1}{6} - n\right)\sigma(n),$$

which is Theorem 12.1 on dividing through by 2.

Example 12.2. Taking $k = 2$ in Theorem 12.3 we obtain (recalling that $B_4 = -1/30$) the following identity:

$$\sum_{m=1}^{n-1}\sigma(m)\sigma_3(n-m) = \sum_{m=1}^{n-1}\sigma(n-m)\sigma_3(m)$$

$$= \frac{7}{80}\sigma_5(n) + \left(\frac{1}{24} - \frac{1}{8}n\right)\sigma_3(n) - \frac{1}{240}\sigma(n).$$

Example 12.3. Taking $k = 3$ in Theorem 12.3 we obtain a linear relation between $\sum_{m=1}^{n-1}\sigma_3(m)\sigma_3(n-m)$ and $\sum_{m=1}^{n-1}\sigma(m)\sigma_5(n-m) = \sum_{m=1}^{n-1}\sigma(n-m)\sigma_5(m)$, see Problem 3. In Chapter 13 we will see how to evaluate each of these two convolution sums individually, see Theorems 13.4 and 13.5.

Example 12.4. We take $f(x) = x^2$ in Problem 10 of Exercises 10. The left hand side is

$$-4\sum_{\substack{m \in \mathbb{N} \\ m < n}}\sigma^*(m)\sigma^*(n-m).$$

The right hand side is

$$-\sigma_3(n) + \sigma_3(n/2) + n\sigma(n) - n\sigma(n/2).$$

Hence

$$\sum_{\substack{m \in \mathbb{N} \\ m < n}} \sigma^*(m)\sigma^*(n-m) = \frac{1}{4}\sigma_3^*(n) - \frac{1}{4}n\sigma^*(n).$$

Our next result follows from the identity of Example 12.4 using Besge's formula (Theorem 12.1).

Theorem 12.4. *Let* $n \in \mathbb{N}$. *Then*

$$\sum_{\substack{m \in \mathbb{N} \\ m < n/2}} \sigma(m)\sigma(n-2m) = \frac{1}{12}\sigma_3(n) + \frac{1}{3}\sigma_3(n/2) + \left(\frac{1}{24} - \frac{1}{8}n\right)\sigma(n)$$

$$+ \left(\frac{1}{24} - \frac{1}{4}n\right)\sigma(n/2).$$

Proof. Recall that for $m \in \mathbb{N}$ we have $\sigma^*(m) = \sigma(m) - \sigma(m/2)$ and $\sigma_3^*(m) = \sigma_3(m) - \sigma_3(m/2)$. Using these in Example 12.4 we obtain

$$\sum_{m=1}^{n-1}(\sigma(m) - \sigma(m/2))(\sigma(n-m) - \sigma((n-m)/2))$$

$$= \frac{1}{4}\sigma_3(n) - \frac{1}{4}\sigma_3(n/2) - \frac{1}{4}n\sigma(n) + \frac{1}{4}\sigma(n/2).$$

Expanding the left hand side, we obtain

$$\sum_{m=1}^{n-1}\sigma(m)\sigma(n-m) - 2\sum_{\substack{m \in \mathbb{N} \\ m < n/2}}\sigma(m)\sigma(n-2m) + \sum_{\substack{m \in \mathbb{N} \\ m < n/2}}\sigma(m)\sigma(n/2-m)$$

$$= \frac{1}{4}\sigma_3(n) - \frac{1}{4}\sigma_3(n/2) - \frac{1}{4}n\sigma(n) + \frac{1}{4}\sigma(n/2).$$

By Besge's formula (Theorem 12.1) we have

$$\sum_{m=1}^{n-1}\sigma(m)\sigma(n-m) = \frac{5}{12}\sigma_3(n) + \left(\frac{1}{12} - \frac{1}{2}n\right)\sigma(n)$$

and

$$\sum_{\substack{m \in \mathbb{N} \\ m < n/2}}\sigma(m)\sigma(n/2-m) = \frac{5}{12}\sigma_3(n/2) + \left(\frac{1}{12} - \frac{1}{4}n\right)\sigma(n/2).$$

Using these evaluations we obtain the assertion of the theorem. $\square$

Glaisher extended Besge's formula by replacing $\sigma(n)$ in the convolution sum in Theorem 12.1 by other sums over the divisors of n. These sums are $d_1(n), \ldots, d_7(n)$, which are defined for $n \in \mathbb{N}$ as follows:

$$d_1(n) := \sum_{\substack{d \in \mathbb{N} \\ d \mid n}} d = \sigma(n),$$

$$d_2(n) := \sum_{\substack{d \in \mathbb{N} \\ d \mid n \\ 2 \nmid n}} d = \sigma(n) - 2\sigma(n/2),$$

$$d_3(n) := \sum_{\substack{d \in \mathbb{N} \\ d \mid n \\ 2 \mid n}} d = 2\sigma(n/2),$$

$$d_4(n) := \sum_{\substack{d \in \mathbb{N} \\ d \mid n \\ 2 \nmid n/d}} d = \sigma(n) - \sigma(n/2),$$

$$d_5(n) := \sum_{\substack{d \in \mathbb{N} \\ d \mid n \\ 2 \mid n/d}} d = \sigma(n/2),$$

$$d_6(n) := \sum_{\substack{d \in \mathbb{N} \\ d \mid n}} (-1)^{d-1} d = \sigma(n) - 4\sigma(n/2),$$

$$d_7(n) := \sum_{\substack{d \in \mathbb{N} \\ d \mid n}} (-1)^{n/d-1} d = \sigma(n) - 2\sigma(n/2).$$

For $r, s \in \{1, 2, 3, 4, 5, 6, 7\}$ and $n \in \mathbb{N}$ we define the convolution sum

$$D(r, s; n) := \sum_{m=1}^{n-1} d_r(m) d_s(n - m).$$

Clearly

$$D(r, s; n) = D(s, r; n).$$

Glaisher evaluated $D(r, s; n)$ in 1885. As $d_3(n) = 2d_5(n)$ and $d_7(n) = d_2(n)$ we can exclude $r, s \in \{3, 7\}$ from the evauation of $D(r, s; n)$.

Theorem 12.5. *Let $n \in \mathbb{N}$. Then*

$$24D(1, 1; n) = 10\sigma_3(n) - 12n\sigma(n) + 2\sigma(n),$$
$$24D(1, 2; n) = 6\sigma_3(n) - 16\sigma_3(n/2) - 6n\sigma(n) + 12n\sigma(n/2) - 2\sigma(n/2),$$

$$24D(1,4;n) = 8\sigma_3(n) - 8\sigma_3(n/2) - 9n\sigma(n) + \sigma(n) + 6n\sigma(n/2) - \sigma(n/2),$$
$$24D(1,5;n) = 2\sigma_3(n) + 8\sigma_3(n/2) - 3n\sigma(n) + \sigma(n) - 6n\sigma(n/2) + \sigma(n/2),$$
$$24D(1,6;n) = 2\sigma_3(n) - 32\sigma_3(n/2) - 2\sigma(n) + 24n\sigma(n/2) - 4\sigma(n/2),$$
$$24D(2,2;n) = 2\sigma_3(n) + 8\sigma_3(n/2) - 2\sigma(n) + 4\sigma(n/2),$$
$$24D(2,4;n) = 4\sigma_3(n) - 4\sigma_3(n/2) - 3n\sigma(n) - \sigma(n) + 6n\sigma(n/2) + \sigma(n/2),$$
$$24D(2,5;n) = 2\sigma_3(n) - 12\sigma_3(n/2) - 3n\sigma(n) + \sigma(n) + 6n\sigma(n/2) - 3\sigma(n/2),$$
$$24D(2,6;n) = -2\sigma_3(n) + 32\sigma_3(n/2) + 6n\sigma(n) - 4\sigma(n) - 12n\sigma(n/2) + 10\sigma(n/2),$$
$$24D(4,4;n) = 6\sigma_3(n) - 6\sigma_3(n/2) - 6n\sigma(n) + 6n\sigma(n/2),$$
$$24D(4,5;n) = 2\sigma_3(n) - 2\sigma_3(n/2) - 3n\sigma(n) + \sigma(n) - \sigma(n/2),$$
$$24D(4,6;n) = 3n\sigma(n) - 3\sigma(n) + 6n\sigma(n/2) + 3\sigma(n/2),$$
$$24D(5,5;n) = 10\sigma_3(n/2) - 6n\sigma(n/2) + 2\sigma(n/2),$$
$$24D(5,6;n) = 2\sigma_3(n) - 32\sigma_3(n/2) - 3n\sigma(n) + \sigma(n) + 18n\sigma(n/2) - 7\sigma(n/2),$$
$$24D(6,6;n) = -6\sigma_3(n) + 96\sigma_3(n/2) + 12n\sigma(n) - 6\sigma(n) - 48n\sigma(n/2) + 24\sigma(n/2).$$

Proof. It suffices to prove one of the formulae of the theorem as the remainder can be proved in a similar manner. We evaluate $D(6,6;n)$. We have

$$D(6,6;n) = \sum_{m=1}^{m-1} d_6(m)d_6(n-m)$$

$$= \sum_{m=1}^{m-1} (\sigma(m) - 4\sigma(m/2))(\sigma(n-m) - 4\sigma((n-m)/2))$$

$$= \sum_{m=1}^{m-1} \sigma(m)\sigma(n-m) - 8 \sum_{\substack{m \in \mathbb{N} \\ m < n/2}} \sigma(m)\sigma(n-2m)$$

$$+ 16 \sum_{\substack{m \in \mathbb{N} \\ m < n/2}} \sigma(m)\sigma(n/2-m)$$

$$= \frac{5}{12}\sigma_3(n) + \left(\frac{1}{12} - \frac{1}{2}n\right)\sigma(n)$$

$$- 8\left(\frac{1}{12}\sigma_3(n) + \frac{1}{3}\sigma_3(n/2) + \left(\frac{1}{24} - \frac{1}{8}n\right)\sigma(n)\right.$$

$$\left. + \left(\frac{1}{24} - \frac{1}{4}n\right)\sigma(n/2)\right)$$

$$+ 16\left(\frac{5}{12}\sigma_3(n/2) + \left(\frac{1}{12} - \frac{1}{4}n\right)\sigma(n/2)\right)$$

$$= -\frac{1}{4}\sigma_3(n) + 4\sigma_3(n/2) + \frac{1}{2}n\sigma(n) - \frac{1}{4}\sigma(n) - 2n\sigma(n/2) + \sigma(n/2),$$

as asserted. $\qquad\square$

Glaisher has observed how the value of the sum $\sum_{m=1}^{n-1} mf(m)f(n-m)$ can be determined from that of $\sum_{m=1}^{n-1} f(m)f(n-m)$.

Theorem 12.6. *Let* $f : \mathbb{N} \to \mathbb{C}$. *Let* $n \in \mathbb{N}$. *Then*

$$\sum_{m=1}^{n-1} mf(m)f(n-m) = \frac{1}{2}n\sum_{m=1}^{n-1} f(m)f(n-m).$$

Proof. Changing the summation variable from m to $n-m$, we obtain

$$\sum_{m=1}^{n-1} mf(m)f(n-m) = \sum_{m=1}^{n-1}(n-m)f(n-m)f(m)$$

$$= n\sum_{m=1}^{n-1} f(m)f(n-m) - \sum_{m=1}^{n-1} mf(m)f(n-m),$$

from which the asserted identity follows. $\qquad\square$

Theorem 12.6 allows us to deduce the value of the sum $\sum_{m=1}^{n-1} m\sigma(m)\sigma(n-m)$ from Besge's formula (Theorem 12.1).

Theorem 12.7. *Let* $n \in \mathbb{N}$. *Then*

$$\sum_{m=1}^{n-1} m\sigma(m)\sigma(n-m) = \frac{5}{24}n\sigma_3(n) + \left(\frac{1}{24}n - \frac{1}{4}n^2\right)\sigma(n).$$

Proof. Taking $f(n) = \sigma(n)$ $(n \in \mathbb{N})$ in Theorem 12.6 and appealing to Besge's formula, we obtain the asserted identity. $\qquad\square$

Our next theorem shows that the sum $\sum_{m=1}^{n-1} m^3 f(m)f(n-m)$ can be obtained from the two sums $\sum_{m=1}^{n-1} f(m)f(n-m)$ and $\sum_{m=1}^{n-1} m^2 f(m)f(n-m)$.

Theorem 12.8. *Let* $f : \mathbb{N} \to \mathbb{C}$. *Let* $n \in \mathbb{N}$. *Then*

$$\sum_{m=1}^{n-1} m^3 f(m)f(n-m) = \frac{3}{2}n\sum_{m=1}^{n-1} m^2 f(m)f(n-m) - \frac{1}{4}n^3\sum_{m=1}^{n-1} f(m)f(n-m).$$

Proof. Changing the summation variable from m to $n-m$, we obtain

$$\sum_{m=1}^{n-1} m^3 f(m)f(n-m) = \sum_{m=1}^{n-1}(n-m)^3 f(n-m)f(m)$$

$$= n^3\sum_{m=1}^{n-1} f(m)f(n-m) - 3n^2\sum_{m=1}^{n-1} mf(m)f(n-m)$$

$$+ 3n\sum_{m=1}^{n-1} m^2 f(m)f(n-m) - \sum_{m=1}^{n-1} m^3 f(m)f(n-m)$$

so that

$$2 \sum_{m=1}^{n-1} m^3 f(m) f(n-m) = n^3 \sum_{m=1}^{n-1} f(m) f(n-m) - 3n^2 \sum_{m=1}^{n-1} m f(m) f(n-m)$$

$$+ 3n \sum_{m=1}^{n-1} m^2 f(m) f(n-m).$$

Appealing to Theorem 12.6 for the value of $\sum_{m=1}^{n-1} m f(m) f(n-m)$, and then dividing by 2, we obtained the required result. $\qquad\square$

It is clear that in general we can express

$$\sum_{m=1}^{n-1} m^{2k+1} f(m) f(n-m), \quad k \in \mathbb{N}_0,$$

in terms of

$$\sum_{m=1}^{n-1} m^{2j} f(m) f(n-m), \quad j = 0, 1, \ldots, k.$$

Exercises 12

1. Let $f : \mathbb{N} \to \mathbb{C}$. Let $n \in \mathbb{N}$. Express

$$\sum_{m=1}^{n-1} m^5 f(m) f(n-m)$$

in terms of

$$\sum_{m=1}^{n-1} m^4 f(m) f(n-m), \quad \sum_{m=1}^{n-1} m^2 f(m) f(n-m) \text{ and } \sum_{m=1}^{n-1} f(m) f(n-m).$$

2. Let $f : \mathbb{N} \to \mathbb{C}$. Let $n \in \mathbb{N}$. Prove that

$$\sum_{\substack{(r,s,t) \in \mathbb{N}^3 \\ r+s+t = n}} r f(r) f(s) f(t) = \frac{1}{3} n \sum_{\substack{(r,s,t) \in \mathbb{N}^3 \\ r+s+t = n}} f(r) f(s) f(t).$$

3. Let $n \in \mathbb{N}$. By taking $k = 3$ in Theorem 12.3, prove that

$$5 \sum_{m=1}^{n-1} \sigma_3(m) \sigma_3(n-m) + 3 \sum_{m=1}^{n-1} \sigma(m) \sigma_5(n-m)$$

$$= \frac{9}{56} \sigma_7(n) + \left(\frac{1}{8} - \frac{1}{4} n \right) \sigma_5(n) - \frac{1}{24} \sigma_3(n) + \frac{1}{168} \sigma(n).$$

4. Let $n \in \mathbb{N}$. Take $k = 4$ in Theorem 12.3 to determine a linear relation

between $\displaystyle\sum_{m=1}^{n-1} \sigma(m)\sigma_7(n-m)$ and $\displaystyle\sum_{m=1}^{n-1} \sigma_3(m)\sigma_5(n-m)$.

5. Let $q \in \mathbb{C}$ be such that $|q| < 1$. Let $k \in \mathbb{N}$. Define

$$G_k(q) := \sum_{n=1}^{\infty} \sigma_k(n)q^n.$$

Prove that

$$G_3(q) = \frac{12}{5} G_1^2(q) - \frac{1}{5} G_1(q) + \frac{6}{5} q G_1'(q).$$

6. Deduce Theorem 12.7 by differentiating the result of Problem 5.

7. Let $n \in \mathbb{N}$. Use $f(x) = (-1)^x x^2$ in Theorem 10.1 to prove Theorem 12.4.

8. Prove that

$$G_1(q)G_3(q) = \frac{7}{80} G_5(q) + \frac{1}{24} G_3(q) - \frac{1}{240} G_1(q) - \frac{1}{8} q G_3'(q).$$

9. Use Theorem 12.4 to determine a formula for $G_1(q)G_1(q^2)$.

Notes on Chapter 12

An extract from a letter from Besge to Liouville giving the value of the sum $\sum_{m=1}^{n-1} \sigma(m)\sigma(n-m)$ is contained in [48]. The erroneous attribution by Dickson occurs in [93, Vol. II, p. 338]. Rankin's attribution to Besgue is given in [236, p. 115]. Lützen's assertion that Besge/Besgue is a pseudonym for Liouville is given in [192, p. 81]. Besge's formula appears in Glaisher [109], [110], [111], Lahiri [159], Lehmer [165], [167, Vol. II, pp. 677–688], Ramanujan [230], [232] and Skoruppa [242]. The proof of Besge's formula (Theorem 12.1) given here is taken from Huard, Spearman, Ou and Williams [137]. Glaisher's theorem (Theorem 12.2) occurs in Glaisher [109, p. 156]. Glaisher [111, p. 33] also determined

$$\left(\sum_{n=1}^{\infty} \sigma(n)q^n\right)^r$$

for $r = 3, 4$ and 5. Theorem 12.3 is based on an assertion of Liouville [170, 1st article, p. 149]. Theorem 12.4 is implicit in the work of Glaisher [110, p. 11] and in the form given here is due to Huard, Spearman, Ou and Williams [137, Theorem 2, p. 247]. Theorem 12.5 is due to Glaisher [110, p. 11]. The first line of Theorem 12.5 is Besge's formula, the fourth line is Theorem 12.4, and the tenth line is the formula of Example 12.4 (as $\sigma^*(n) = d_4(n)$). Theorem 12.6 is an observation of Glaisher [111, p. 36].

Example 12.2 is a result attributed to Glaisher by MacMahon [193, p. 101], [194, Vol. II, p. 329]. It also appears in Ramanujan [230, Table IV], [232, p. 146], Lahiri [159, formula (5.1), p. 198] and Huard, Spearman, Ou and Williams [137, formula (3.12), p. 237].

Hahn [121, p. 3] has defined for $s, n \in \mathbb{N}$

$$\tilde{\sigma}_s(n) := \sum_{\substack{d \in \mathbb{N} \\ d \mid n}} (-1)^{d-1} d^s$$

and

$$\hat{\sigma}_s(n) := \sum_{\substack{d \in \mathbb{N} \\ d \mid n}} (-1)^{n/d-1} d^s.$$

By Problem 2 of Exercises 3 we have

$$\tilde{\sigma}_s(n) = \sigma_s(n) - 2^{s+1}\sigma_s(n/2) \tag{12.1}$$

and by Problem 1 of Exercises 3

$$\hat{\sigma}_s(n) = \sigma_s(n) - 2\sigma_s(n/2). \tag{12.2}$$

Further set

$$\tilde{\sigma}(n) := \tilde{\sigma}_1(n) = d_6(n) = \sigma(n) - 4\sigma(n/2),$$
$$\hat{\sigma}(n) := \hat{\sigma}_1(n) = d_7(n) = \sigma(n) - 2\sigma(n/2)$$

and

$$\tilde{\sigma}_s(n) = \hat{\sigma}_s(n) = 0, \quad n \notin \mathbb{N}.$$

The functions $\tilde{\sigma}_s$ and $\hat{\sigma}_s$ are related by

$$\tilde{\sigma}_s(n) - 2\tilde{\sigma}_s(n/2) - \hat{\sigma}_s(n) + 2^{s+1}\hat{\sigma}_s(n/2) = 0, \quad n \in \mathbb{N}.$$

Hahn [121, Theorem 2.7.2, pp. 55–56] has shown that

$$\sum_{m=1}^{n-1} \tilde{\sigma}(m)\tilde{\sigma}(n-m) = -\frac{1}{4}\tilde{\sigma}_3(n) + \left(\frac{1}{2}n - \frac{1}{4}\right)\tilde{\sigma}(n), \tag{12.3}$$

$$\sum_{m=1}^{n-1} \hat{\sigma}(m)\tilde{\sigma}(n-m) = -\frac{1}{12}\tilde{\sigma}_3(n) + \left(\frac{1}{4}n - \frac{1}{8}\right)\hat{\sigma}(n) - \frac{1}{24}\tilde{\sigma}(n). \tag{12.4}$$

Equation (12.3) is a reformulation of Glaisher's formula (Theorem 12.5)

$$\sum_{m=1}^{n-1} d_6(m)d_6(n-m) = -\frac{1}{4}\sigma_3(n) + 4\sigma_3(n/2) - \left(\frac{1}{4} - \frac{1}{2}n\right)\sigma(n) + (1 - 2n)\sigma(n/2)$$

and equation (12.4) of Glaisher's formula

$$\sum_{m=1}^{n-1} d_7(m)d_6(n-m)$$

$$= -\frac{1}{12}\sigma_3(n) + \frac{4}{3}\sigma_3(n/2) - \left(\frac{1}{6} - \frac{1}{4}n\right)\sigma(n) + \left(\frac{5}{12} - \frac{1}{2}n\right)\sigma(n/2).$$

Hahn did not consider the sum $\sum_{m=1}^{n-1} \hat{\sigma}(m)\hat{\sigma}(n-m)$. We have

$$\sum_{m=1}^{n-1} \hat{\sigma}(m)\hat{\sigma}(n-m) = \sum_{m=1}^{n-1} d_2(m)d_2(n-m)$$

$$= \frac{1}{12}\sigma_3(n) + \frac{1}{3}\sigma_3(n/2) - \frac{1}{12}\sigma(n) + \frac{1}{6}\sigma(n/2)$$

$$= \frac{5}{42}\hat{\sigma}_3(n) - \frac{1}{28}\tilde{\sigma}_3(n) - \frac{1}{12}\hat{\sigma}(n),$$

by Theorem 12.5. Hahn [121, pp. 56, 58, 60] has also evaluated other convolution sums involving $\hat{\sigma}$ and $\tilde{\sigma}$.

Huber [141, Section 4.8, pp. 135–141] has extended ideas of Ramanujan to obtain some formulae similar to Theorem 12.3. Unfortunately some of his results contain errors. For example Huber's equation (4.211) fails for $k = 1$ and $k = 4$.

Cheng [68, Theorem 2.9.1, p. 102] has evaluated a number of sums similar to that of Theorem 12.4.

13

An Identity of Huard, Ou, Spearman and Williams

In 2000 Huard, Ou, Spearman and Williams proved a far reaching generalization of Liouville's identity (Theorem 10.1).

Theorem 13.1. *Let $f : \mathbb{Z}^4 \to \mathbb{C}$ be such that*

$$f(a, b, x, y) - f(x, y, a, b) = f(-a, -b, x, y) - f(x, y, -a, -b) \quad (13.1)$$

for all integers a, b, x and y. Let $n \in \mathbb{N}$. Then

$$\sum_{\substack{(a, b, x, y) \in \mathbb{N}^4 \\ ax + by = n}} (f(a, b, x, -y) - f(a, -b, x, y) + f(a, a - b, x + y, y)$$

$$- f(a, a + b, y - x, y) + f(b - a, b, x, x + y) - f(a + b, b, x, x - y))$$

$$= \sum_{\substack{d \in \mathbb{N} \\ d \mid n}} \sum_{\substack{x \in \mathbb{N} \\ x < d}} (f(0, n/d, x, d) + f(n/d, 0, d, x) + f(n/d, n/d, d - x, -x)$$

$$- f(x, x - d, n/d, n/d) - f(x, d, 0, n/d) - f(d, x, n/d, 0)).$$

Proof. We set

$$g(a, b, x, y) = f(a, b, x, y) - f(x, y, a, b)$$

so that

$$g(a, -b, x, y) = g(-a, b, x, y)$$

and

$$g(a, b, x, y) = -g(x, y, a, b).$$

137

Then

$$\sum_{\substack{(a,b,x,y) \in \mathbb{N}^4 \\ ax+by=n}} (f(a,b,x,-y) - f(a,-b,x,y) + f(a,a-b,x+y,y)$$

$$- f(a,a+b,y-x,y) + f(b-a,b,x,x+y) - f(a+b,b,x,x-y))$$

$$= \sum_{\substack{(a,b,x,y) \in \mathbb{N}^4 \\ ax+by=n}} (f(a,b,x,-y) - f(x,-y,a,b) + f(a,a-b,x+y,y)$$

$$- f(y,x+y,a-b,a) + f(a-b,a,y,x+y) - f(x+y,y,a,a-b))$$

$$= \sum_{\substack{(a,b,x,y) \in \mathbb{N}^4 \\ ax+by=n}} (g(a,a-b,x+y,y) + g(a-b,a,y,x+y) + g(a,b,x,-y))$$

and

$$\sum_{\substack{d \in \mathbb{N} \\ d\,|\,n}} \sum_{\substack{x \in \mathbb{N} \\ x<d}} (f(0,n/d,x,d) + f(n/d,0,d,x) + f(n/d,n/d,d-x,-x)$$

$$- f(x,x-d,n/d,n/d) - f(x,d,0,n/d) - f(d,x,n/d,0))$$

$$= \sum_{\substack{d \in \mathbb{N} \\ d\,|\,n}} \sum_{\substack{t \in \mathbb{N} \\ t<d}} (f(0,n/d,t,d) + f(n/d,0,d,t) + f(n/d,n/d,d-t,-t)$$

$$- f(d-t,-t,n/d,n/d) - f(t,d,0,n/d) - f(d,t,n/d,0))$$

$$= \sum_{\substack{d \in \mathbb{N} \\ d\,|\,n}} \sum_{\substack{t \in \mathbb{N} \\ t<d}} (g(n/d,0,d,t) + g(0,n/d,t,d) + g(n/d,n/d,d-t,-t))$$

so we must prove that

$$\sum_{\substack{(a,b,x,y) \in \mathbb{N}^4 \\ ax+by=n}} (g(a,a-b,x+y,y) + g(a-b,a,y,x+y) + g(a,b,x,-y))$$

$$= \sum_{\substack{d \in \mathbb{N} \\ d\,|\,n}} \sum_{\substack{t \in \mathbb{N} \\ t<d}} (g(n/d,0,d,t) + g(0,n/d,t,d) + g(n/d,n/d,d-t,-t)).$$

First we consider the terms with $a = b$ in the left hand sum. We have

$$\sum_{\substack{(a, b, x, y) \in \mathbb{N}^4 \\ ax + by = n \\ a = b}} (g(a, a - b, x + y, y) + g(a - b, a, y, x + y) + g(a, b, x, -y))$$

$$= \sum_{\substack{(a, x, y) \in \mathbb{N}^3 \\ a(x + y) = n}} (g(a, 0, x + y, y) + g(0, a, y, x + y) + g(a, a, x, -y))$$

$$= \sum_{\substack{d \in \mathbb{N} \\ d \mid n}} \sum_{\substack{t \in \mathbb{N} \\ t < d}} (g(n/d, 0, d, t) + g(0, n/d, t, d) + g(n/d, n/d, d - t, -t)).$$

Secondly we consider the terms with $a < b$. We have

$$\sum_{\substack{(a, b, x, y) \in \mathbb{N}^4 \\ ax + by = n \\ a < b}} (g(a, a - b, x + y, y) + g(a - b, a, y, x + y) + g(a, b, x, -y))$$

$$= \sum_{\substack{(a, b, x, y) \in \mathbb{N}^4 \\ a(x + y) + (b - a)y = n \\ a < b}} (g(a, a - b, x + y, y) + g(a - b, a, y, x + y))$$

$$+ \sum_{\substack{(a, b, x, y) \in \mathbb{N}^4 \\ ax + by = n \\ a < b}} g(a, b, x, -y)$$

$$= \sum_{\substack{(a, b, x, y) \in \mathbb{N}^4 \\ ax + by = n \\ x > y}} (g(a, -b, x, y) + g(-b, a, y, x))$$

$$+ \sum_{\substack{(a, b, x, y) \in \mathbb{N}^4 \\ ax + by = n \\ x < y}} g(x, y, a, -b)$$

$$= \sum_{\substack{(a, b, x, y) \in \mathbb{N}^4 \\ ax + by = n \\ x > y}} g(a, -b, x, y) + \sum_{\substack{(a, b, x, y) \in \mathbb{N}^4 \\ ax + by = n \\ x > y}} g(b, -a, y, x)$$

$$- \sum_{\substack{(a, b, x, y) \in \mathbb{N}^4 \\ ax + by = n \\ x < y}} g(a, -b, x, y)$$

$$= - \sum_{\substack{(a,b,x,y) \in \mathbb{N}^4 \\ ax+by=n \\ x>y}} g(x,y,a,-b) + \sum_{\substack{(a,b,x,y) \in \mathbb{N}^4 \\ ax+by=n \\ x<y}} g(a,-b,x,y)$$

$$- \sum_{\substack{(a,b,x,y) \in \mathbb{N}^4 \\ ax+by=n \\ x<y}} g(a,-b,x,y)$$

$$= - \sum_{\substack{(a,b,x,y) \in \mathbb{N}^4 \\ ax+by=n \\ x>y}} g(x,y,a,-b).$$

Thirdly we consider the terms with $a > b$. We have

$$\sum_{\substack{(a,b,x,y) \in \mathbb{N}^4 \\ ax+by=n \\ a>b}} (g(a,a-b,x+y,y)$$

$$+ g(a-b,a,y,x+y) + g(a,b,x,-y)) = S_1 + S_2,$$

where

$$S_1 = \sum_{\substack{(a,b,x,y) \in \mathbb{N}^4 \\ ax+by=n \\ a>b}} (g(a,a-b,x+y,y) + g(a-b,a,y,x+y))$$

$$= \sum_{\substack{(a,b,x,y) \in \mathbb{N}^4 \\ ax+b(x+y)=n}} (g(a+b,a,x+y,y) + g(a,a+b,y,x+y))$$

$$= \sum_{\substack{(a,b,x,y) \in \mathbb{N}^4 \\ ax+by=n \\ y>x}} (g(a+b,a,y,y-x) + g(a,a+b,y-x,y))$$

$$= \sum_{\substack{(a,b,x,y) \in \mathbb{N}^4 \\ ax+by=n \\ a>b}} (g(x+y,y,a,a-b) + g(y,x+y,a-b,a))$$

$$= -S_1,$$

so that $S_1 = 0$, and

$$S_2 = \sum_{\substack{(a,b,x,y) \in \mathbb{N}^4 \\ ax+by=n \\ a>b}} g(a,b,x,-y) = \sum_{\substack{(a,b,x,y) \in \mathbb{N}^4 \\ ax+by=n \\ x>y}} g(x,y,a,-b).$$

This completes the proof of the theorem. $\qquad\qquad\qquad\qquad\qquad\qquad$ □

We remark that if f satisfies

$$f(a, -b, x, y) = f(-a, b, x, y), \quad f(a, b, x, -y) = f(a, b, -x, y),$$

for all integers a, b, x and y then f satisfies (13.1).

We now deduce Theorem 10.3 from Theorem 13.1.

Proof of Theorem 10.3. Let $f : \mathbb{Z} \times \mathbb{Z} \to \mathbb{C}$ satisfy

$$f(x, y) = f(-x, y) = f(x, -y)$$

for all integers x and y. Hence $f(x, y) = f(-x, -y)$ for all $x, y \in \mathbb{Z}$. Define

$$F(a, b, x, y) = f(a - b, x - y), \quad a, b, x, y \in \mathbb{Z}.$$

Then

$$
\begin{aligned}
F(a, b, x, y) - F(x, y, a, b) &= f(a - b, x - y) - f(x - y, a - b) \\
&= f(-a + b, x - y) - f(x - y, -a + b) \\
&= F(-a, -b, x, y) - F(x, y, -a, -b)
\end{aligned}
$$

so that F satisfies (13.1).

With this choice of F, the left hand side of Theorem 13.1 reduces to

$$\sum_{\substack{(a, b, x, y) \in \mathbb{N}^4 \\ ax + by = n}} (f(a - b, x + y) - f(a + b, x - y))$$

and the right hand side is

$$\sum_{\substack{d \in \mathbb{N} \\ d \mid n}} \sum_{\substack{x \in \mathbb{N} \\ x < d}} (f(-n/d, x - d) + f(n/d, d - x) + f(0, d)$$

$$- f(d, 0) - f(x - d, -n/d) - f(d - x, n/d))$$

$$= \sum_{\substack{d \in \mathbb{N} \\ d \mid n}} (d - 1)(f(0, d) - f(d, 0))$$

$$+ 2 \sum_{\substack{d \in \mathbb{N} \\ d \mid n}} \sum_{\substack{x \in \mathbb{N} \\ x < d}} (f(n/d, d - x) - f(d - x, n/d)).$$

Now

$$\sum_{\substack{d \in \mathbb{N} \\ d \mid n}} \sum_{\substack{x \in \mathbb{N} \\ x < d}} (f(n/d, d - x) - f(d - x, n/d))$$

$$= \sum_{\substack{d \in \mathbb{N} \\ d \mid n}} \sum_{\substack{x \in \mathbb{N} \\ x < n/d}} (f(d, n/d - x) - f(n/d - x, d))$$

$$= \sum_{\substack{d \in \mathbb{N} \\ d \mid n}} \sum_{\substack{e \in \mathbb{N} \\ e < n/d}} (f(d, e) - f(e, d)).$$

Equating the left and right hand sides, we obtain Theorem 10.3.

In 1993 Skoruppa proved an identity of Liouville type. We show that Skoruppa's identity is also a special case of Theorem 13.1.

Theorem 13.2. *Let $h : \mathbb{Z} \times \mathbb{Z} \to \mathbb{C}$ be such that*

$$h(y, y - x) = h(x, y), \quad x, y \in \mathbb{Z}. \tag{13.2}$$

Let $n \in \mathbb{N}$. Then

$$\sum_{\substack{(a, b, x, y) \in \mathbb{N}^4 \\ ax + by = n}} (h(a, b) - h(a, -b)) = \sum_{\substack{d \in \mathbb{N} \\ d \mid n}} \left(\frac{n}{d} h(d, 0) - \sum_{j=0}^{d-1} h(d, j) \right).$$

Proof. Appealing to (13.2) we obtain

$$h(a, b) = h(b, b - a) = h(b - a, -a) = h(-a, -b)$$
$$= h(-b, a - b) = h(a - b, a)$$

for all $a, b \in \mathbb{Z}$. In particular we have

$$h(a, 0) = h(0, -a) = h(-a, -a) = h(-a, 0) = h(0, a) = h(a, a)$$

for all $a \in \mathbb{Z}$. We choose

$$f(a, b, x, y) = h(a, b).$$

Then

$$f(a, b, x, y) - f(x, y, a, b) = h(a, b) - h(x, y)$$
$$= h(-a, -b) - h(x, y)$$
$$= f(-a, -b, x, y) - f(x, y, -a, -b)$$

so that (13.1) is satisfied. We note that

$$f(a, -b, x, y) = h(a, -b),$$
$$f(a, a - b, x + y, y) = h(a, a - b) = h(b, a),$$
$$f(a, a + b, y - x, y) = h(a, a + b) = h(-b, a),$$
$$f(b - a, b, x, x + y) = h(b - a, b) = h(b, a),$$
$$f(a + b, b, x, x - y) = h(a + b, b) = h(-b, a).$$

Before applying Theorem 13.1 with the specified choice of f, we make two simple but useful observations. First, mapping $(a, b, x, y) \rightarrow (b, a, y, x)$, we see that

$$\sum_{\substack{(a, b, x, y) \in \mathbb{N}^4 \\ ax + by = n}} h(a, b) = \sum_{\substack{(a, b, x, y) \in \mathbb{N}^4 \\ ax + by = n}} h(b, a).$$

Secondly we note that

$$\sum_{\substack{x \in \mathbb{N} \\ x < d}} h(x, d) = \sum_{\substack{x \in \mathbb{N} \\ x < d}} h(d - x, d) = \sum_{\substack{x \in \mathbb{N} \\ x < d}} h(d, x).$$

Thus the left hand side of Theorem 13.1 is

$$\sum_{\substack{(a, b, x, y) \in \mathbb{N}^4 \\ ax + by = n}} (h(a, b) - h(a, -b) + h(b, a) - h(-b, a) + h(b, a) - h(-b, a))$$

$$= 3 \sum_{\substack{(a, b, x, y) \in \mathbb{N}^4 \\ ax + by = n}} (h(a, b) - h(a, -b)),$$

and the right hand side is

$$\sum_{\substack{d \in \mathbb{N} \ x \in \mathbb{N} \\ d \mid n \ x < d}} (h(0, n/d) + h(n/d, 0) + h(n/d, n/d)$$

$$- h(x, x - d) - h(x, d) - h(d, x))$$

$$= 3 \sum_{\substack{d \in \mathbb{N} \ x \in \mathbb{N} \\ d \mid n \ x < d}} (h(n/d, 0) - h(d, x))$$

$$= 3 \sum_{\substack{d \in \mathbb{N} \\ d \mid n}} (d - 1)h(n/d, 0) - 3 \sum_{\substack{d \in \mathbb{N} \ x \in \mathbb{N} \\ d \mid n \ x < d}} h(d, x)$$

$$= 3 \sum_{\substack{d \in \mathbb{N} \\ d \mid n}} \left(\frac{n}{d} - 1 \right) h(d, 0) - 3 \sum_{\substack{d \in \mathbb{N} \\ d \mid n}} \sum_{\substack{x \in \mathbb{N} \\ x < d}} h(d, x)$$

$$= 3 \sum_{\substack{d \in \mathbb{N} \\ d \mid n}} \frac{n}{d} h(d, 0) - 3 \sum_{\substack{d \in \mathbb{N} \\ d \mid n}} \sum_{x=0}^{d-1} h(d, x).$$

Equating the left and right hand sides, we obtain the assertion of Theorem 13.2. □

Example 13.1. We show that Liouville's identity given in Chapter 2 (see (2.7)) follows from Theorem 13.1. For $\ell \in \mathbb{Z}$ we define

$$G_2(\ell) := \begin{cases} 0, & \text{if } 2 \mid \ell, \\ 1, & \text{if } 2 \nmid \ell. \end{cases} \tag{13.3}$$

The function G_2 has the property

$$G_2(\ell + 2m) = G_2(\ell), \quad \ell, m \in \mathbb{Z}. \tag{13.4}$$

Replacing ℓ by $\ell - m$ in (13.4), we see that

$$G_2(\ell + m) = G_2(\ell - m), \quad \ell, m \in \mathbb{Z}, \tag{13.5}$$

and taking $m = -\ell$ in (13.4), we obtain

$$G_2(\ell) = G_2(-\ell), \quad \ell \in \mathbb{Z}. \tag{13.6}$$

The function G_2 is related to the function F_2 by

$$G_2(\ell) = 1 - F_2(\ell), \quad \ell \in \mathbb{Z}, \tag{13.7}$$
$$G_2(\ell + 1) = F_2(\ell), \quad \ell \in \mathbb{Z}, \tag{13.8}$$
$$G_2(\ell) = F_2(\ell + 1), \quad \ell \in \mathbb{Z}, \tag{13.9}$$
$$F_2(\ell) G_2(\ell) = 0, \quad \ell \in \mathbb{Z}, \tag{13.10}$$
$$F_2(\ell + m) G_2(m) = G_2(\ell) G_2(m), \quad \ell, m \in \mathbb{Z}, \tag{13.11}$$
$$F_2(m) G_2(\ell + m) = F_2(m) G_2(\ell), \quad \ell, m \in \mathbb{Z}. \tag{13.12}$$

Let $f : \mathbb{Z} \longrightarrow \mathbb{C}$ be an even function. We choose

$$f(a, b, x, y) = f(a) F_2(a) G_2(b) G_2(x) F_2(y), \quad a, b, x, y \in \mathbb{Z}.$$

For all $a, b, x, y \in \mathbb{Z}$ we have

$$
\begin{aligned}
f(a, &b, x, y) - f(x, y, a, b) \\
&= f(a)F_2(a)G_2(b)G_2(x)F_2(y) - f(x)F_2(x)G_2(y)G_2(a)F_2(b) \\
&= f(-a)F_2(-a)G_2(-b)G_2(x)F_2(y) - f(x)F_2(x)G_2(y)G_2(-a)F_2(-b) \\
&= f(-a, -b, x, y) - f(x, y, -a, -b)
\end{aligned}
$$

so that (13.1) is satisfied. With this choice of f, and with n replaced by $2n$, the left hand side of Theorem 13.1 becomes using properties (13.6) and (13.11)

$$
\begin{aligned}
\sum_{\substack{(a, b, x, y) \in \mathbb{N}^4 \\ ax + by = 2n}} & (f(b - a)F_2(b - a)G_2(b)G_2(x)F_2(x + y) \\
& - f(a + b)F_2(a + b)G_2(b)G_2(x)F_2(x - y)) \\
= \sum_{\substack{(a, b, x, y) \in \mathbb{N}^4 \\ ax + by = 2n}} & (f(a - b) - f(a + b))G_2(a)G_2(b)G_2(x)G_2(y) \\
= \sum_{\substack{(a, b, x, y) \in \mathbb{N}^4 \\ ax + by = 2n \\ a, b, x, y \text{ odd}}} & (f(a - b) - f(a + b)).
\end{aligned}
$$

Putting this choice of f into the right hand side of Theorem 13.1, we see that the second and fifth sums vanish as $G_2(0) = 0$, and that the third and fourth sums vanish in view of property (13.10). Thus the right hand side is $A - B$, where

$$
\begin{aligned}
A := \sum_{\substack{d \in \mathbb{N} \\ d \mid 2n \\ 2n/d \text{ odd}, d \text{ even}}} \sum_{\substack{x = 1 \\ x \text{ odd}}}^{d-1} f(0) \\
= f(0) \sum_{\substack{d \in \mathbb{N} \\ d \mid n \\ n/d \text{ odd}}} \sum_{\substack{x = 1 \\ x \text{ odd}}}^{2d-1} 1 \\
= f(0) \sum_{\substack{d \in \mathbb{N} \\ d \mid n \\ n/d \text{ odd}}} d
\end{aligned}
$$

and

$$B := \sum_{\substack{d \in \mathbb{N} \\ d \mid 2n \\ 2n/d \text{ odd}, d \text{ even}}} \sum_{\substack{x = 1 \\ x \text{ odd}}}^{d-1} f(d)$$

$$= \sum_{\substack{d \in \mathbb{N} \\ d \mid n \\ n/d \text{ odd}}} f(2d) \sum_{\substack{x = 1 \\ x \text{ odd}}}^{2d-1} 1$$

$$= \sum_{\substack{d \in \mathbb{N} \\ d \mid n \\ n/d \text{ odd}}} df(2d).$$

This completes the proof.

For $e, f, n \in \mathbb{N}$ we define the convolution sum

$$S_{e,f}(n) := \sum_{m=1}^{n-1} \sigma_e(m)\sigma_f(n - m). \tag{13.13}$$

Clearly we have

$$S_{e,f}(n) = S_{f,e}(n)$$

for all $e, f, n \in \mathbb{N}$. Ramanujan showed that the sum $S_{e,f}(n)$ can be evaluated in terms $\sigma_{e+f+1}(n)$, $\sigma_{e+f-1}(n)$, ..., $\sigma_3(n)$, $\sigma(n)$ for the nine pairs $(e, f) \in \mathbb{N}^2$ satisfying

$$e + f = 2, 4, 6, 8, 12, \quad e \le f, \quad e \equiv f \equiv 1 \;(\text{mod } 2).$$

Huard, Ou, Spearman and Williams proved that all nine of Ramanujan's evaluations follow from Theorem 13.1 and thus all can be proved in an elementary arithmetic manner in the spirit of Liouville. The value of $S_{1,1}(n)$ is given in Theorem 12.1. We now give the remaining eight of Ramanujan's results leaving their verification to the reader (see Exercises 13).

Theorem 13.3. *Let $n \in \mathbb{N}$. Then*

$$\sum_{m=1}^{n-1} \sigma(m)\sigma_3(n - m) = \frac{7}{80}\sigma_5(n) + \left(\frac{1}{24} - \frac{1}{8}n\right)\sigma_3(n) - \frac{1}{240}\sigma(n).$$

Proof. Take $f(a, b, x, y) = xy^3 + x^3y$ in Theorem 13.1. $\square$

We remark that the identity of Theorem 13.3 was proved in a different way in Example 12.2.

Theorem 13.4. *Let $n \in \mathbb{N}$. Then*

$$\sum_{m=1}^{n-1} \sigma(m)\sigma_5(n-m) = \frac{5}{126}\sigma_7(n) + \left(\frac{1}{24} - \frac{1}{12}n\right)\sigma_5(n) + \frac{1}{504}\sigma(n).$$

Proof. Take $f(a, b, x, y) = xy^5 - 20x^3y^3 + x^5y$ in Theorem 13.1.   □

Theorem 13.5. *Let $n \in \mathbb{N}$. Then*

$$\sum_{m=1}^{n-1} \sigma_3(m)\sigma_3(n-m) = \frac{1}{120}\sigma_7(n) - \frac{1}{120}\sigma_3(n).$$

Proof. Take $f(a, b, x, y) = xy^5 - 2x^3y^3 + x^5y$ in Theorem 13.1.   □

Theorem 13.6. *Let $n \in \mathbb{N}$. Then*

$$\sum_{m=1}^{n-1} \sigma(m)\sigma_7(n-m) = \frac{11}{480}\sigma_9(n) + \left(\frac{1}{24} - \frac{1}{16}n\right)\sigma_7(n) - \frac{1}{480}\sigma(n).$$

Proof. Take

$$f(a, b, x, y) = 11xy^7 - 56x^3y^5 - 56x^5y^3 + 11x^7y$$

in Theorem 13.1.   □

Theorem 13.7. *Let $n \in \mathbb{N}$. Then*

$$\sum_{m=1}^{n-1} \sigma_3(m)\sigma_5(n-m) = \frac{11}{5040}\sigma_9(n) - \frac{1}{240}\sigma_5(n) + \frac{1}{504}\sigma_3(n).$$

Proof. Take

$$f(a, b, x, y) = xy^7 - x^3y^5 - x^5y^3 + x^7y$$

in Theorem 13.1.   □

Theorem 13.8. *Let $n \in \mathbb{N}$. Then*

$$\sum_{m=1}^{n-1} \sigma(m)\sigma_{11}(n-m) = \frac{691}{65520}\sigma_{13}(n) + \left(\frac{1}{24} - \frac{1}{24}n\right)\sigma_{11}(n) - \frac{691}{65520}\sigma(n).$$

Proof. Take

$$f(a, b, x, y) = 271xy^{11} - 1540x^3y^9 + 1584x^5y^7 + 1584x^7y^5$$
$$- 1540x^9y^3 + 271x^{11}y$$

in Theorem 13.1.   □

Theorem 13.9. *Let $n \in \mathbb{N}$. Then*

$$\sum_{m=1}^{n-1} \sigma_3(m)\sigma_9(n-m) = \frac{1}{2640}\sigma_{13}(n) - \frac{1}{240}\sigma_9(n) + \frac{1}{264}\sigma_3(n).$$

Proof. Take

$$f(a, b, x, y) = 2xy^{11} - 11x^3y^9 + 9x^5y^7 + 9x^7y^5 - 11x^9y^3 + 2x^{11}y$$

in Theorem 13.1. ☐

Theorem 13.10. *Let $n \in \mathbb{N}$. Then*

$$\sum_{m=1}^{n-1} \sigma_5(m)\sigma_7(n-m) = \frac{1}{10080}\sigma_{13}(n) + \frac{1}{504}\sigma_7(n) - \frac{1}{480}\sigma_5(n).$$

Proof. Take

$$f(a, b, x, y) = 8xy^{11} - 35x^3y^9 + 27x^5y^7 + 27x^7y^5 - 35x^9y^3 + 8x^{11}y$$

in Theorem 13.1. ☐

Example 13.2. Let $n \in \mathbb{N}$. We use Theorems 12.1, 12.7 and 13.3 to show that

$$\sum_{\substack{(r,s,t) \in \mathbb{N}^3 \\ r+s+t=n}} \sigma(r)\sigma(s)\sigma(t) = \frac{7}{192}\sigma_5(n) + \left(\frac{5}{96} - \frac{5}{32}n\right)\sigma_3(n)$$

$$+ \left(\frac{1}{192} - \frac{1}{16}n + \frac{1}{8}n^2\right)\sigma(n).$$

By Theorem 12.1 we have

$$\sum_{\substack{(r,s,t) \in \mathbb{N}^3 \\ r+s+t=n}} \sigma(r)\sigma(s)\sigma(t) = \sum_{t=1}^{n-2} \sigma(t) \sum_{\substack{(r,s) \in \mathbb{N}^2 \\ r+s=n-t}} \sigma(r)\sigma(s)$$

$$= \sum_{t=1}^{n-2} \sigma(t)\left(\frac{5}{12}\sigma_3(n-t) + \left(\frac{1}{12} - \frac{1}{2}(n-t)\right)\sigma(n-t)\right)$$

$$= \sum_{t=1}^{n-1} \sigma(t)\left(\frac{5}{12}\sigma_3(n-t) + \left(\frac{1}{12} - \frac{1}{2}(n-t)\right)\sigma(n-t)\right)$$

$$= \frac{5}{12}\sum_{t=1}^{n-1} \sigma(t)\sigma_3(n-t) + \frac{1}{12}\sum_{t=1}^{n-1} \sigma(t)\sigma(n-t)$$

$$- \frac{1}{2}\sum_{t=1}^{n-1}(n-t)\sigma(t)\sigma(n-t).$$

Appealing to Theorem 13.3 for the value of the first sum, Theorem 12.1 for that of the second sum, and Theorem 12.7 for that of the third sum, we obtain the asserted result.

For $e, f, g, n \in \mathbb{N}$ we define the twisted convolution sum

$$T_{e,f,g}(n) := \sum_{\substack{m \in \mathbb{N} \\ m < n/g}} \sigma_e(m)\sigma_f(n - gm). \tag{13.14}$$

The integer g is called the twist. We observe that

$$T_{e,f,1}(n) = S_{e,f}(n).$$

The sum

$$T_{1,1,2}(n) = \sum_{\substack{m \in \mathbb{N} \\ m < n/2}} \sigma(m)\sigma(n - 2m)$$

was evaluated in Theorem 12.4. We now evaluate $T_{1,1,2}(n)$ using Theorem 13.1.

Example 13.3. We choose $f(a, b, x, y) = a^2 F_2(x)$ in Theorem 13.1. Condition (13.1) is satisfied for this choice of f. The left hand side is

$$\sum_{\substack{(a, b, x, y) \in \mathbb{N}^4 \\ ax + by = n}} ((b - a)^2 - (a + b)^2)F_2(x) = -4 \sum_{\substack{(a, b, x, y) \in \mathbb{N}^4 \\ 2ax + by = n}} ab$$

$$= -4 \sum_{\substack{m \in \mathbb{N} \\ m < n/2}} \sigma(m)\sigma(n - 2m)$$

$$= -4T_{1,1,2}(n).$$

The right hand side is

$$\sum_{\substack{d \in \mathbb{N} \\ d \mid n}} \sum_{\substack{x \in \mathbb{N} \\ x < d}} \left((n/d)^2 F_2(d) + (n/d)^2 F_2(d - x) - x^2 F_2(n/d) - x^2 - d^2 F_2(n/d)\right).$$

Easy calculations show that

$$\sum_{\substack{d \in \mathbb{N} \\ d \mid n}} \sum_{\substack{x \in \mathbb{N} \\ x < d}} (n/d)^2 F_2(d) = n\sigma(n/2) - \sigma_2(n/2),$$

$$\sum_{\substack{d \in \mathbb{N} \\ d \mid n}} \sum_{\substack{x \in \mathbb{N} \\ x < d}} (n/d)^2 F_2(d - x) = \frac{1}{2}n\sigma(n) - \frac{1}{2}\sigma_2(n) - \frac{1}{2}\sigma_2(n/2),$$

$$\sum_{\substack{d \in \mathbb{N} \ x \in \mathbb{N} \\ d \mid n \ x < d}} \sum x^2 F_2(n/d) = \frac{1}{3}\sigma_3(n/2) - \frac{1}{2}\sigma_2(n/2) + \frac{1}{6}\sigma(n/2),$$

$$\sum_{\substack{d \in \mathbb{N} \ x \in \mathbb{N} \\ d \mid n \ x < d}} \sum x^2 = \frac{1}{3}\sigma_3(n) - \frac{1}{2}\sigma_2(n) + \frac{1}{6}\sigma(n),$$

$$\sum_{\substack{d \in \mathbb{N} \ x \in \mathbb{N} \\ d \mid n \ x < d}} \sum d^2 F_2(n/d) = \sigma_3(n/2) - \sigma_2(n/2).$$

Hence the right hand side is

$$-\frac{1}{3}\sigma_3(n) - \frac{4}{3}\sigma_3(n/2) + \left(\frac{1}{2}n - \frac{1}{6}\right)\sigma(n) + \left(n - \frac{1}{6}\right)\sigma(n/2).$$

Equating the left and right hand sides of Theorem 13.1, we deduce

$$T_{1,1,2}(n) = \sum_{\substack{m \in \mathbb{N} \\ m < n/2}} \sigma(m)\sigma(n - 2m)$$

$$= \frac{1}{12}\sigma_3(n) + \frac{1}{3}\sigma_3(n/2) + \left(\frac{1}{24} - \frac{1}{8}n\right)\sigma(n) + \left(\frac{1}{24} - \frac{1}{4}n\right)\sigma(n/2).$$

Example 13.4. As a consequence of Example 13.3 we have

$$\sum_{k=0}^{n-1} \sigma(2k + 1)\sigma(n - k) = \sum_{m=1}^{n} \sigma(2n - 2m + 1)\sigma(m)$$

$$= \sum_{\substack{m \in \mathbb{N} \\ m < (2n + 1)/2}} \sigma((2n + 1) - 2m)\sigma(m)$$

$$= \frac{1}{12}\sigma_3(2n + 1) + \left(\frac{1}{24} - \frac{1}{8}(2n + 1)\right)\sigma(2n + 1)$$

so that

$$\sum_{k=0}^{n-1} \sigma(2k + 1)\sigma(n - k) = \frac{1}{12}\sigma_3(2n + 1) - \left(\frac{1}{12} + \frac{1}{4}n\right)\sigma(2n + 1).$$

In the next theorem we use Theorem 13.1 to evaluate the twisted convolution sums

$$T_{3,1,2}(n) = \sum_{\substack{m \in \mathbb{N} \\ m < n/2}} \sigma_3(m)\sigma(n - 2m)$$

and

$$T_{1,3,2}(n) = \sum_{\substack{m \in \mathbb{N} \\ m < n/2}} \sigma(m)\sigma_3(n - 2m).$$

Theorem 13.11. *Let $n \in \mathbb{N}$. Then*

$$\sum_{\substack{m \in \mathbb{N} \\ m < n/2}} \sigma_3(m)\sigma(n - 2m)$$

$$= \frac{1}{240}\sigma_5(n) + \frac{1}{12}\sigma_5(n/2) + \left(\frac{1}{24} - \frac{1}{8}n\right)\sigma_3(n/2) - \frac{1}{240}\sigma(n)$$

and

$$\sum_{\substack{m \in \mathbb{N} \\ m < n/2}} \sigma(m)\sigma_3(n - 2m)$$

$$= \frac{1}{48}\sigma_5(n) + \frac{1}{15}\sigma_5(n/2) + \left(\frac{1}{24} - \frac{1}{16}n\right)\sigma_3(n) - \frac{1}{240}\sigma(n/2).$$

Proof. We set

$$X := \sum_{\substack{m \in \mathbb{N} \\ m < n/2}} \sigma_3(m)\sigma(n - 2m), \quad Y := \sum_{\substack{m \in \mathbb{N} \\ m < n/2}} \sigma(m)\sigma_3(n - 2m).$$

With $f(a, b, x, y) = a^4 F_2(x)$ the left hand side of Theorem 13.1 is

$$\sum_{\substack{(a, b, x, y) \in \mathbb{N}^4 \\ ax + by = n}} ((b - a)^4 - (b + a)^4)F_2(x) = \sum_{\substack{(a, b, x, y) \in \mathbb{N}^4 \\ 2ax + by = n}} (-8a^3b - 8ab^3)$$

$$= -8X - 8Y$$

and the right hand side is (after some calculation)

$$-\frac{1}{5}\sigma_5(n) - \frac{6}{5}\sigma_5(n/2) + \left(\frac{n}{2} - \frac{1}{3}\right)\sigma_3(n)$$

$$+ \left(n - \frac{1}{3}\right)\sigma_3(n/2) + \frac{1}{30}\sigma(n) - \frac{1}{30}\sigma(n/2).$$

Next, with $f(a, b, x, y) = b^4 F_2(a)$ the left hand side of Theorem 13.1 is

$$\sum_{\substack{(a, b, x, y) \in \mathbb{N}^4 \\ ax + by = n}} ((a - b)^4 - (a + b)^4)F_2(a) = \sum_{\substack{(a, b, x, y) \in \mathbb{N}^4 \\ 2ax + by = n}} (-64a^3b - 16ab^3)$$

$$= -64X - 16Y$$

and the right hand side is (after some calculation)

$$-\frac{3}{5}\sigma_5(n) - \frac{32}{5}\sigma_5(n/2) + \left(n - \frac{2}{3}\right)\sigma_3(n)$$

$$+ \left(8n - \frac{8}{3}\right)\sigma_3(n/2) + \frac{4}{15}\sigma(n) + \frac{1}{15}\sigma(n/2).$$

Solving the two linear equations for X and Y resulting from Theorem 13.1, we obtain the two identities of the theorem. $\qquad\square$

As a consequence of Theorem 13.11 we have the following result.

Theorem 13.12. *Let $n \in \mathbb{N}$. Then*

$$\sum_{k=0}^{n-1} \sigma(2k+1)\sigma_3(n-k) = \frac{1}{240}\sigma_5(2n+1) - \frac{1}{240}\sigma(2n+1)$$

and

$$\sum_{k=0}^{n-1} \sigma_3(2k+1)\sigma(n-k) = \frac{1}{48}\sigma_5(2n+1) - \left(\frac{1}{48} + \frac{1}{8}n\right)\sigma_3(2n+1).$$

Proof. We have

$$\sum_{k=0}^{n-1} \sigma(2k+1)\sigma_3(n-k) = \sum_{m=1}^{n} \sigma(2n - 2m + 1)\sigma_3(m)$$

$$= \sum_{\substack{m \in \mathbb{N} \\ m < (2n+1)/2}} \sigma((2n+1) - 2m)\sigma_3(m)$$

$$= \frac{1}{240}\sigma_5(2n+1) - \frac{1}{240}\sigma(2n+1),$$

by Theorem 13.11. The second identity can be proved similarly. $\qquad\square$

As a further consequence of Theorem 13.11, we prove the following identity of Liouville.

Theorem 13.13. *Let M be a positive odd integer. Then*

$$\sum_{m=0}^{M-1} \sigma(2m+1)\sigma_3(2M - 2m - 1) = \sigma_5(M).$$

Proof. Recalling the identities (see Theorem 3.1(ii))

$$\sigma(2k) = 3\sigma(k) - 2\sigma(k/2), \quad \sigma_3(2k) = 9\sigma_3(k) - 8\sigma_3(k/2), \quad k \in \mathbb{N},$$

we have

$$\sum_{m=1}^{M-1} \sigma(2m)\sigma_3(2M - 2m) = 27S_1 - 18S_2 - 24S_3 + 16S_4,$$

where

$$S_1 = \sum_{m=1}^{M-1} \sigma(m)\sigma_3(M - m),$$

$$S_2 = \sum_{\substack{m = 1 \\ 2 \mid m}}^{M-1} \sigma(m/2)\sigma_3(M - m),$$

$$S_3 = \sum_{\substack{m = 1 \\ 2 \mid M - m}}^{M-1} \sigma(m)\sigma_3((M - m)/2),$$

$$S_4 = \sum_{\substack{m = 1 \\ 2 \mid m \\ 2 \mid M - m}}^{M-1} \sigma(m/2)\sigma_3((M - m)/2).$$

By Theorem 13.3 we have

$$S_1 = \frac{7}{80}\sigma_5(M) + \left(\frac{1}{24} - \frac{1}{8}M\right)\sigma_3(M) - \frac{1}{240}\sigma(M).$$

By the second identity in Theorem 13.11, we have

$$S_2 = \sum_{\substack{m \in \mathbb{N} \\ n < M/2}} \sigma(m)\sigma_3(M - 2m) = \frac{1}{48}\sigma_5(M) + \left(\frac{1}{24} - \frac{1}{16}M\right)\sigma_3(M).$$

Further, by Theorem 13.12, we have

$$S_3 = \sum_{\substack{m = 1 \\ 2 \mid M - m}}^{M-1} \sigma(m)\sigma_3((M - m)/2)$$

$$= \sum_{k=0}^{(M-3)/2} \sigma(2k + 1)\sigma_3\left(\frac{M - 1}{2} - k\right)$$

$$= \frac{1}{240}\sigma_5(M) - \frac{1}{240}\sigma(M).$$

Finally, as M is odd, m and $M - m$ are of opposite parity, so that $S_4 = 0$. Putting these evaluations together, we obtain

$$\sum_{m=1}^{M-1} \sigma(2m)\sigma_3(2M - 2m) = \frac{151}{80}\sigma_5(M) + \left(\frac{3}{8} - \frac{9}{4}M\right)\sigma_3(M) - \frac{1}{80}\sigma(M).$$

Then, appealing to Theorem 13.3, we obtain

$$\sum_{m=0}^{M-1} \sigma(2m + 1)\sigma_3(2M - 2m - 1)$$

$$= \sum_{m=1}^{2M-1} \sigma(m)\sigma_3(2M - m) - \sum_{m=1}^{M-1} \sigma(2m)\sigma_3(2M - 2m)$$

$$= \left(\frac{7}{80}\sigma_5(2M) + \left(\frac{1}{24} - \frac{1}{4}M\right)\sigma_3(2M) - \frac{1}{240}\sigma(2M)\right)$$

$$\quad - \left(\frac{151}{80}\sigma_5(M) + \left(\frac{3}{8} - \frac{9}{4}M\right)\sigma_3(M) - \frac{1}{80}\sigma(M)\right)$$

$$= \sigma_5(M),$$

which is the asserted result. $\square$

We conclude this chapter by proving the following identity of Giraud.

Theorem 13.14. *Let* $n \in \mathbb{N}$. *Then*

$$\sum_{\substack{(a, b, x, y) \in \mathbb{N}^4 \\ ax + by = n}} \min(a, b)^2 = \frac{1}{6}\sigma_3(n) - \frac{1}{6}\sigma(n).$$

Proof. We choose $f(a, b, x, y) = \frac{1}{2}(ab - |ab|)$. Clearly

$$f(a, -b, x, y) = f(-a, b, x, y) \quad \text{and} \quad f(a, b, x, -y) = f(a, b, -x, y)$$

for all $a, b, x, y \in \mathbb{Z}$ so that we can apply Theorem 13.1. With this choice the left hand side of Theorem 13.1 is

$$\sum_{\substack{(a, b, x, y) \in \mathbb{N}^4 \\ ax + by = n}} \left(\frac{1}{2}a^2 + \frac{1}{2}b^2 - \frac{1}{2}a|a - b| - \frac{1}{2}b|a - b|\right) = \sum_{\substack{(a, b, x, y) \in \mathbb{N}^4 \\ ax + by = n}} \min(a, b)^2.$$

The right hand side is

$$\sum_{\substack{d \in \mathbb{N} \\ d \mid n}} \sum_{\substack{x \in \mathbb{N} \\ x < d}} (dx - x^2) = \frac{1}{6}\sum_{\substack{d \in \mathbb{N} \\ d \mid n}} (d^3 - d) = \frac{1}{6}\sigma_3(n) - \frac{1}{6}\sigma(n).$$

Theorem 13.1 now gives the asserted formula. $\square$

The problems of Exercises 13 contain many other applications of Theorem 13.1.

Exercises 13

1. Let $n \in \mathbb{N}$. By taking $f(a, b, x, y) = b^2 y^4 - b^2 x y^3$ in Theorem 13.1, prove that

$$\sum_{m=1}^{n-1} m^2 \sigma(m) \sigma(n-m) = \frac{1}{8} n^2 \sigma_3(n) + \left(\frac{1}{24} n^2 - \frac{1}{6} n^3 \right) \sigma(n).$$

2. Let $n \in \mathbb{N}$. Deduce the identity

$$\sum_{m=1}^{n-1} m \sigma(m) \sigma(n-m) = \frac{5}{24} n \sigma_3(n) + \left(\frac{1}{24} n - \frac{1}{4} n^2 \right) \sigma(n)$$

   from Problem 1. This gives another proof of Theorem 12.7.

3. By taking

$$f(a, b, x, y) = abx^3 y - b^2 x y^3$$

   in Theorem 13.1, prove that

$$\sum_{m=1}^{n-1} m(n-m) \sigma(m) \sigma(n-m) = \frac{1}{12} n^2 \sigma_3(n) - \frac{1}{12} n^3 \sigma(n).$$

4. Let $n \in \mathbb{N}$. Prove that

$$\sum_{m=1}^{n-1} m \sigma_3(m) \sigma_3(n-m) = \frac{1}{240} n \sigma_7(n) - \frac{1}{240} n \sigma_3(n).$$

5. Let $n \in \mathbb{N}$. Prove that

$$\sum_{m=1}^{n-1} m(n-m)^2 \sigma(m) \sigma(n-m) = \frac{1}{24} n^3 \sigma_3(n) - \frac{1}{24} n^4 \sigma(n).$$

6. Let $n \in \mathbb{N}$. Prove that

$$\sum_{m=1}^{n-1} m^3 \sigma(m) \sigma(n-m) = \frac{1}{12} n^3 \sigma_3(n) + \left(\frac{1}{24} n^3 - \frac{1}{8} n^4 \right) \sigma(n).$$

7. Let $n \in \mathbb{N}$. Prove that

$$\sum_{m=1}^{n-1} m^2 \sigma_3(m) \sigma(n-m) = \frac{1}{24} n^2 \sigma_5(n) + \left(\frac{1}{24} n^2 - \frac{1}{12} n^3 \right) \sigma_3(n).$$

8. Let $n \in \mathbb{N}$. Prove that

$$\sum_{m=1}^{n-1} m^2 \sigma(m) \sigma_3(n-m) = \frac{1}{80} n^2 \sigma_5(n) - \frac{1}{120} n^3 \sigma_3(n) - \frac{1}{240} n^2 \sigma(n).$$

9. Let $n \in \mathbb{N}$. Prove that

$$\sum_{m=1}^{n-1} m(n-m) \sigma(m) \sigma_3(n-m) = \frac{1}{60} n^2 \sigma_5(n) - \frac{1}{60} n^3 \sigma_3(n).$$

10. Let $n \in \mathbb{N}$. Prove that

$$\sum_{m=1}^{n-1} m \sigma(m) \sigma_5(n-m) = \frac{5}{504} n \sigma_7(n) - \frac{1}{84} n^2 \sigma_5(n) + \frac{1}{504} n \sigma(n).$$

11. Let $n \in \mathbb{N}$. Prove that

$$\sum_{m=1}^{n-1} m \sigma_5(m) \sigma(n-m) = \frac{5}{168} n \sigma_7(n) + \left(\frac{1}{24} n - \frac{1}{14} n^2 \right) \sigma_5(n).$$

12. Let $k, n \in \mathbb{N}$. By taking

$$f(a, b, x, y) = (a^2 - ab) F_k(x)$$

in Theorem 13.1, prove that

$$\sum_{\substack{m \in \mathbb{N} \\ m < n/k}} \sigma(m) \sigma(n-km) = -\frac{1}{24} \sigma_3(n) + \frac{1}{24} \sigma(n) + \frac{1}{4} \sigma_3(n/k) - \frac{n}{4} \sigma(n/k)$$

$$+ \frac{1}{4} \sum_{\substack{(a, b, x, y) \in \mathbb{N}^4 \\ ax + by = n \\ x \equiv y \pmod{k}}} ab + \frac{1}{4} \sum_{\substack{(a, b, x, y) \in \mathbb{N}^4 \\ ax + by = n \\ x \equiv -y \pmod{k}}} ab.$$

13. Deduce Besge's formula (Theorem 12.1) from Problem 12.

14. Let $n \in \mathbb{N}$ and $g : \mathbb{N}^2 \to \mathbb{C}$. Prove that

$$\sum_{\substack{(a, b, x, y) \in \mathbb{N}^4 \\ ax + by = n \\ x \equiv y \pmod{2}}} g(a, b) = \sum_{\substack{(a, b, x, y) \in \mathbb{N}^4 \\ ax + by = n}} g(a, b) + 2 \sum_{\substack{(a, b, x, y) \in \mathbb{N}^4 \\ ax + by = n \\ x \equiv y \equiv 0 \pmod{2}}} g(a, b)$$

$$- \sum_{\substack{(a, b, x, y) \in \mathbb{N}^4 \\ ax + by = n \\ x \equiv 0 \pmod{2}}} g(a, b) - \sum_{\substack{(a, b, x, y) \in \mathbb{N}^4 \\ ax + by = n \\ y \equiv 0 \pmod{2}}} g(a, b).$$

15. Let $n \in \mathbb{N}$. From Problem 12 with $k = 2$ and Problem 14 with $g(a, b) = ab$, deduce that

$$\sum_{\substack{m \in \mathbb{N} \\ m < n/2}} \sigma(m)\sigma(n - 2m) = \frac{1}{12}\sigma_3(n) + \frac{1}{3}\sigma_3(n/2) + \left(\frac{1}{24} - \frac{1}{8}n\right)\sigma(n)$$

$$+ \left(\frac{1}{24} - \frac{1}{4}n\right)\sigma(n/2).$$

This gives another proof of Theorem 12.4, see also Example 13.3.

16. Use Theorem 13.3 to prove

$$\sum_{m=1}^{n-1} m\sigma(m)\sigma_3(n - m) + \sum_{m=1}^{n-1} m\sigma(n - m)\sigma_3(m)$$

$$= \frac{7}{80}n\sigma_5(n) + \left(\frac{1}{24}n - \frac{1}{8}n^2\right)\sigma_3(n) - \frac{1}{240}n\sigma(n).$$

17. By taking

$$f(a, b, x, y) = 3b^2xy^3 + 5abx^3y - 3abx^2y^2 - 2b^2x^3y$$

in Theorem 13.1, prove that

$$2\sum_{m=1}^{n-1} m\sigma(m)\sigma_3(n - m) - \sum_{m=1}^{n-1} m\sigma(n - m)\sigma_3(m)$$

$$= \left(\frac{1}{20}n^2 - \frac{1}{24}n\right)\sigma_3(n) - \frac{1}{120}n\sigma(n).$$

18. Deduce from Problems 16 and 17 that

$$\sum_{m=1}^{n-1} m\sigma(m)\sigma_3(n - m) = \frac{7}{240}n\sigma_5(n) - \frac{1}{40}n^2\sigma_3(n) - \frac{1}{240}n\sigma(n)$$

and

$$\sum_{m=1}^{n-1} m\sigma(n - m)\sigma_3(m) = \frac{7}{120}n\sigma_5(n) + \left(\frac{1}{24}n - \frac{1}{10}n^2\right)\sigma_3(n).$$

19. Prove Theorem 13.3.
20. Prove Theorem 13.4.
21. Prove Theorem 13.5.
22. Prove Theorem 13.6.
23. Prove Theorem 13.7.
24. Prove Theorem 13.8.
25. Prove Theorem 13.9.

26. Prove Theorem 13.10.

27. Let $n \in \mathbb{N}$. Use Theorems 12.1, 13.3, 13.4 and Problem 18 to prove that

$$2880 \sum_{\substack{(m_1, m_2, m_3) \in \mathbb{N}^3 \\ m_1 + m_2 + m_3 = n}} \sigma(m_1)\sigma(m_2)\sigma_3(m_3)$$

$$= 10\sigma_7(n) + (21 - 42n)\sigma_5(n) - (30n - 36n^2)\sigma_3(n) - (1 - 6n)\sigma(n).$$

28. Let $n \in \mathbb{N}$. Use Example 13.2, Theorems 12.1, 13.3 and 13.4, and Problems 1,2 and 18 to prove that

$$3456 \sum_{\substack{(m_1, m_2, m_3, m_4) \in \mathbb{N}^4 \\ m_1 + m_2 + m_3 + m_4 = n}} \sigma(m_1)\sigma(m_2)\sigma(m_3)\sigma(m_4) = 5\sigma_7(n) + (21 - 42n)\sigma_5(n)$$

$$+ (15 - 90n + 108n^2)\sigma_3(n) + (1 - 18n + 72n^2 - 72n^3)\sigma(n).$$

29. Let $n \in \mathbb{N}$. Let $f : \mathbb{Z} \to \mathbb{C}$, $g : \mathbb{Z} \to \mathbb{C}$ and $h : \mathbb{Z} \to \mathbb{C}$ be even functions. Set

$$F(a, b) := (f(a - b) - f(a + b))g(a)h(b) + (g(a - b)$$
$$- g(a + b))h(a)f(b) + (h(a - b) - h(a + b))f(a)g(b),$$
$$G(d) := f(0)g(d)h(d) + g(0)h(d)f(d) + h(0)f(d)g(d),$$
$$H(d, k) := f(d - k)g(d)h(k) + g(d - k)h(d)f(k) + h(d - k)f(d)g(k).$$

Prove that

$$\sum_{\substack{(a, b, x, y) \in \mathbb{N}^4 \\ ax + by = n}} F(a, b) = \sum_{\substack{d \in \mathbb{N} \\ d \mid n}} \frac{n}{d}G(d) - \sum_{\substack{d \in \mathbb{N} \\ d \mid n}} \sum_{k=1}^{d} H(d, k).$$

30. Deduce Theorem 10.1 from Problem 29.

31. Let $n \in \mathbb{N}$. Let $f : \mathbb{Z} \to \mathbb{C}$ be an even function. Prove that

$$\sum_{\substack{(a, b, x, y) \in \mathbb{N}^4 \\ ax + by = n}} f(a)f(b)(f(a - b) - f(a + b))$$

$$= f(0) \sum_{\substack{d \in \mathbb{N} \\ d \mid n}} \frac{n}{d}f^2(d) - \sum_{\substack{d \in \mathbb{N} \\ d \mid n}} f(d) \left(\sum_{k=1}^{d} f(k)f(d - k) \right).$$

32. Use Problem 31 to prove Theorem 13.5.

33. Let $f : \mathbb{Z}^2 \to \mathbb{C}$, $g : \mathbb{Z}^2 \to \mathbb{C}$ and $h : \mathbb{Z}^2 \to \mathbb{C}$ be such that

$$f(x, y) = f(-x, y) = f(x, -y) = f(-x, -y),$$
$$g(x, y) = g(-x, y) = g(x, -y) = g(-x, -y)$$

and

$$h(x, y) = h(-x, y) = h(x, -y) = h(-x, -y),$$

for all $(x, y) \in \mathbb{Z}^2$. Let

$$
\begin{aligned}
F(a, b, x, y) :=\ & (f(a - b, x + y) - f(a + b, x - y))g(a, y)h(b, x) \\
& + (g(a - b, x + y) - g(a + b, x - y))h(a, y)f(b, x) \\
& + (h(a - b, x + y) - h(a + b, x - y))f(a, y)g(b, x),
\end{aligned}
$$

$$
\begin{aligned}
G_n(d, a) :=\ & f(0, d)g(n/d, a)h(n/d, d - a) \\
& + f(d, 0)g(a, n/d)h(d - a, n/d),
\end{aligned}
$$

$$H_n(d, e) := f(d, e)(g(d, n/d - e)h(0, n/d) + g(0, n/d)h(d, n/d - e)),$$

$$I_n(d, e) := f(d, e)(g(n/e, 0)h(n/e - d, e) + g(n/e - d, d)h(n/e, 0)).$$

Prove that

$$
\sum_{\substack{(a, b, x, y) \in \mathbb{N}^4 \\ ax + by = n}} F(a, b, x, y) = \sum_{\substack{d \in \mathbb{N} \\ d \mid n}} \sum_{a=1}^{d-1} G_n(d, a) + \sum_{\substack{d \in \mathbb{N} \\ d \mid n}} \sum_{\substack{e \in \mathbb{N} \\ e < n/d}} H_n(d, e)
$$

$$
- \sum_{\substack{e \in \mathbb{N} \\ e \mid n}} \sum_{\substack{d \in \mathbb{N} \\ d < n/e}} I_n(d, e).
$$

34. Deduce the result of Problem 29 from that of Problem 33.

35. Use Problem 2 of Exercises 12 and Example 13.2 to prove that for all $n \in \mathbb{N}$ the following identity holds

$$
\sum_{\substack{(r, s, t) \in \mathbb{N}^3 \\ r + s + t = n}} r\sigma(r)\sigma(s)\sigma(t) = \frac{7}{576}n\sigma_5(n) + \left(\frac{5}{288}n - \frac{5}{96}n^2\right)\sigma_3(n)
$$

$$
+ \left(\frac{1}{576}n - \frac{1}{48}n^2 + \frac{1}{24}n^3\right)\sigma(n).
$$

Notes on Chapter 13

Theorem 13.1 is due to Huard, Ou, Spearman and Williams [137, Theorem 1, p. 230]. The proof that we present is taken from their paper. Giraud [108] has used the Huard-Ou-Spearman-Williams identity in his solution to a problem in mathematical physics. Skoruppa gave his identity (Theorem 13.2) in [242, Theorem, p. 69] and used it to prove identities between Eisenstein series.

Theorem 13.3 is attributed to Glaisher by MacMahon [193, p. 101], [194, Vol. II, p. 329]. It also appears in Ramanujan [230], [232, Table IV, p. 146], is formula (5.1) of Lahiri [159], and is formula (3.12) in Huard, Ou, Spearman and Williams [137, p. 237]. Theorems 13.4, 13.6 and 13.8 are due to Ramanujan [230], [232, Table IV, p. 146]. Theorems 13.5, 13.7, 13.9 and 13.10 are due to Glaisher [111, p. 35]. Theorems 13.4, 13.5, 13.6, 13.7, 13.8, 13.9 and 13.10 are formulae (7.2), (7.1), (9.2), (9.1), (13.4), (13.2) and (13.1) respectively of Lahiri [159], and formulae (3.18), (3.17), (3.28), (3.27), (3.31), (3.30) and (3.29) respectively of Huard, Ou, Spearman and Williams [137]. Theorem 13.4 is also given in the work of MacMahon [193, p. 103], [194, Vol. II, p. 331].

Example 13.2 is identity (5.2) of Lahiri [159]. We remark that the identity of Example 13.2 is implicit in the work of Glaisher [111, p. 33] and is proved in Bambah and Chowla [35, eq. (28), p. 145], [72, Vol. II, p. 649], see also Chowla [71], [72, Vol. II, p. 669], as well as in Huard, Ou, Spearman and Williams [137, pp. 244–246].

The result of Example 13.3 is implicit in the work of Glaisher [110, p. 11] and in the form given here is due to Huard, Ou, Spearman and Williams [137, Theorem 2, p. 247]. The special case of it when n is odd is due to Melfi [202, eq. (8)].

The identity of Example 13.4 does not appear to have been stated before. A related formula is given in Huard, Ou, Spearman and Williams [137, eq. (4.5), p. 248].

Theorem 13.11 is a result of Huard, Ou, Spearman and Williams [137, Theorem 6, p. 250]. The special case when n is odd is due to Melfi [202, eqs. (9), (10)].

The first identity of Theorem 13.12 was explicitly stated but never proved by Ramanujan [230], [232, p. 146]. A result equivalent to Theorem 13.12 was first proved by Masser, see Berndt [44, Part II, p. 329] and Berndt and Evans [46, p. 136], with later proofs by Atkin (see Berndt [44, Part II, p. 329]) and Ramamani [229]. None of these proofs is elementary. The proof presented here is the first elementary proof of Ramanujan's identity and is due to Huard, Ou, Spearman and Williams [137, Corollary 2, p. 251].

Theorem 13.13 was stated by Liouville in [170, 1st article, p. 147]. The proof we give is taken from Huard, Ou, Spearman and Williams [137, Corollary 3, pp. 252–253].

Theorem 13.14 is due to Giraud [108].

Problem 1 is implicit in the work of Glaisher [111] and is formula (3.4) in Lahiri's paper [159] and formula (3.16) in Huard, Ou, Spearman and Williams [137].

The formula of Problem 2 is due to Glaisher [111, p. 36]. Note that the multiplier 12 on the left hand side of Glaisher's formula should be replaced by 24. It is also formula (3.2) of Lahiri [159] and formula (3.11) of Huard, Ou, Spearman and Williams [137].

Problem 3 is formula (3.3) of Lahiri [159] and formula (3.15) of Huard, Spearman and Williams [137]. It is due to Glaisher [111, p. 35].

Problem 4 is formula (7.5) of Lahiri [159] and formula (3.19) of Huard, Ou, Spearman and Williams [137]. It is implicit in the work of Glaisher [111, pp. 35–36].

Problem 5 is due to Glaisher [111, p. 36]. It is formula (3.6) of Lahiri [159] and formula (3.20) of Huard, Ou, Spearman and Williams [137].

Problem 6 is due to Lahiri [159, formula (3.6)] and is formula (3.21) of Huard, Ou, Spearman and Williams [137].

Problem 7 is due to Lahiri [159, formula (5.7)] and is formula (3.22) of Huard, Ou, Spearman and Williams [137].

Problem 8 is due to Lahiri [159, formula (5.8)] and is formula (3.23) of Huard, Ou, Spearman and Williams [137].

Problem 9 is formula (5.6) of Lahiri [159] and formula (3.26) of Huard, Ou, Spearman and Williams [137].

Problem 10 is formula (7.7) of Lahiri [159] and formula (3.24) of Huard, Ou, Spearman and Williams [137].

Problem 11 is due to Lahiri [159, formula (7.6)] and is formula (3.25) of Huard, Ou, Spearman and Williams [137].

The identity of Problem 12 is due to Huard, Ou, Spearman and Williams [137, Lemma 1, p. 247].

Problems 14 and 15 give another proof of the formula for $T_{1,1,2}(n)$ given in Theorem 12.7.

Problems 16, 17 and 18 are taken from Huard, Ou, Spearman and Williams [137, pp. 237–238].

Problems 19–28 are taken from Huard, Ou, Spearman and Williams [137]. The identities of Problems 29 and 33 are due to Ou [218].

The identity of Problem 35 is formula (5.5) of Lahiri [159].

Lahiri [159] has given 37 sums of the form

$$\sum_{\substack{(m_1,\ldots,m_r) \in \mathbb{N}^r \\ m_1 + \cdots + m_r = n}} m_1{}^{a_1} \cdots m_r{}^{a_r} \sigma_{b_1}(m_1) \cdots \sigma_{b_r}(m_r), \ a_1, \ldots, a_r \in \mathbb{N}_0, b_1, \ldots, b_r \in \mathbb{N},$$

each of which can be expressed as a finite linear combination of $\sigma(n)$, $\sigma_3(n)$, $\sigma_5(n), \ldots, \sigma_{b_1+\cdots+b_r+r-1}(n)$ with coefficients which are polynomials in n of degree at most $a_1 + \cdots + a_r + r - 1$ with rational coefficients. For example

$$\sum_{\substack{(m_1, m_2, m_3) \in \mathbb{N}^3 \\ m_1 + m_2 + m_3 = n}} m_1 m_2 \sigma(m_1) \sigma(m_2) \sigma(m_3)$$

$$= \frac{1}{288} \left(n^2 \sigma_5(n) + (n^2 - 4n^3) \sigma_3(n) - (n^3 - 3n^4) \sigma(n) \right),$$

see Lahiri [159, formula (5.9)]. Huard, Ou, Spearman and Williams [137] have verified that each of these 37 identities is a consequence of Theorem 13.1. Lahiri [160] has also exhibited 41 sums, each of which can be expressed as a finite linear combination of $\sigma(n)$, $\sigma_3(n)$, $\sigma_5(n)$, ... and $\tau(n)$ with coefficients which are polynomials in n, where $\tau(n)$ is the Ramanujan tau function defined by

$$q \prod_{n=1}^{\infty} (1 - q^n)^{24} = \sum_{n=1}^{\infty} \tau(n) q^n, \quad q \in \mathbb{C}, \ |q| < 1. \tag{13.15}$$

For example

$$\sum_{m=1}^{n-1} m(n - m) \sigma_3(m) \sigma_3(n - m) = \frac{1}{540} n^2 \sigma_7(n) - \frac{1}{540} \tau(n).$$

The first few values of $\tau(n)$ are $\tau(1) = 1$, $\tau(2) = -24$, $\tau(3) = 252$, $\tau(4) = -1472$, $\tau(5) = 4830$, $\tau(6) = -6048$, $\tau(7) = -16744$ and $\tau(8) = 84480$.

Levitt [169] has used the theory of modular forms to show in a certain sense that the nine arithmetic evaluations of $S_{e,f}(n)$ with e and f odd (Theorems 12.1 and 13.3–13.10) are the only ones. Grosjean [117], [118] and Radoux have given recurrence relations for $S_{e,f}(n)$. O'Sullivan [217] has proved some of the identities in this chapter using modular forms.

14

Four Elementary Arithmetic Formulae

Let $n \in \mathbb{N}$. Let $f : \mathbb{Z} \to \mathbb{C}$ be an even function. For positive integers A, B, C and D, we consider the sum

$$S(A, B, C, D, f;n) := \sum_{\substack{(a,b,x,y) \in \mathbb{N}^4 \\ Cax + Dby = n}} (f(Aa - Bb) - f(Aa + Bb)). \quad (14.1)$$

This sum has the following three properties:

$$S(A, B, C, D, f;n) = S(B, A, D, C, f;n), \quad (14.2)$$

$$S(A, B, C, D, f;n) = S(A, B, C/E, D/E, f;n/E), \quad (14.3)$$

where $E = \gcd(C, D)$, and

$$S(A, B, C, D, f;n) = S(A/F, B/F, C, D, g;n), \quad (14.4)$$

where $F = \gcd(A, B)$ and $g(x) = f(Fx)$. Thus we may suppose that

$$A \geq B, \quad C \geq D \text{ if } A = B, \quad \gcd(A, B) = \gcd(C, D) = 1.$$

We know from Theorem 10.1 that the sum $S(1, 1, 1, 1, f;n)$ has an elementary evaluation in terms of sums over the divisors of n. In this chapter we show that four other such sums have similar evaluations. They are the sums given by

$$(A, B, C, D) = (1, 1, 2, 1), \ (2, 1, 1, 1), \ (2, 1, 2, 1) \text{ and } (2, 1, 4, 1),$$

see Theorems 14.1, 14.2, 14.3 and 14.4 respectively. We deduce these theorems from Theorem 13.1. It is not known if the sum $S(A, B, C, D, f;n)$ can be evaluated in a similar manner for any other values of (A, B, C, D) for arbitrary even functions f and positive integers n. We first treat the sum (14.1) with $(A, B, C, D) = (1, 1, 2, 1)$.

Theorem 14.1. *Let $n \in \mathbb{N}$. Let $f : \mathbb{Z} \to \mathbb{C}$ be an even function. Then*

$$\sum_{\substack{(a,b,x,y)\in\mathbb{N}^4 \\ 2ax+by=n}} (f(a-b) - f(a+b)) = \frac{1}{2} f(0)\, (\sigma(n) - d(n) - d(n/2))$$

$$+ \frac{1}{2} \sum_{\substack{d \in \mathbb{N} \\ d \mid n}} \left(1 + \frac{n}{d}\right) f(d) + \frac{1}{2} \sum_{\substack{d \in \mathbb{N} \\ d \mid n/2}} \left(1 - 2d + \frac{2n}{d}\right) f(d)$$

$$- \sum_{\substack{d \in \mathbb{N} \\ d \mid n}} \left(\sum_{\ell=1}^{d} f(\ell)\right) - \sum_{\substack{d \in \mathbb{N} \\ d \mid n/2}} \left(\sum_{\ell=1}^{d} f(\ell)\right).$$

Proof. Let $f : \mathbb{Z} \to \mathbb{C}$ be an even function. We choose

$$f(a, b, x, y) = f(a)F_2(x), \quad (a, b, x, y) \in \mathbb{Z}^4,$$

in Theorem 13.1. We are using f in two different ways but this should not be confusing to the reader. It is easily checked that the condition (13.1) is satisfied. Let $n \in \mathbb{N}$. The left hand side of Theorem 13.1 becomes with this choice

$$\sum_{\substack{(a, b, x, y) \in \mathbb{N}^4 \\ ax + by = n}} \left(f(a)F_2(x) - f(a)F_2(x) + f(a)F_2(x + y) \right.$$

$$\left. - f(a)F_2(y - x) + f(b - a)F_2(x) - f(a + b)F_2(x) \right)$$

$$= \sum_{\substack{(a, b, x, y) \in \mathbb{N}^4 \\ 2ax + by = n}} (f(a-b) - f(a+b)).$$

The right hand side of Theorem 13.1 with this choice is

$$\sum_{\substack{d \in \mathbb{N}\, x \in \mathbb{N} \\ d \mid n\ \ x < d}} \sum \left(f(0)F_2(x) + f(n/d)F_2(d) + f(n/d)F_2(d - x) \right.$$

$$\left. - f(x)F_2(n/d) - f(x)F_2(0) - f(d)F_2(n/d) \right).$$

The first sum is

$$f(0) \sum_{\substack{d \in \mathbb{N}\, x \in \mathbb{N} \\ d \mid n\ \ x < d}} \sum F_2(x) = \frac{1}{2} f(0)(\sigma(n) - d(n) - d(n/2)),$$

by Problem 31 of Exercises 3. The second sum is

$$\sum_{\substack{d \in \mathbb{N} \\ d \mid n}} \sum_{\substack{x \in \mathbb{N} \\ x < d}} f(n/d) F_2(d) = \sum_{\substack{d \in \mathbb{N} \\ d \mid n/2}} \left(\frac{n}{d} - 1\right) f(d),$$

by Problem 32 of Exercises 3. The third sum is

$$\sum_{\substack{d \in \mathbb{N} \\ d \mid n}} \sum_{\substack{x \in \mathbb{N} \\ x < d}} f(n/d) F_2(d - x) = \frac{1}{2} \sum_{\substack{d \in \mathbb{N} \\ d \mid n}} \left(\frac{n}{d} - 1\right) f(d) - \frac{1}{2} \sum_{\substack{d \in \mathbb{N} \\ d \mid n/2}} f(d),$$

by Problem 37 of Exercises 3. The fourth sum is

$$\sum_{\substack{d \in \mathbb{N} \\ d \mid n}} \sum_{\substack{x \in \mathbb{N} \\ x < d}} f(x) F_2(n/d) = \sum_{\substack{d \in \mathbb{N} \\ d \mid n/2}} \left(\sum_{x=1}^{d} f(x)\right) - \sum_{\substack{d \in \mathbb{N} \\ d \mid n/2}} f(d),$$

by Problem 36 of Exercises 3. The fifth sum is

$$\sum_{\substack{d \in \mathbb{N} \\ d \mid n}} \sum_{\substack{x \in \mathbb{N} \\ x < d}} f(x) F_2(0) = \sum_{\substack{d \in \mathbb{N} \\ d \mid n}} \left(\sum_{x=1}^{d} f(x)\right) - \sum_{\substack{d \in \mathbb{N} \\ d \mid n}} f(d).$$

The sixth sum is

$$\sum_{\substack{d \in \mathbb{N} \\ d \mid n}} \sum_{\substack{x \in \mathbb{N} \\ x < d}} f(d) F_2(n/d) = \sum_{\substack{d \in \mathbb{N} \\ d \mid n/2}} (d - 1) f(d),$$

by Problem 33 of Exercises 3.
    The asserted formula now follows by Theorem 13.1. $\qquad\square$

Before continuing we give an application of Theorem 14.1.

Example 14.1. Let $k, n \in \mathbb{N}$. We take $f(x) = F_k(x)$ $(x \in \mathbb{Z})$ in Theorem 14.1 to determine the sum

$$\sum_{\substack{(a, b, x, y) \in \mathbb{N}^4 \\ 2ax + by = n}} (F_k(a - b) - F_k(a + b)).$$

By Problems 6, 7 and 8 of Exercises 3 we have

$$\sum_{\substack{d \in \mathbb{N} \\ d \mid n}} \left(1 + \frac{n}{d}\right) F_k(d) = d(n/k) + \sigma(n/k)$$

and

$$\sum_{\substack{d \in \mathbb{N} \\ d \,|\, n/2}} \left(1 - 2d + \frac{2n}{d}\right) F_k(d) = d(n/2k) - 2k\sigma(n/2k) + 4\sigma(n/2k).$$

By Example 3.7 we have

$$\sum_{\substack{d \in \mathbb{N} \\ d \,|\, n}} \left(\sum_{\ell=1}^{d} F_k(\ell)\right) = \frac{1}{k}\sigma(n) - \frac{1}{k}\sum_{\ell=1}^{k-1} \ell d_{\ell,k}(n)$$

and

$$\sum_{\substack{d \in \mathbb{N} \\ d \,|\, n/2}} \left(\sum_{\ell=1}^{d} F_k(\ell)\right) = \frac{1}{k}\sigma(n/2) - \frac{1}{k}\sum_{\ell=1}^{k-1} \ell d_{\ell,k}(n/2).$$

Taking $f(x) = F_k(x)$ in Theorem 14.1, and appealing to these identities, we obtain

$$\sum_{\substack{(a, b, x, y) \in \mathbb{N}^4 \\ 2ax + by = n}} (F_k(a - b) - F_k(a + b))$$

$$= \left(\frac{1}{2} - \frac{1}{k}\right)\sigma(n) - \frac{1}{k}\sigma(n/2) + \frac{1}{2}\sigma(n/k) + (2 - k)\sigma(n/2k)$$

$$- \frac{1}{2}d(n) - \frac{1}{2}d(n/2) + \frac{1}{2}d(n/k) + \frac{1}{2}d(n/2k)$$

$$+ \frac{1}{k}\sum_{\ell=1}^{k-1} \ell d_{\ell,k}(n) + \frac{1}{k}\sum_{\ell=1}^{k-1} \ell d_{\ell,k}(n/2),$$

which is the formula we sought.

Next we treat the sum (14.1) with $(A, B, C, D) = (2, 1, 1, 1)$.

Theorem 14.2. *Let $n \in \mathbb{N}$. Let $f : \mathbb{Z} \to \mathbb{C}$ be an even function. Then*

$$\sum_{\substack{(a, b, x, y) \in \mathbb{N}^4 \\ ax + by = n}} (f(2a - b) - f(2a + b)) = \frac{1}{2}f(0)\left(\sigma(n) - d(n) - d(n/2)\right)$$

$$+ \frac{1}{2}\sum_{\substack{d \in \mathbb{N} \\ d \,|\, n}} \left(1 - d + \frac{4n}{d}\right)f(d) + \frac{1}{2}\sum_{\substack{d \in \mathbb{N} \\ d \,|\, n}} \left(1 + \frac{n}{d}\right)f(2d)$$

$$- \sum_{\substack{d \in \mathbb{N} \\ d \,|\, n}} \sum_{\ell=1}^{2d} f(\ell) - \sum_{\substack{d \in \mathbb{N} \\ d \,|\, n}} \sum_{\substack{\ell = 1 \\ \ell \equiv d \,(\mathrm{mod}\, 2)}}^{d} f(\ell).$$

Proof. In Theorem 13.1 we replace n by $2n$ and choose $f(a, b, x, y) = F_2(a)f(b)F_2(y)$, where $f : \mathbb{Z} \to \mathbb{C}$ is an even function. With this choice the condition (13.1) is satisfied. The left hand side of Theorem 13.1 becomes

$$\sum_{\substack{(a, b, x, y) \in \mathbb{N}^4 \\ ax + by = 2n}} \left(F_2(a)f(b)F_2(-y) - F_2(a)f(-b)F_2(y) \right.$$

$$+ F_2(a)f(a-b)F_2(y) - F_2(a)f(a+b)F_2(y)$$

$$\left. + F_2(b-a)f(b)F_2(x+y) - F_2(a+b)f(b)F_2(x-y) \right)$$

$$= \sum_{\substack{(a, b, x, y) \in \mathbb{N}^4 \\ 2ax + 2by = 2n}} (f(2a-b) - f(2a+b))$$

$$= \sum_{\substack{(a, b, x, y) \in \mathbb{N}^4 \\ ax + by = n}} (f(2a-b) - f(2a+b)).$$

The right hand side of Theorem 13.1 is

$$S_1 + S_2 + S_3 - S_4 - S_5 - S_6,$$

where

$$S_1 := \sum_{\substack{d \in \mathbb{N} \\ d \mid 2n}} \sum_{x=1}^{d-1} F_2(0)f(2n/d)F_2(d),$$

$$S_2 := \sum_{\substack{d \in \mathbb{N} \\ d \mid 2n}} \sum_{x=1}^{d-1} F_2(2n/d)f(0)F_2(x),$$

$$S_3 := \sum_{\substack{d \in \mathbb{N} \\ d \mid 2n}} \sum_{x=1}^{d-1} F_2(2n/d)f(2n/d)F_2(-x),$$

$$S_4 := \sum_{\substack{d \in \mathbb{N} \\ d \mid 2n}} \sum_{x=1}^{d-1} F_2(x)f(x-d)F_2(2n/d),$$

$$S_5 := \sum_{\substack{d \in \mathbb{N} \\ d \mid 2n}} \sum_{x=1}^{d-1} F_2(x)f(d)F_2(2n/d),$$

$$S_6 := \sum_{\substack{d \in \mathbb{N} \\ d \mid 2n}} \sum_{x=1}^{d-1} F_2(d)f(x)F_2(0).$$

Straightforward calculations show that

$$S_1 = \sum_{\substack{d \in \mathbb{N} \\ d \mid n}} \left(\frac{2n}{d} - 1\right) f(d), \tag{14.5}$$

$$S_2 = \frac{1}{2} f(0) \left(\sigma(n) - d(n) - d(n/2)\right), \tag{14.6}$$

$$S_3 = \frac{1}{2} \sum_{\substack{d \in \mathbb{N} \\ d \mid n}} \left(\frac{n}{d} - 1\right) f(2d) - \frac{1}{2} \sum_{\substack{d \in \mathbb{N} \\ d \mid n/2}} f(2d), \tag{14.7}$$

$$S_4 = \sum_{\substack{d \in \mathbb{N} \\ d \mid n \ \ k \equiv d \ (\mathrm{mod} \ 2)}} \sum_{k=1}^{d} f(k) - \sum_{\substack{d \in \mathbb{N} \\ d \mid n}} f(d), \tag{14.8}$$

$$S_5 = \frac{1}{2} \sum_{\substack{d \in \mathbb{N} \\ d \mid n}} (d - 1) f(d) - \frac{1}{2} \sum_{\substack{d \in \mathbb{N} \\ d \mid n/2}} f(2d), \tag{14.9}$$

$$S_6 = \sum_{\substack{d \in \mathbb{N} \\ d \mid n}} \sum_{k=1}^{2d} f(k) - \sum_{\substack{d \in \mathbb{N} \\ d \mid n}} f(2d). \tag{14.10}$$

Equating the left and right hand sides of Theorem 13.1, we obtain the identity of Theorem 14.2. □

Example 14.2. Let $k, n \in \mathbb{N}$ with k odd. We take $f(x) = F_k(x)$ ($x \in \mathbb{Z}$) in Theorem 14.2 to determine the sum

$$\sum_{\substack{(a, b, x, y) \in \mathbb{N}^4 \\ ax + by = n}} \left(F_k(2a - b) - F_k(2a + b)\right).$$

By Problems 6, 7 and 8 of Exercises 3, we have

$$\sum_{\substack{d \in \mathbb{N} \\ d \mid n}} \left(1 - d + \frac{4n}{d}\right) F_k(d) = d(n/k) + (4 - k)\sigma(n/k).$$

Further, by Problems 6 and 8 of Exercises 3, we have as k is odd

$$\sum_{\substack{d \in \mathbb{N} \\ d \mid n}} \left(1 + \frac{n}{d}\right) F_k(2d) = \sum_{\substack{d \in \mathbb{N} \\ d \mid n}} \left(1 + \frac{n}{d}\right) F_k(d) = d(n/k) + \sigma(n/k).$$

By Problem 30 of Exercises 3, we have as k is odd

$$\sum_{\substack{d \in \mathbb{N} \\ d \mid n}} \sum_{\ell=1}^{2d} F_k(\ell) = \frac{2}{k}\sigma(n) - \frac{1}{k}\sum_{\ell=1}^{k-1} \ell d_{(k+1)\ell/2,k}(n).$$

By Problem 25 of Exercises 3, we have as k is odd

$$\sum_{\substack{d \in \mathbb{N} \\ d \mid n}} \sum_{\substack{\ell=1 \\ \ell \equiv d \,(\mathrm{mod}\, 2)}}^{d} F_k(\ell) = \frac{1}{2k}\sigma(n) - \frac{1}{k}\sum_{\ell=1}^{k-1} \ell d_{\ell,k}(n/2) - \frac{1}{2k}\sum_{\substack{\ell=-k \\ 2\nmid\ell}}^{k-1} \ell d_{\ell,2k}(n).$$

Taking $f(x) = F_k(x)$ in Theorem 14.2, and appealing to these identities, we obtain

$$\sum_{\substack{(a,b,x,y) \in \mathbb{N}^4 \\ ax+by=n}} (F_k(2a-b) - F_k(2a+b))$$

$$= \left(\frac{1}{2} - \frac{2}{k} - \frac{1}{2k}\right)\sigma(n) + \left(\frac{5}{2} - \frac{k}{2}\right)\sigma(n/k)$$

$$- \frac{1}{2}d(n) - \frac{1}{2}d(n/2) + d(n/k)$$

$$+ \frac{1}{k}\sum_{\ell=1}^{k-1} \ell d_{(k+1)\ell/2,k}(n) + \frac{1}{k}\sum_{\ell=1}^{k-1} \ell d_{\ell,k}(n/2) + \frac{1}{2k}\sum_{\substack{\ell=-k \\ 2\nmid\ell}}^{k-1} \ell d_{\ell,2k}(n),$$

which is the formula we sought.

Our next theorem evaluates the sum (14.1) for $(A, B, C, D) = (2, 1, 2, 1)$.

Theorem 14.3. *Let $n \in \mathbb{N}$. Let $f : \mathbb{Z} \to \mathbb{C}$ be an even function. Then*

$$\sum_{\substack{(a,b,x,y) \in \mathbb{N}^4 \\ 2ax+by=n}} (f(2a-b) - f(2a+b)) = f(0)(\sigma(n/2) - d(n/2))$$

$$+ \frac{1}{2}\sum_{\substack{d \in \mathbb{N} \\ d \mid n}} \left(1 - d + \frac{2n}{d}\right)f(d) + \frac{1}{2}\sum_{\substack{d \in \mathbb{N} \\ d \mid n/2}} \left(1 + \frac{n}{d}\right)f(2d)$$

$$- \sum_{\substack{d \in \mathbb{N} \\ d \mid n}} \sum_{\substack{\ell=1 \\ \ell \equiv d \,(\mathrm{mod}\, 2)}}^{d} f(\ell) - \sum_{\substack{d \in \mathbb{N} \\ d \mid n/2}} \sum_{\ell=1}^{2d} f(\ell).$$

Proof. In Theorem 13.1 we replace n by $2n$ and choose

$$f(a, b, x, y) = F_2(a) f(b) F_2(x) F_2(y),$$

where $f : \mathbb{Z} \to \mathbb{C}$ is an even function. With this choice condition (13.1) is satisfied. The left hand side of Theorem 13.1 becomes

$$\sum_{\substack{(a, b, x, y) \in \mathbb{N}^4 \\ ax + by = 2n}} \Big(F_2(a) f(b) F_2(x) F_2(-y) - F_2(a) f(-b) F_2(x) F_2(y)$$

$$+ F_2(a) f(a - b) F_2(x + y) F_2(y) - F_2(a) f(a + b) F_2(y - x) F_2(y)$$

$$+ F_2(b - a) f(b) F_2(x) F_2(x + y) - F_2(a + b) f(b) F_2(x) F_2(x - y) \Big)$$

$$= \sum_{\substack{(a, b, x, y) \in \mathbb{N}^4 \\ ax + by = 2n}} \Big(F_2(a) f(a - b) F_2(x + y) F_2(y) - F_2(a) f(a + b) F_2(y - x) F_2(y) \Big)$$

$$= \sum_{\substack{(a, b, x, y) \in \mathbb{N}^4 \\ ax + by = 2n}} \Big(F_2(a) f(a - b) F_2(x) F_2(y) - F_2(a) f(a + b) F_2(x) F_2(y) \Big)$$

$$= \sum_{\substack{(a, b, x, y) \in \mathbb{N}^4 \\ ax + by = 2n}} F_2(a) F_2(x) F_2(y) (f(a - b) - f(a + b))$$

$$= \sum_{\substack{(a, b, x, y) \in \mathbb{N}^4 \\ 4ax + 2by = 2n}} (f(2a - b) - f(2a + b))$$

$$= \sum_{\substack{(a, b, x, y) \in \mathbb{N}^4 \\ 2ax + by = n}} (f(2a - b) - f(2a + b)).$$

The right hand side of Theorem 13.1 is

$$T_1 + T_2 + T_3 - T_4 - T_5 - T_6,$$

where

$$T_1 := \sum_{\substack{d \in \mathbb{N} \\ d \mid 2n}} \sum_{x=1}^{d-1} F_2(0) f(2n/d) F_2(x) F_2(d),$$

$$T_2 := \sum_{\substack{d \in \mathbb{N} \\ d \mid 2n}} \sum_{x=1}^{d-1} F_2(2n/d) f(0) F_2(d) F_2(x),$$

$$T_3 := \sum_{\substack{d \in \mathbb{N} \\ d \,|\, 2n}} \sum_{x=1}^{d-1} F_2(2n/d) f(2n/d) F_2(d-x) F_2(-x),$$

$$T_4 := \sum_{\substack{d \in \mathbb{N} \\ d \,|\, 2n}} \sum_{x=1}^{d-1} F_2(x) f(x-d) F_2(2n/d) F_2(2n/d),$$

$$T_5 := \sum_{\substack{d \in \mathbb{N} \\ d \,|\, 2n}} \sum_{x=1}^{d-1} F_2(x) f(d) F_2(0) F_2(2n/d),$$

$$T_6 := \sum_{\substack{d \in \mathbb{N} \\ d \,|\, 2n}} \sum_{x=1}^{d-1} F_2(d) f(x) F_2(2n/d) F_2(0).$$

We leave it to the reader to show that

$$T_1 = \sum_{\substack{d \in \mathbb{N} \\ d \,|\, n}} \left(\frac{n}{d} - 1 \right) f(d), \tag{14.11}$$

$$T_2 = f(0) \left(\sigma(n/2) - d(n/2) \right), \tag{14.12}$$

$$T_3 = \frac{1}{2} \sum_{\substack{d \in \mathbb{N} \\ d \,|\, n/2}} \left(\frac{n}{d} - 2 \right) f(2d), \tag{14.13}$$

$$T_4 = \sum_{\substack{d \in \mathbb{N} \\ d \,|\, n \ \ k \equiv d \,(\mathrm{mod}\, 2)}} \sum_{k=1}^{d} f(k) - \sum_{\substack{d \in \mathbb{N} \\ d \,|\, n}} f(d), \tag{14.14}$$

$$T_5 = \frac{1}{2} \sum_{\substack{d \in \mathbb{N} \\ d \,|\, n}} (d-1) f(d) - \frac{1}{2} \sum_{\substack{d \in \mathbb{N} \\ d \,|\, n/2}} f(2d), \tag{14.15}$$

$$T_6 = \sum_{\substack{d \in \mathbb{N} \\ d \,|\, n/2}} \sum_{k=1}^{2d} f(k) - \sum_{\substack{d \in \mathbb{N} \\ d \,|\, n/2}} f(2d). \tag{14.16}$$

Theorem 14.3 now follows by equating the left and right hand sides. $\qquad \Box$

By taking $f(x) = F_k(x)$ in Theorem 14.3, a formula for the sum

$$\sum_{\substack{(a, b, x, y) \in \mathbb{N}^4 \\ 2ax + by = n}} (F_k(2a - b) - F_k(2a + b)), \quad k, n \in \mathbb{N},$$

can be determined, see Problem 28.

Finally we evaluate the sum (14.1) for $(A, B, C, D) = (2, 1, 4, 1)$.

Theorem 14.4. *Let $n \in \mathbb{N}$. Let $f : \mathbb{Z} \to \mathbb{C}$ be an even function. Then*

$$\sum_{\substack{(a, b, x, y) \in \mathbb{N}^4 \\ 4ax + by = n}} (f(2a - b) - f(2a + b)) = \frac{1}{2} f(0) (\sigma(n/2) - d(n/2) - d(n/4))$$

$$+ \frac{1}{2} \sum_{\substack{d \in \mathbb{N} \\ d \mid n}} \left(1 + \frac{n}{d}\right) f(d) - \frac{1}{2} \sum_{\substack{d \in \mathbb{N} \\ d \mid n/2}} df(d) + \frac{1}{2} \sum_{\substack{d \in \mathbb{N} \\ d \mid n/4}} \left(1 + \frac{n}{d}\right) f(2d)$$

$$- \sum_{\substack{d \in \mathbb{N} \\ d \mid n}} \sum_{\substack{\ell = 1 \\ \ell \equiv d \,(\mathrm{mod}\ 2)}}^{d} f(\ell) - \sum_{\substack{d \in \mathbb{N} \\ d \mid n/4}} \sum_{\ell=1}^{2d} f(\ell).$$

Proof. The proof of this theorem appears to require a more subtle choice of the function $f(a, b, x, y)$ in Theorem 13.1. A choice that works is

$$f(a, b, x, y) = f(a - b)F_2(ax),$$

where $f : \mathbb{Z} \to \mathbb{C}$ is an even function. First we note that

$$\begin{aligned}
f(a, b, x, y) - f(x, y, a, b) &= f(a - b)F_2(ax) - f(x - y)F_2(xa) \\
&= f(-a + b)F_2(-ax) - f(x - y)F_2(-xa) \\
&= f(-a, -b, x, y) - f(x, y, -a, -b)
\end{aligned}$$

so that our choice satisfies (13.1). Next we observe that

$$\begin{aligned}
f(a, b, x, -y) - f(a, -b, x, y) &= f(a - b)F_2(ax) - f(a + b)F_2(ax) \\
&= (f(a - b) - f(a + b))F_2(ax),
\end{aligned}$$

as well as

$$\begin{aligned}
f(a, a - b, x + y, y) &- f(a, a + b, y - x, y) \\
&= f(b)F_2(ax + ay) - f(-b)F_2(ay - ax) \\
&= f(b)F_2(ax + ay) - f(b)F_2(ax + ay) \\
&= 0
\end{aligned}$$

and

$$f(b - a, b, x, x + y) - f(a + b, b, x, x - y)$$
$$= f(-a)F_2(bx - ax) - f(a)F_2(ax + bx)$$
$$= f(a)F_2(ax + bx) - f(a)F_2(ax + bx)$$
$$= 0.$$

Thus the left hand side of Theorem 13.1 with this choice of the function $f(a, b, x, y)$ is

$$\sum_{\substack{(a, b, x, y) \in \mathbb{N}^4 \\ ax + by = n}} (f(a - b) - f(a + b))F_2(ax)$$

$$= \sum_{\substack{(a, b, x, y) \in \mathbb{N}^4 \\ ax + by = n \\ 2 \,|\, ax}} (f(a - b) - f(a + b))$$

$$= \sum_{\substack{(a, b, x, y) \in \mathbb{N}^4 \\ ax + by = n \\ 2 \,|\, a}} (f(a - b) - f(a + b))$$

$$+ \sum_{\substack{(a, b, x, y) \in \mathbb{N}^4 \\ ax + by = n \\ 2 \,|\, x}} (f(a - b) - f(a + b))$$

$$- \sum_{\substack{(a, b, x, y) \in \mathbb{N}^4 \\ ax + by = n \\ 2 \,|\, a, \, 2 \,|\, x}} (f(a - b) - f(a + b))$$

$$= \sum_{\substack{(a, b, x, y) \in \mathbb{N}^4 \\ 2ax + by = n}} (f(2a - b) - f(2a + b))$$

$$+ \sum_{\substack{(a, b, x, y) \in \mathbb{N}^4 \\ 2ax + by = n}} (f(a - b) - f(a + b))$$

$$- \sum_{\substack{(a, b, x, y) \in \mathbb{N}^4 \\ 4ax + by = n}} (f(2a - b) - f(2a + b)).$$

The value of the first sum is given in Theorem 14.3 and the value of the second sum in Theorem 14.1. The third sum is the sum whose evaluation we seek.

We now turn to the evaluation of the right hand side of Theorem 13.1 for our choice of the function $f(a, b, x, y)$. The right hand side is

$$U_1 + U_2 + U_3 - U_4 - U_5 - U_6,$$

where

$$U_1 := \sum_{\substack{d \in \mathbb{N} \\ d \mid n}} \sum_{x=1}^{d-1} f(-n/d)F_2(0),$$

$$U_2 := \sum_{\substack{d \in \mathbb{N} \\ d \mid n}} \sum_{x=1}^{d-1} f(n/d)F_2(n),$$

$$U_3 := \sum_{\substack{d \in \mathbb{N} \\ d \mid n}} \sum_{x=1}^{d-1} f(0)F_2(n(d-x)/d),$$

$$U_4 := \sum_{\substack{d \in \mathbb{N} \\ d \mid n}} \sum_{x=1}^{d-1} f(d)F_2(nx/d),$$

$$U_5 := \sum_{\substack{d \in \mathbb{N} \\ d \mid n}} \sum_{x=1}^{d-1} f(x-d)F_2(0),$$

$$U_6 := \sum_{\substack{d \in \mathbb{N} \\ d \mid n}} \sum_{x=1}^{d-1} f(d-x)F_2(n).$$

We leave it to the reader to carry out the straightforward calculations needed to show that

$$U_1 = \sum_{\substack{d \in \mathbb{N} \\ d \mid n}} \left(\frac{n}{d} - 1 \right) f(d), \tag{14.17}$$

$$U_2 = F_2(n) \sum_{\substack{d \in \mathbb{N} \\ d \mid n}} \left(\frac{n}{d} - 1 \right) f(d), \tag{14.18}$$

$$U_3 = \frac{1}{2} f(0)(\sigma(n) + \sigma(n/2) - d(n) - 2d(n/2) + d(n/4)), \tag{14.19}$$

$$U_4 = \frac{1}{2} \sum_{\substack{d \in \mathbb{N} \\ d \mid n}} (d-1)f(d) + \frac{1}{2} \sum_{\substack{d \in \mathbb{N} \\ d \mid n/2}} (d-1)f(d)$$

$$- \frac{1}{2} \sum_{\substack{d \in \mathbb{N} \\ d \mid n/2}} f(2d) + \frac{1}{2} \sum_{\substack{d \in \mathbb{N} \\ d \mid n/4}} f(2d), \tag{14.20}$$

$$U_5 = \sum_{\substack{d \in \mathbb{N} \\ d \mid n}} \sum_{\ell=1}^{d} f(\ell) - \sum_{\substack{d \in \mathbb{N} \\ d \mid n}} f(d), \tag{14.21}$$

$$U_6 = F_2(n) \sum_{\substack{d \in \mathbb{N} \\ d \mid n}} \sum_{\ell=1}^{d} f(\ell) - F_2(n) \sum_{\substack{d \in \mathbb{N} \\ d \mid n}} f(d). \tag{14.22}$$

Appealing to Problem 34 of Exercises 3, we obtain

$$U_2 = \sum_{\substack{d \in \mathbb{N} \\ d \mid n/2}} \left(\frac{n}{d} - 1 \right) f(d) + \sum_{\substack{d \in \mathbb{N} \\ d \mid n/2}} \left(\frac{n}{2d} - 1 \right) f(2d) - \sum_{\substack{d \in \mathbb{N} \\ d \mid n/4}} \left(\frac{n}{2d} - 1 \right) f(2d)$$

and

$$U_6 = \sum_{\substack{d \in \mathbb{N} \\ d \mid n/2}} \left(\sum_{\ell=1}^{d} f(\ell) \right) - \sum_{\substack{d \in \mathbb{N} \\ d \mid n/2}} f(d)$$

$$+ \sum_{\substack{d \in \mathbb{N} \\ d \mid n/2}} \left(\sum_{\ell=1}^{2d} f(\ell) \right) - \sum_{\substack{d \in \mathbb{N} \\ d \mid n/2}} f(2d)$$

$$- \sum_{\substack{d \in \mathbb{N} \\ d \mid n/4}} \left(\sum_{\ell=1}^{2d} f(\ell) \right) + \sum_{\substack{d \in \mathbb{N} \\ d \mid n/4}} f(2d).$$

Theorem 14.4 now follows by equating the left and right hand sides. $\square$

By taking $f(x) = F_k(x)$ in Theorem 14.4, a formula for the sum

$$\sum_{\substack{(a, b, x, y) \in \mathbb{N}^4 \\ 4ax + by = n}} (F_k(2a - b) - F_k(2a + b)), \quad k, n \in \mathbb{N},$$

can be determined, see Problem 29.

The identity of the next example will be used in the proof of Theorem 14.5.

Example 14.3. We use Example 14.2 to determine a formula for

$$\sum_{\substack{(a, b, x, y) \in \mathbb{N}^4 \\ ax + by = n}} (F_7(2a - b) - F_7(2a + b)).$$

For $n \in \mathbb{N}$ we set

$$e_i := d_{i,14}(n), \quad i = 0, 1, 2, \ldots, 13,$$

so that

$$d_{i,7}(n) = e_i + e_{i+7}, \quad i = 0, 1, \ldots, 6.$$

With $k = 7$ we have

$$\left(\frac{1}{2} - \frac{2}{k} - \frac{1}{2k}\right)\sigma(n) + \left(\frac{5}{2} - \frac{k}{2}\right)\sigma(n/k) = \frac{1}{7}\sigma(n) - \sigma(n/7)$$

and

$$-\frac{1}{2}d(n) - \frac{1}{2}d(n/2) + d(n/k) = -\frac{1}{2}d(n) - \frac{1}{2}d(n/2) + d(n/7)$$

$$= -\frac{1}{2}(d_{0,14}(n) + d_{1,14}(n) + \cdots + d_{13,14}(n))$$

$$- \frac{1}{2}(d_{0,14}(n) + d_{2,14}(n) + \cdots + d_{12,14}(n))$$

$$+ d_{0,14}(n) + d_{7,14}(n)$$

$$= -\frac{1}{2}e_1 - e_2 - \frac{1}{2}e_3 - e_4 - \frac{1}{2}e_5 - e_6 + \frac{1}{2}e_7$$

$$- e_8 - \frac{1}{2}e_9 - e_{10} - \frac{1}{2}e_{11} - e_{12} - \frac{1}{2}e_{13}.$$

Also

$$\frac{1}{k}\sum_{\ell=1}^{k-1} \ell d_{(k+1)\ell/2,k}(n) = \frac{1}{7}\sum_{\ell=1}^{6} \ell d_{4\ell,7}(n)$$

$$= \frac{1}{7}d_{4,7}(n) + \frac{2}{7}d_{1,7}(n) + \frac{3}{7}d_{5,7}(n)$$

$$+ \frac{4}{7}d_{2,7}(n) + \frac{5}{7}d_{6,7}(n) + \frac{6}{7}d_{3,7}(n)$$

$$= \frac{2}{7}e_1 + \frac{4}{7}e_2 + \frac{6}{7}e_3 + \frac{1}{7}e_4 + \frac{3}{7}e_5 + \frac{5}{7}e_6$$

$$+ \frac{2}{7}e_8 + \frac{4}{7}e_9 + \frac{6}{7}e_{10} + \frac{1}{7}e_{11} + \frac{3}{7}e_{12} + \frac{5}{7}e_{13},$$

$$\frac{1}{2k}\sum_{\substack{\ell=-k \\ 2\nmid\ell}}^{k-1} \ell d_{\ell,2k}(n) = \frac{1}{14}\sum_{\substack{\ell=-7 \\ 2\nmid\ell}}^{6} \ell d_{\ell,14}(n)$$

$$= \frac{1}{14}e_1 + \frac{3}{14}e_3 + \frac{5}{14}e_5 - \frac{1}{2}e_7 - \frac{5}{14}e_9 - \frac{3}{14}e_{11} - \frac{1}{14}e_{13},$$

and

$$\frac{1}{k}\sum_{\ell=1}^{k-1}\ell d_{\ell,k}(n/2) = \frac{1}{7}\sum_{\ell=1}^{6}\ell d_{\ell,7}(n/2)$$

$$= \frac{1}{7}d_{1,7}(n/2) + \frac{2}{7}d_{2,7}(n/2) + \frac{3}{7}d_{3,7}(n/2)$$

$$+ \frac{4}{7}d_{4,7}(n/2) + \frac{5}{7}d_{5,7}(n/2) + \frac{6}{7}d_{6,7}(n/2)$$

$$= \frac{1}{7}e_2 + \frac{2}{7}e_4 + \frac{3}{7}e_6 + \frac{4}{7}e_8 + \frac{5}{7}e_{10} + \frac{6}{7}e_{12}.$$

By Example 14.2 we obtain

$$\sum_{\substack{(a,b,x,y)\in\mathbb{N}^4 \\ ax+by=n}} (F_7(2a-b) - F_7(2a+b))$$

$$= \frac{1}{7}\sigma(n) - \sigma(n/7) - \frac{1}{7}e_1 - \frac{2}{7}e_2 + \frac{4}{7}e_3 - \frac{4}{7}e_4 + \frac{2}{7}e_5$$

$$+ \frac{1}{7}e_6 - \frac{1}{7}e_8 - \frac{2}{7}e_9 + \frac{4}{7}e_{10} - \frac{4}{7}e_{11} + \frac{2}{7}e_{12} + \frac{1}{7}e_{13}$$

$$= \frac{1}{7}\sigma(n) - \sigma(n/7) - \frac{1}{7}d_{1,7}(n) - \frac{2}{7}d_{2,7}(n)$$

$$+ \frac{4}{7}d_{3,7}(n) - \frac{4}{7}d_{4,7}(n) + \frac{2}{7}d_{5,7}(n) + \frac{1}{7}d_{6,7}(n),$$

which is the required formula.

The proof of the final theorem of this chapter makes use of Examples 10.3 and 14.3. For $n \in \mathbb{N}_0$ we set

$$w(n) := \text{card}\{(x_1, x_2) \in \mathbb{Z}^2 \mid x_1^2 + x_1 x_2 + 2x_2^2 = n\}$$

so that $w(0) = 1$. We need the following evaluation of $w(n)$ ($n \in \mathbb{N}$), namely

$$w(n) = 2\sum_{\substack{m \in \mathbb{N} \\ m \mid n}} \left(\frac{-7}{m}\right), \tag{14.23}$$

where $\left(\dfrac{-7}{m}\right)$ ($m \in \mathbb{N}$) is the Legendre-Jacobi-Kronecker symbol for discriminant -7, that is

$$\left(\frac{-7}{m}\right) = \begin{cases} +1, & \text{if } m \equiv 1, 2, 4 \pmod{7}, \\ -1, & \text{if } m \equiv 3, 5, 6 \pmod{7}, \\ 0, & \text{if } m \equiv 0 \pmod{7}, \end{cases}$$

see the notes at the end of this chapter. It is convenient to set for $n \in \mathbb{N}$

$$d_i := d_{i,7}(n), \quad i = 0, 1, 2, 3, 4, 5, 6,$$

so that by (14.23)

$$w(n) = 2d_1 + 2d_2 - 2d_3 + 2d_4 - 2d_5 - 2d_6. \tag{14.24}$$

Theorem 14.5. *Let $n \in \mathbb{N}$. Then the number of $(x_1, x_2, x_3, x_4) \in \mathbb{Z}^4$ such that*

$$n = x_1{}^2 + x_1 x_2 + 2x_2{}^2 + x_3{}^2 + x_3 x_4 + 2x_4{}^2$$

is

$$4\sigma(n) - 28\sigma(n/7).$$

Proof. The number of $(x_1, x_2, x_3, x_4) \in \mathbb{Z}^4$ such that

$$n = x_1{}^2 + x_1 x_2 + 2x_2{}^2 + x_3{}^2 + x_3 x_4 + 2x_4{}^2$$

is

$$\sum_{\substack{(k, \ell) \in \mathbb{N}_0^2 \\ k + \ell = n}} w(k)w(l) = 2w(n) + \sum_{k=1}^{n-1} w(k)w(n - k).$$

Now, by (14.23), we have

$$\sum_{k=1}^{n-1} w(k)w(n - k) = \sum_{k=1}^{n-1} \left(2 \sum_{\substack{a \in \mathbb{N} \\ a \mid k}} \left(\frac{-7}{a} \right) \right) \left(2 \sum_{\substack{b \in \mathbb{N} \\ b \mid n - k}} \left(\frac{-7}{b} \right) \right)$$

$$= 4 \sum_{\substack{(a, b, x, y) \in \mathbb{N}^4 \\ ax + by = n}} \left(\frac{-7}{ab} \right).$$

Appealing to the second part of Problem 5 of Exercises 3, we obtain

$$\sum_{\substack{(a, b, x, y) \in \mathbb{N}^4 \\ ax + by = n}} \left(\frac{-7}{ab} \right) = \sum_{\substack{(a, b, x, y) \in \mathbb{N}^4 \\ ax + by = n}} (F_7(a - b) - F_7(a + b))$$

$$+ \sum_{\substack{(a, b, x, y) \in \mathbb{N}^4 \\ ax + by = n}} (F_7(a - 2b) - F_7(a + 2b))$$

$$+ \sum_{\substack{(a, b, x, y) \in \mathbb{N}^4 \\ ax + by = n}} (F_7(2a - b) - F_7(2a + b))$$

$$= \sum_{\substack{(a, b, x, y) \in \mathbb{N}^4 \\ ax + by = n}} (F_7(a - b) - F_7(a + b))$$

$$+ 2 \sum_{\substack{(a, b, x, y) \in \mathbb{N}^4 \\ ax + by = n}} (F_7(2a - b) - F_7(2a + b)).$$

By Example 10.3 we have

$$\sum_{\substack{(a, b, x, y) \in \mathbb{N}^4 \\ ax + by = n}} (F_7(a - b) - F_7(a + b))$$

$$= \frac{5}{7}\sigma(n) - 5\sigma(n/7) - \frac{5}{7}d_1 - \frac{3}{7}d_2 - \frac{1}{7}d_3 + \frac{1}{7}d_4 + \frac{3}{7}d_5 + \frac{5}{7}d_6.$$

By Example 14.3 we have

$$\sum_{\substack{(a, b, x, y) \in \mathbb{N}^4 \\ ax + by = n}} (F_7(2a - b) - F_7(2a + b))$$

$$= \frac{1}{7}\sigma(n) - \sigma(n/7) - \frac{1}{7}d_1 - \frac{2}{7}d_2 + \frac{4}{7}d_3 - \frac{4}{7}d_4 + \frac{2}{7}d_5 + \frac{1}{7}d_6.$$

Hence, by (14.24), we obtain

$$\sum_{\substack{(a, b, x, y) \in \mathbb{N}^4 \\ ax + by = n}} \left(\frac{-7}{ab}\right) = \sigma(n) - 7\sigma(n/7) - d_1 - d_2 + d_3 - d_4 + d_5 + d_6$$

$$= \sigma(n) - 7\sigma(n/7) - \frac{1}{2}w(n).$$

Thus the required number is

$$2w(n) + 4\left(\sigma(n) - 7\sigma(n/7) - \frac{1}{2}w(n)\right) = 4\sigma(n) - 28\sigma(n/7)$$

as claimed. $\square$

Exercises 14

1. Prove (14.5).
2. Prove (14.6).
3. Prove (14.7).
4. Prove (14.8).
5. Prove (14.9).

6. Prove (14.10).
7. Prove (14.11).
8. Prove (14.12).
9. Prove (14.13).
10. Prove (14.14).
11. Prove (14.15).
12. Prove (14.16).
13. Prove (14.17).
14. Prove (14.18).
15. Prove (14.19).
16. Prove (14.20).
17. Prove (14.21).
18. Prove (14.22).
19. Prove (14.23).
20. Let $n \in \mathbb{N}$. Set $n = 7^{\alpha} N$, where $\alpha \in \mathbb{N}_0, n \in \mathbb{N}$ and $\gcd(N, 7) = 1$. Deduce from Theorem 14.5 that the number of $(x_1, x_2, x_3, x_4) \in \mathbb{Z}^4$ such that

$$n = x_1^2 + x_1 x_2 + 2x_2^2 + x_3^2 + x_3 x_4 + 2x_4^2$$

   is

$$4\sigma(N) = 4\sum_{\substack{d \mid n \\ 7 \nmid d}} d.$$

   Is every positive integer of the form $x_1^2 + x_1 x_2 + 2x_2^2 + x_3^2 + x_3 x_4 + 2x_4^2$ for some integers x_1, x_2, x_3, x_4?

21. Let $k, n \in \mathbb{N}$. Extend the ideas of the proof of Theorem 10.2 to prove that

$$\sum_{\substack{(a, b, x, y) \in \mathbb{N}^4 \\ 2ax + by = n \\ a - b = k}} 1 + \sum_{\substack{(a, b, x, y) \in \mathbb{N}^4 \\ 2ax + by = n \\ a - b = -k}} 1 - \sum_{\substack{(a, b, x, y) \in \mathbb{N}^4 \\ 2ax + by = n \\ a + b = k}} 1$$

$$= -\sum_{\substack{d \in \mathbb{N} \\ d \mid n \\ d \geq k}} 1 - \sum_{\substack{d \in \mathbb{N} \\ d \mid n/2 \\ d \geq k}} 1 + \frac{1}{2}\left(1 + \frac{n}{k}\right) F_k(n)$$

$$+ \frac{1}{2}\left(1 - 2k + \frac{2n}{k}\right) F_2(k) F_{k/2}(n).$$

   Deduce Theorem 14.1 from this identity.

22. Let $k, n \in \mathbb{N}$. If $2 \nmid k$ extend the ideas of the proof of Theorem 10.2 to prove that

$$\sum_{\substack{(a,b,x,y) \in \mathbb{N}^4 \\ ax+by=n \\ 2a-b=k}} 1 + \sum_{\substack{(a,b,x,y) \in \mathbb{N}^4 \\ ax+by=n \\ 2a-b=-k}} 1 - \sum_{\substack{(a,b,x,y) \in \mathbb{N}^4 \\ ax+by=n \\ 2a+b=k}} 1$$

$$= -\sum_{\substack{d \in \mathbb{N} \\ d \mid n \\ d > k/2}} 1 - \sum_{\substack{d \in \mathbb{N} \\ d \mid n \\ d > k \\ d \text{ odd}}} 1 + \left(\frac{2n}{k} - \frac{k}{2} - \frac{1}{2}\right) F_k(n).$$

Similarly if $2 \mid k$ prove that

$$\sum_{\substack{(a,b,x,y) \in \mathbb{N}^4 \\ ax+by=n \\ 2a-b=k}} 1 + \sum_{\substack{(a,b,x,y) \in \mathbb{N}^4 \\ ax+by=n \\ 2a-b=-k}} 1 - \sum_{\substack{(a,b,x,y) \in \mathbb{N}^4 \\ ax+by=n \\ 2a+b=k}} 1$$

$$= -\sum_{\substack{d \in \mathbb{N} \\ d \mid n \\ d > k/2}} 1 - \sum_{\substack{d \in \mathbb{N} \\ d \mid n/2 \\ d > k/2}} 1 + \left(\frac{2n}{k} - \frac{k}{2} - \frac{1}{2}\right) F_k(n) + \left(\frac{n}{k} - \frac{1}{2}\right) F_{k/2}(n).$$

Deduce Theorem 14.2 from these identities.

23. Let $k, n \in \mathbb{N}$. Extend the ideas of the proof of Theorem 10.2 to prove that

$$\sum_{\substack{(a,b,x,y) \in \mathbb{N}^4 \\ 2ax+by=n \\ 2a-b=k}} 1 + \sum_{\substack{(a,b,x,y) \in \mathbb{N}^4 \\ 2ax+by=n \\ 2a-b=-k}} 1 - \sum_{\substack{(a,b,x,y) \in \mathbb{N}^4 \\ 2ax+by=n \\ 2a+b=k}} 1 = -\sum_{\substack{d \in \mathbb{N} \\ d \mid n \\ d \geq k \\ d \equiv k \,(\mathrm{mod}\, 2)}} 1 - \sum_{\substack{d \in \mathbb{N} \\ d \mid n/2 \\ d \geq k/2}} 1$$

$$+ \frac{1}{2}\left(1 - k + \frac{2n}{k}\right) F_k(n) + \frac{1}{2}\left(1 + \frac{2n}{k}\right) F_2(k) F_k(n).$$

Deduce Theorem 14.3 from this identity.

24. Let $k, n \in \mathbb{N}$. Extend the ideas of the proof of Theorem 10.2 to prove that

$$\sum_{\substack{(a,b,x,y) \in \mathbb{N}^4 \\ 4ax+by=n \\ 2a-b=k}} 1 + \sum_{\substack{(a,b,x,y) \in \mathbb{N}^4 \\ 4ax+by=n \\ 2a-b=-k}} 1 - \sum_{\substack{(a,b,x,y) \in \mathbb{N}^4 \\ 4ax+by=n \\ 2a+b=k}} 1 = -\sum_{\substack{d \in \mathbb{N} \\ d \mid n \\ d \geq k \\ d \equiv k \,(\mathrm{mod}\, 2)}} 1 - \sum_{\substack{d \in \mathbb{N} \\ d \mid n/4 \\ d \geq k/2}} 1$$

$$+ \frac{1}{2}\left(1 + \frac{n}{k}\right) F_k(n) - \frac{1}{2} k F_k(n/2) + \frac{1}{2}\left(1 + \frac{2n}{k}\right) F_2(k) F_{k/2}(n/4).$$

Deduce Theorem 14.4 from this identity.

25. Let $n \in \mathbb{N}$. Deduce from Example 14.1 that

$$\sum_{\substack{(a,b,x,y) \in \mathbb{N}^4 \\ 2ax + by = n}} (F_3(a-b) - F_3(a+b)) = \frac{1}{6}\sigma(n) - \frac{1}{3}\sigma(n/2) + \frac{1}{2}\sigma(n/3) - \sigma(n/6)$$

$$- \frac{1}{6}(d_{1,3}(n) - d_{2,3}(n)) - \frac{1}{6}(d_{1,3}(n/2) - d_{2,3}(n/2)).$$

26. Let $n \in \mathbb{N}$. Deduce from Example 14.1 that

$$\sum_{\substack{(a,b,x,y) \in \mathbb{N}^4 \\ 2ax + by = n}} (F_4(a-b) - F_4(a+b)) = \frac{1}{4}\sigma(n) - \frac{1}{4}\sigma(n/2) + \frac{1}{2}\sigma(n/4) - 2\sigma(n/8)$$

$$- \frac{1}{4}(d_{1,4}(n) - d_{3,4}(n)) - \frac{1}{4}(d_{1,4}(n/2) - d_{3,4}(n/2)).$$

27. Let $k, n \in \mathbb{N}$. Determine the sum

$$\sum_{\substack{(a,b,x,y) \in \mathbb{N}^4 \\ ax + by = n}} (F_k(2a-b) - F_k(2a+b))$$

in the case when k is even. The case k odd is treated in Example 14.2.

28. Let $k, n \in \mathbb{N}$. Take $f(x) = F_k(x)$ $(x \in \mathbb{Z})$ in Theorem 14.3 to determine a formula for the sum

$$\sum_{\substack{(a,b,x,y) \in \mathbb{N}^4 \\ 2ax + by = n}} (F_k(2a-b) - F_k(2a+b)).$$

29. Let $k, n \in \mathbb{N}$. Take $f(x) = F_k(x)$ $(x \in \mathbb{Z})$ in Theorem 14.4 to determine a formula for the sum

$$\sum_{\substack{(a,b,x,y) \in \mathbb{N}^4 \\ 4ax + by = n}} (F_k(2a-b) - F_k(2a+b)).$$

Notes on Chapter 14

Theorems 14.1, 14.2, 14.3 and 14.4 were discovered by Huard, Ou, Spearman and Williams in their research leading up to their paper [137]. Theorem 14.1 is given in Williams [266, Proposition 2, p. 236].

We indicate how (14.24) can be proved. Let d be a negative integer with $d \equiv 0$ or $1 \pmod 4$. Let $\{ f_i(x, y) = a_i x_i^2 + b_i xy + c_i y^2 \mid i = 1, 2, \ldots, h \}$ be a representative set of inequivalent, primitive, positive-definite, integral, binary quadratic forms of discriminant d. Let $N(n, d)$ denote the number of representations of $n \in \mathbb{N}$ by the forms $f_i(x, y)$ $(i = 1, 2, \ldots, h)$. Let f be the largest positive integer such that $f^2 \mid d$ and $d/f^2 \equiv 0$ or $1 \pmod 4$. The integer f is called the conductor of the discriminant d. Kaplan and Williams [152] have proved the following extension of a theorem of Dirichlet, see [94, Theorem 64, p. 78]:

$$N(n, d) = \lambda(d) \sum_{\substack{m \in \mathbb{N} \\ m \mid n}} \left(\frac{d}{m} \right), \quad \gcd(n, f) = 1, \qquad (14.25)$$

where

$$\lambda(d) := \begin{cases} 2, & \text{if } d < -4, \\ 4, & \text{if } d = -4, \\ 6, & \text{if } d = -3. \end{cases}$$

When $d = -7$ we have $f = 1, h = 1$ and

$$\{f_i(x, y) \mid i = 1, 2, \ldots, h\} = \{x^2 + xy + 2y^2\}.$$

Thus by (14.26) we have for all $n \in \mathbb{N}$

$$w(n) = N(n, -7) = 2 \sum_{\substack{m \in \mathbb{N} \\ m \mid n}} \left(\frac{-7}{m} \right).$$

Theorem 14.5 is given in Lomadze [189]. The first elementary arithmetic proof of this result is due to Williams [269, Theorem 1.1, p. 793]. This is the proof given here.

When $d = -11$ we have $f = 1, h = 1$ and

$$\{f_i(x, y) \mid i = 1, 2, \ldots, h\} = \{x^2 + xy + 3y^2\}.$$

Thus, by (14.25), we have for all $n \in \mathbb{N}$

$$\text{card}\{(x_1, x_2) \in \mathbb{Z}^2 \mid n = x_1{}^2 + x_1 x_2 + 3x_2{}^2\} = 2 \sum_{\substack{m \in \mathbb{N} \\ m \mid n}} \left(\frac{-11}{m} \right).$$

The number of representations of $n \in \mathbb{N}$ by the quadratic form $x_1{}^2 + x_1 x_2 + 3x_2{}^2 + x_3{}^2 + x_3 x_4 + 3x_4{}^2$ is given in the book by Petersson [222, p. 80]. Recently Shavgulidze [241] has determined the number of representations of a positive integer by the quadratic form

$$x_1{}^2 + x_1 x_2 + 3x_2{}^2 + \cdots + x_{2k-1}{}^2 + x_{2k-1} x_{2k} + 3x_{2k}{}^2$$

in the cases $k = 3, 4$ and 5.

15

Some Twisted Convolution Sums

For $n \in \mathbb{N}$ and $s \in \mathbb{N}$ we define the twisted convolution sum $A_s(n)$ by

$$A_s(n) := T_{1,1,s}(n) = \sum_{\substack{m \in \mathbb{N} \\ m < n/s}} \sigma(m)\sigma(n - sm), \qquad (15.1)$$

where the sum $T_{e,f,g}(n)$ was defined in (13.14). By Besge's formula (Theorem 12.1) we know that

$$A_1(n) = \frac{5}{12}\sigma_3(n) + \left(\frac{1}{12} - \frac{1}{2}n\right)\sigma(n). \qquad (15.2)$$

By Theorem 12.4 we have

$$A_2(n) = \frac{1}{12}\sigma_3(n) + \frac{1}{3}\sigma_3(n/2) + \left(\frac{1}{24} - \frac{1}{8}n\right)\sigma(n) + \left(\frac{1}{24} - \frac{1}{4}n\right)\sigma(n/2),$$

see also Example 13.3 and Problem 15 of Exercises 13. In this chapter we begin by giving another proof of Theorem 12.4, this one using Theorem 14.1, and then we determine $A_3(n)$ and $A_4(n)$.

Theorem 15.1. *Let $n \in \mathbb{N}$. Then*

$$A_2(n) = \frac{1}{12}\sigma_3(n) + \frac{1}{3}\sigma_3(n/2) + \left(\frac{1}{24} - \frac{1}{8}n\right)\sigma(n) + \left(\frac{1}{24} - \frac{1}{4}n\right)\sigma(n/2).$$

Proof. We take $f(x) = x^2$ in Theorem 14.1. The left hand side is

$$\sum_{\substack{(a,b,x,y) \in \mathbb{N}^4 \\ 2ax + by = n}} \left((a-b)^2 - (a+b)^2\right) = -4 \sum_{\substack{(a,b,x,y) \in \mathbb{N}^4 \\ 2ax + by = n}} ab$$

$$= -4 \sum_{\substack{(\ell,m) \in \mathbb{N}^2 \\ 2\ell + m = n}} \sum_{\substack{a \in \mathbb{N} \\ a \mid \ell}} a \sum_{\substack{b \in \mathbb{N} \\ b \mid m}} b$$

$$= -4 \sum_{\substack{(\ell, m) \in \mathbb{N}^2 \\ 2\ell + m = n}} \sigma(\ell)\sigma(m)$$

$$= -4 \sum_{\substack{\ell \in \mathbb{N} \\ 1 \le \ell < n/2}} \sigma(\ell)\sigma(n - 2\ell)$$

$$= -4A_2(n).$$

The right hand side is

$$\frac{1}{2} \sum_{\substack{d \in \mathbb{N} \\ d \mid n}} \left(1 + \frac{n}{d}\right) d^2 + \frac{1}{2} \sum_{\substack{d \in \mathbb{N} \\ d \mid n/2}} \left(1 - 2d + \frac{2n}{d}\right) d^2$$

$$- \sum_{\substack{d \in \mathbb{N} \\ d \mid n}} \left(\sum_{v=1}^{d} v^2\right) - \sum_{\substack{d \in \mathbb{N} \\ d \mid n/2}} \left(\sum_{v=1}^{d} v^2\right).$$

Clearly

$$\sum_{\substack{d \in \mathbb{N} \\ d \mid n}} \left(1 + \frac{n}{d}\right) d^2 = \sigma_2(n) + n\sigma(n)$$

and

$$\sum_{\substack{d \in \mathbb{N} \\ d \mid n/2}} \left(1 - 2d + \frac{2n}{d}\right) d^2 = \sigma_2(n/2) - 2\sigma_3(n/2) + 2n\sigma(n/2).$$

From Problem 28 of Exercises 3 we have

$$\sum_{\substack{d \in \mathbb{N} \\ d \mid n}} \left(\sum_{v=1}^{d} v^2\right) = \frac{1}{3}\sigma_3(n) + \frac{1}{2}\sigma_2(n) + \frac{1}{6}\sigma(n)$$

and

$$\sum_{\substack{d \in \mathbb{N} \\ d \mid n/2}} \left(\sum_{v=1}^{d} v^2\right) = \frac{1}{3}\sigma_3(n/2) + \frac{1}{2}\sigma_2(n/2) + \frac{1}{6}\sigma(n/2).$$

Thus the right hand side is

$$-\frac{1}{3}\sigma_3(n) - \frac{4}{3}\sigma_3(n/2) - \left(\frac{1}{6} - \frac{1}{2}n\right)\sigma(n) - \left(\frac{1}{6} - n\right)\sigma(n/2).$$

Equating the left and right hand sides, we obtain the evaluation of $A_2(n)$. $\square$

Next we use Theorem 14.4 to determine $A_4(n)$.

Theorem 15.2. *Let $n \in \mathbb{N}$. Then*

$$A_4(n) = \frac{1}{48}\sigma_3(n) + \frac{1}{16}\sigma_3(n/2) + \frac{1}{3}\sigma_3(n/4) + \left(\frac{1}{24} - \frac{1}{16}n\right)\sigma(n)$$

$$+ \left(\frac{1}{24} - \frac{1}{4}n\right)\sigma(n/4).$$

Proof. We take $f(x) = x^2$ in Theorem 14.4. The left hand side is

$$\sum_{\substack{(a,b,x,y) \in \mathbb{N}^4 \\ 4ax + by = n}} \left((2a - b)^2 - (2a + b)^2\right) = -8 \sum_{\substack{(a,b,x,y) \in \mathbb{N}^4 \\ 4ax + by = n}} ab$$

$$= -8 \sum_{\substack{(\ell,m) \in \mathbb{N}^2 \\ 4\ell + m = n}} \sum_{\substack{a \in \mathbb{N} \\ a \mid \ell}} a \sum_{\substack{b \in \mathbb{N} \\ b \mid m}} b$$

$$= -8 \sum_{\substack{(\ell,m) \in \mathbb{N}^2 \\ 4\ell + m = n}} \sigma(\ell)\sigma(m)$$

$$= -8 \sum_{\substack{\ell \in \mathbb{N} \\ 1 \le \ell < n/4}} \sigma(\ell)\sigma(n - 4\ell)$$

$$= -8A_4(n).$$

The right hand side is

$$\frac{1}{2} \sum_{\substack{d \in \mathbb{N} \\ d \mid n}} \left(1 + \frac{n}{d}\right)d^2 - \frac{1}{2} \sum_{\substack{d \in \mathbb{N} \\ d \mid n/2}} d^3 + 2 \sum_{\substack{d \in \mathbb{N} \\ d \mid n/4}} \left(1 + \frac{n}{d}\right)d^2$$

$$- \sum_{\substack{d \in \mathbb{N} \\ d \mid n}} \sum_{\substack{k = 1 \\ k \equiv d \;(\mathrm{mod}\; 2)}}^{d} k^2 - \sum_{\substack{d \in \mathbb{N} \\ d \mid n/4}} \sum_{k=1}^{2d} k^2.$$

We have

$$\frac{1}{2} \sum_{\substack{d \in \mathbb{N} \\ d \mid n}} \left(1 + \frac{n}{d}\right)d^2 = \frac{1}{2}\sigma_2(n) + \frac{1}{2}n\sigma(n),$$

$$\frac{1}{2} \sum_{\substack{d \in \mathbb{N} \\ d \mid n/2}} d^3 = \frac{1}{2}\sigma_3(n/2),$$

and

$$2 \sum_{\substack{d \in \mathbb{N} \\ d \mid n/4}} \left(1 + \frac{n}{d}\right) d^2 = 2\sigma_2(n/4) + 2n\sigma(n/4).$$

From Problem 29 of Exercises 3 we have

$$\sum_{\substack{d \in \mathbb{N} \\ d \mid n}} \sum_{\substack{k=1 \\ k \equiv d \,(\mathrm{mod}\, 2)}}^{d} k^2 = \frac{1}{6}\sigma_3(n) + \frac{1}{2}\sigma_2(n) + \frac{1}{3}\sigma(n)$$

and from Problem 35 of the same exercises

$$\sum_{\substack{d \in \mathbb{N} \\ d \mid n/4}} \sum_{k=1}^{2d} k^2 = \frac{8}{3}\sigma_3(n/4) + 2\sigma_2(n/4) + \frac{1}{3}\sigma(n/4).$$

Thus the right hand side is

$$-\frac{1}{6}\sigma_3(n) - \frac{1}{2}\sigma_3(n/2) - \frac{8}{3}\sigma_3(n/4) + \left(-\frac{1}{3} + \frac{1}{2}n\right)\sigma(n) + \left(-\frac{1}{3} + 2n\right)\sigma(n/4).$$

Equating the left and right hand sides, we obtain the evaluation of $A_4(n)$. □

Example 15.1. We use Theorems 12.1, 15.1 and 15.2 to prove three formulae due to Hahn. We recall that for $m \in \mathbb{N}$

$$\tilde{\sigma}(m) = \sigma(m) - 4\sigma(m/2), \quad \hat{\sigma}(m) = \sigma(m) - 2\sigma(m/2),$$

and

$$\tilde{\sigma}_3(m) = \sigma_3(m) - 16\sigma_3(m/2), \quad \hat{\sigma}_3(m) = \sigma_3(m) - 2\sigma_3(m/2).$$

Thus

$$\sum_{\substack{m \in \mathbb{N} \\ m < n/2}} \tilde{\sigma}(m)\tilde{\sigma}(n - 2m) = A_2(n) - 4A_4(n) - 4A_1(n/2) + 16A_2(n/2).$$

Appealing to Theorems 12.1, 15.1 and 15.2 for the evaluations of $A_1(n/2)$, $A_2(n)$, $A_2(n/2)$ and $A_4(n)$, we obtain

$$\sum_{\substack{m \in \mathbb{N} \\ m < n/2}} \tilde{\sigma}(m)\tilde{\sigma}(n - 2m) = -\frac{1}{4}\sigma_3(n/2) + 4\sigma_3(n/4) + \left(\frac{1}{8}n - \frac{1}{8}\right)\sigma(n)$$

$$- \left(\frac{1}{4}n - \frac{3}{8}\right)\sigma(n/2) - \left(n - \frac{1}{2}\right)\sigma(n/4).$$

Thus

$$\sum_{\substack{m \in \mathbb{N} \\ m < n/2}} \tilde{\sigma}(m)\tilde{\sigma}(n - 2m) = -\frac{1}{4}\tilde{\sigma}_3(n/2) + \left(\frac{1}{8}n - \frac{1}{8}\right)\tilde{\sigma}(n) + \left(\frac{1}{4}n - \frac{1}{8}\right)\tilde{\sigma}(n/2),$$

which is the first formula of Hahn. The second and third formulae of Hahn, namely,

$$\sum_{\substack{m \in \mathbb{N} \\ m < n/2}} \tilde{\sigma}(m)\hat{\sigma}(n - 2m) = -\frac{1}{12}\tilde{\sigma}_3(n/2) + \left(\frac{1}{12}n - \frac{1}{8}\right)\hat{\sigma}(n) + \frac{1}{6}n\hat{\sigma}(n/2)$$

$$+ \frac{1}{24}n\tilde{\sigma}(n) - \left(\frac{1}{12}n + \frac{1}{24}\right)\tilde{\sigma}(n/2),$$

$$\sum_{\substack{m \in \mathbb{N} \\ m < n/2}} \hat{\sigma}(m)\tilde{\sigma}(n - 2m) = \frac{1}{24}\tilde{\sigma}_3(n) - \frac{1}{8}\tilde{\sigma}_3(n/2) + \left(\frac{1}{4}n - \frac{1}{8}\right)\hat{\sigma}(n/2) - \frac{1}{24}\tilde{\sigma}(n),$$

can be proved in a similar manner. Although not considered by Hahn the formula

$$\sum_{\substack{m \in \mathbb{N} \\ m < n/2}} \hat{\sigma}(m)\hat{\sigma}(n - 2m) = \frac{5}{168}\hat{\sigma}_3(n) + \frac{1}{84}\tilde{\sigma}_3(n) - \frac{1}{24}\tilde{\sigma}_3(n/2)$$

$$- \frac{5}{96}\hat{\sigma}(n) + \frac{1}{96}\tilde{\sigma}(n) - \frac{1}{48}\tilde{\sigma}(n/2)$$

can be proved in exactly the same way.

Our next goal in this chapter is to determine $A_3(n)$. Since none of Theorems 14.1–14.4 contains the constraint $3ax + by = n$, we cannot use any of these theorems with $f(x) = x^2$ to determine $A_3(n)$. Instead we make use of the Huard-Ou-Spearman-Williams identity (Theorem 13.1) with a suitable choice of $f(a, b, x, y)$.

Theorem 15.3. *Let $n \in \mathbb{N}$. Then*

$$A_3(n) = \frac{1}{24}\sigma_3(n) + \frac{3}{8}\sigma_3(n/3) + \left(\frac{1}{24} - \frac{1}{12}n\right)\sigma(n) + \left(\frac{1}{24} - \frac{1}{4}n\right)\sigma(n/3).$$

Proof. We apply Theorem 13.1 with

$$f(a, b, x, y) = (a^2 - ab)F_3(x).$$

It is easily checked that this choice of f satisfies (13.1). With this choice the left hand side of Theorem 13.1 is

$$\sum_{\substack{(a,b,x,y) \in \mathbb{N}^4 \\ ax+by=n}} \left((a^2 - ab)F_3(x) - (a^2 + ab)F_3(x) + (a^2 - a(a-b))F_3(x+y) \right.$$

$$- (a^2 - a(a+b))F_3(y-x) + ((b-a)^2 - (b-a)b)F_3(x)$$
$$\left. - ((a+b)^2 - (a+b)b)F_3(x) \right)$$

$$= - \sum_{\substack{(a,b,x,y) \in \mathbb{N}^4 \\ ax+by=n}} 4ab\,F_3(x) + \sum_{\substack{(a,b,x,y) \in \mathbb{N}^4 \\ ax+by=n}} ab\,F_3(x+y)$$

$$+ \sum_{\substack{(a,b,x,y) \in \mathbb{N}^4 \\ ax+by=n}} ab\,F_3(y-x)$$

$$= -4 \sum_{\substack{(a,b,x,y) \in \mathbb{N}^4 \\ 3ax+by=n}} ab + \sum_{\substack{(a,b,x,y) \in \mathbb{N}^4 \\ ax+by=n \\ x \equiv -y \,(\mathrm{mod}\,3)}} ab + \sum_{\substack{(a,b,x,y) \in \mathbb{N}^4 \\ ax+by=n \\ x \equiv y \,(\mathrm{mod}\,3)}} ab.$$

The first sum is

$$-4 \sum_{\substack{(a,b,x,y) \in \mathbb{N}^4 \\ 3ax+by=n}} ab = -4 \sum_{\substack{(\ell,m) \in \mathbb{N}^2 \\ 3\ell+m=n}} \sum_{\substack{a \in \mathbb{N} \\ a \mid \ell}} a \sum_{\substack{b \in \mathbb{N} \\ b \mid m}} b = -4 \sum_{\substack{(\ell,m) \in \mathbb{N}^2 \\ 3\ell+m=n}} \sigma(\ell)\sigma(m)$$

$$= -4 \sum_{\substack{\ell \in \mathbb{N} \\ 1 \le \ell < n/3}} \sigma(\ell)\sigma(n-3\ell) = -4A_3(n),$$

which is the sum we wish to evaluate. By consideration of the residues of x and y modulo 3, we obtain

$$\sum_{\substack{(a,b,x,y) \in \mathbb{N}^4 \\ ax+by=n \\ x \equiv -y \,(\mathrm{mod}\,3)}} ab + \sum_{\substack{(a,b,x,y) \in \mathbb{N}^4 \\ ax+by=n \\ x \equiv y \,(\mathrm{mod}\,3)}} ab$$

$$= 3 \sum_{\substack{(a,b,x,y) \in \mathbb{N}^4 \\ ax+by=n \\ x \equiv 0 \,(\mathrm{mod}\,3) \\ y \equiv 0 \,(\mathrm{mod}\,3)}} ab + \sum_{\substack{(a,b,x,y) \in \mathbb{N}^4 \\ ax+by=n}} ab$$

$$- \sum_{\substack{(a,b,x,y) \in \mathbb{N}^4 \\ ax+by=n \\ x \equiv 0 \,(\mathrm{mod}\,3)}} ab - \sum_{\substack{(a,b,x,y) \in \mathbb{N}^4 \\ ax+by=n \\ y \equiv 0 \,(\mathrm{mod}\,3)}} ab$$

$$= 3 \sum_{\substack{(a, b, x, y) \in \mathbb{N}^4 \\ ax + by = n/3}} ab + \sum_{\substack{(a, b, x, y) \in \mathbb{N}^4 \\ ax + by = n}} ab - 2 \sum_{\substack{(a, b, x, y) \in \mathbb{N}^4 \\ 3ax + by = n}} ab$$

$$= 3A_1(n/3) + A_1(n) - 2A_3(n).$$

Hence the left hand side of Theorem 13.1 for our choice of f is

$$3A_1(n/3) + A_1(n) - 6A_3(n).$$

By Besge's formula (Theorem 12.1) we have

$$A_1(n) = \frac{5}{12}\sigma_3(n) + \left(\frac{1}{12} - \frac{1}{2}n\right)\sigma(n)$$

and

$$A_1(n/3) = \frac{5}{12}\sigma_3(n/3) + \left(\frac{1}{12} - \frac{1}{6}n\right)\sigma(n/3)$$

so that the left hand side is

$$\frac{5}{12}\sigma_3(n) + \frac{5}{4}\sigma_3(n/3) + \left(\frac{1}{12} - \frac{1}{2}n\right)\sigma(n) + \left(\frac{1}{4} - \frac{1}{2}n\right)\sigma(n/3) - 6A_3(n).$$

We now turn to the evaluation of the right hand side of Theorem 13.1 for our choice of f. The right hand side is

$$\sum_{\substack{d \in \mathbb{N} \\ d \mid n}} \sum_{\substack{x \in \mathbb{N} \\ x < d}} \left(\left(\frac{n}{d}\right)^2 F_3(d) - (x^2 - x(x-d))F_3(n/d) \right.$$

$$\left. - (x^2 - xd)F_3(0) - (d^2 - dx)F_3(n/d) \right)$$

$$= \sum_{\substack{d \in \mathbb{N} \\ d \mid n \\ 3 \mid d}} \left(\frac{n}{d}\right)^2 (d - 1) - \sum_{\substack{d \in \mathbb{N} \\ d \mid n}} \sum_{x=1}^{d-1}(x^2 - xd) - \sum_{\substack{d \in \mathbb{N} \\ d \mid n \\ 3 \mid n/d}} d^2(d - 1).$$

Next we evaluate each of these three sums. The first sum is

$$\sum_{\substack{d \in \mathbb{N} \\ d \mid n \\ 3 \mid d}} \left(\frac{n}{d}\right)^2 (d - 1) = \sum_{\substack{d \in \mathbb{N} \\ d \mid n \\ 3 \mid n/d}} d^2 \left(\frac{n}{d} - 1\right)$$

$$= \sum_{\substack{d \in \mathbb{N} \\ d \mid n/3}} d^2 \left(\frac{n}{d} - 1\right)$$

$$= n\sigma(n/3) - \sigma_2(n/3).$$

The second sum is

$$\sum_{\substack{d \in \mathbb{N} \\ d \mid n}} \sum_{x=1}^{d-1} (x^2 - xd) = \sum_{\substack{d \in \mathbb{N} \\ d \mid n}} \left(\sum_{x=1}^{d-1} x^2 - d \sum_{x=1}^{d-1} x \right)$$

$$= \sum_{\substack{d \in \mathbb{N} \\ d \mid n}} \left(\frac{1}{6}(d-1)d(2d-1) - \frac{1}{2}d^2(d-1) \right)$$

$$= \frac{1}{6} \sum_{\substack{d \in \mathbb{N} \\ d \mid n}} (d - d^3)$$

$$= \frac{1}{6}\sigma(n) - \frac{1}{6}\sigma_3(n).$$

The third sum is

$$\sum_{\substack{d \in \mathbb{N} \\ d \mid n \\ 3 \mid n/d}} d^2(d-1) = \sum_{\substack{d \in \mathbb{N} \\ d \mid n \\ 3 \mid n/d}} (d^3 - d^2)$$

$$= \sum_{\substack{d \in \mathbb{N} \\ d \mid n/3}} (d^3 - d^2)$$

$$= \sigma_3(n/3) - \sigma_2(n/3).$$

Thus the right hand side is

$$\frac{1}{6}\sigma_3(n) - \sigma_3(n/3) - \frac{1}{6}\sigma(n) + n\sigma(n/3).$$

Equating the left and right hand sides, we obtain the asserted formula for $A_3(n)$. $\square$

As an application of Theorems 15.1 and 15.2, we obtain the following result.

Theorem 15.4. *Let $n \in \mathbb{N}$. Then*

$$\sum_{\substack{m \in \mathbb{N} \\ m < n \\ m \text{ even}}} \sigma(m)\sigma(n-m) = \frac{5}{24}\sigma_3(n) + \frac{7}{8}\sigma_3(n/2) - \frac{2}{3}\sigma_3(n/4) + \left(\frac{1}{24} - \frac{1}{4}n \right)\sigma(n)$$

$$+ \left(\frac{1}{8} - \frac{3}{4}n \right)\sigma(n/2) - \left(\frac{1}{12} - \frac{1}{2}n \right)\sigma(n/4).$$

Proof. From Theorem 3.1 (ii) (with $k = 1$ and $p = 2$) we have

$$\sigma(2m) = 3\sigma(m) - 2\sigma(m/2), \quad m \in \mathbb{N}.$$

Hence

$$\sum_{\substack{m \in \mathbb{N} \\ m < n \\ m \text{ even}}} \sigma(m)\sigma(n-m)$$

$$= \sum_{\substack{m \in \mathbb{N} \\ m < n/2}} \sigma(2m)\sigma(n-2m)$$

$$= 3 \sum_{\substack{m \in \mathbb{N} \\ m < n/2}} \sigma(m)\sigma(n-2m) - 2 \sum_{\substack{m \in \mathbb{N} \\ m < n/2}} \sigma(m/2)\sigma(n-2m)$$

$$= 3 \sum_{\substack{m \in \mathbb{N} \\ m < n/2}} \sigma(m)\sigma(n-2m) - 2 \sum_{\substack{m \in \mathbb{N} \\ m < n/4}} \sigma(m)\sigma(n-4m)$$

$$= 3A_2(n) - 2A_4(n).$$

The theorem now follows from Theorems 15.1 and 15.2. □

The next result follows from Besge's formula (Theorem 12.1) and Theorem 15.4. It will be used in the proof of the formula for the number of representations of a nonnegative integer as the sum of eight triangular numbers (Theorem 16.13).

Theorem 15.5. *Let $n \in \mathbb{N}$. Then*

$$\sum_{\substack{m \in \mathbb{N} \\ m < n \\ m \text{ odd}}} \sigma(m)\sigma(n-m) = \frac{5}{24}\sigma_3(n) - \frac{7}{8}\sigma_3(n/2) + \frac{2}{3}\sigma_3(n/4) + \left(\frac{1}{24} - \frac{1}{4}n\right)\sigma(n)$$

$$- \left(\frac{1}{8} - \frac{3}{4}n\right)\sigma(n/2) + \left(\frac{1}{12} - \frac{1}{2}n\right)\sigma(n/4).$$

Proof. We have

$$\sum_{\substack{m \in \mathbb{N} \\ m < n \\ m \text{ odd}}} \sigma(m)\sigma(n-m) = \sum_{\substack{m \in \mathbb{N} \\ m < n}} \sigma(m)\sigma(n-m) - \sum_{\substack{m \in \mathbb{N} \\ m < n \\ m \text{ even}}} \sigma(m)\sigma(n-m).$$

The value of the first sum on the right hand side is given by Besge's formula and that of the second by Theorem 15.4. □

Example 15.2. We use Theorem 15.5 to reprove (2.8). For $n \in \mathbb{N}$, by Theorem 3.1(ii) (with $k = 3$ and $p = 2$), we have

$$\sigma_3(2n) = 9\sigma_3(n) - 8\sigma_3(n/2)$$

so that

$$\frac{5}{24}\sigma_3(2n) - \frac{7}{8}\sigma_3(n) + \frac{2}{3}\sigma_3(n/2) = \frac{1}{8}\sigma_3(2n) - \frac{1}{8}\sigma_3(n) = \frac{1}{8}\sigma_3^*(2n).$$

Also, as

$$\sigma(2n) = 3\sigma(n) - 2\sigma(n/2),$$

we have

$$\left(\frac{1}{24} - \frac{1}{2}n\right)\sigma(2n) - \left(\frac{1}{8} - \frac{3}{2}n\right)\sigma(n) + \left(\frac{1}{12} - n\right)\sigma(n/2) = 0.$$

Hence, by Theorem 15.5, we have

$$\sum_{\substack{m \in \mathbb{N} \\ m < 2n \\ m \text{ odd}}} \sigma(m)\sigma(2n - m) = \frac{1}{8}\sigma_3^*(2n) = \sigma_3^*(n),$$

which is (2.8).

We next use Theorem 15.5 to prove a theorem of Liouville about binary quadratic forms.

Theorem 15.6. *Let p be a prime with $p \equiv 3 \pmod 8$. Then there exists a prime $q \equiv 5 \pmod 8$ and integers x and y such that*

$$2p = x^2 + qy^2.$$

Proof. Taking $n = 2p$ in Theorem 15.5 we obtain

$$\sum_{m=1}^{p} \sigma(2m - 1)\sigma(2p - 2m + 1) = \sigma_3(p).$$

Now

$$\sigma_3(p) = 1^3 + p^3 \equiv 1 + 3^3 \equiv 4 \pmod 8$$

so that

$$\sum_{m=1}^{p} \sigma(2m - 1)\sigma(2p - 2m + 1) \equiv 4 \pmod 8.$$

Hence

$$2\sum_{m=1}^{(p-1)/2} \sigma(2m - 1)\sigma(2p - 2m + 1) + \sigma(p)^2 \equiv 4 \pmod 8.$$

Now

$$\sigma(p)^2 = (1+p)^2 \equiv 0 \pmod{16}$$

so

$$\sum_{m=1}^{(p-1)/2} \sigma(2m-1)\sigma(2p-2m+1) \equiv 2 \pmod{4}.$$

Suppose there exists $m \in \{1, 2, \ldots, (p-1)/2\}$ such that

$$\sigma(2m-1)\sigma(2p-2m+1) \equiv 1 \pmod{2}.$$

Then

$$\sigma(2m-1) \equiv \sigma(2p-2m+1) \equiv 1 \pmod{2}.$$

Hence there exist odd integers x and y such that

$$2m-1 = x^2, \quad 2p-2m+1 = y^2,$$

so

$$2p = x^2 + y^2 \equiv 2 \pmod{8},$$

contradicting $p \equiv 3 \pmod{8}$. Hence for all $m \in \{1, 2, \ldots, (p-1)/2\}$ we have

$$\sigma(2m-1)\sigma(2p-2m+1) \equiv 0 \pmod{2}.$$

If

$$\sigma(2m-1)\sigma(2p-2m+1) \equiv 0 \pmod{4}$$

for all $m \in \{1, 2, \ldots, (p-1)/2\}$ then

$$\sum_{m=1}^{(p-1)/2} \sigma(2m-1)\sigma(2p-2m+1) \equiv 0 \pmod{4},$$

a contradiction. Hence there exists $m \in \{1, 2, \ldots, (p-1)/2\}$ such that

$$\sigma(2m-1)\sigma(2p-2m+1) \equiv 2 \pmod{4}.$$

Thus there exists $m \in \{1, 2, \ldots, p-1\}$ such that

$$\sigma(2m-1) \equiv 1 \pmod{2}, \quad \sigma(2p-2m+1) \equiv 2 \pmod{4}.$$

Hence there exist an odd integer x, an odd integer y, and a prime $q \equiv 1 \pmod 4$ such that

$$2m - 1 = x^2, \quad 2p - 2m + 1 = qy^2,$$

so that

$$2p = x^2 + qy^2,$$

as asserted. Finally we have

$$1 + q \equiv x^2 + qy^2 \equiv 2p \equiv 6 \pmod 8,$$

so that

$$q \equiv 5 \pmod 8$$

as claimed. $\square$

For $a \in \mathbb{Z}$ and $b, n \in \mathbb{N}$ we define

$$W_{a,b}(n) := \sum_{\substack{m \in \mathbb{N} \\ m < n \\ m \equiv a \,(\mathrm{mod}\, b)}} \sigma(m)\sigma(n - m).$$

As $W_{a,b}(n) = W_{a',b}(n)$ if $a \equiv a' \pmod b$ we usually restrict a to satisfy $0 \le a \le b - 1$.

Theorem 15.7. *For $a \in \mathbb{Z}$ and $b, n \in \mathbb{N}$ we have*

$$W_{a,b}(n) = W_{n-a,b}(n).$$

Proof. We have

$$W_{a,b}(n) = \sum_{\substack{m = 1 \\ m \equiv a \,(\mathrm{mod}\, b)}}^{n-1} \sigma(m)\sigma(n - m)$$

$$= \sum_{\substack{m = 1 \\ n - m \equiv a \,(\mathrm{mod}\, b)}}^{n-1} \sigma(n - m)\sigma(m)$$

$$= \sum_{\substack{m = 1 \\ m \equiv n - a \,(\mathrm{mod}\, b)}}^{n-1} \sigma(m)\sigma(n - m)$$

$$= W_{n-a,b}(n),$$

as asserted. $\square$

Theorem 15.8. *For $b, n \in \mathbb{N}$ we have*

$$\sum_{a=0}^{b-1} W_{a,b}(n) = \frac{5}{12}\sigma_3(n) + \left(\frac{1}{12} - \frac{1}{2}n\right)\sigma(n).$$

Proof. We have

$$\sum_{a=0}^{b-1} W_{a,b}(n) = \sum_{a=0}^{b-1} \sum_{\substack{m=1 \\ m \equiv a \,(\mathrm{mod}\, b)}}^{n-1} \sigma(m)\sigma(n-m)$$

$$= \sum_{m=1}^{n-1} \sigma(m)\sigma(n-m)$$

$$= \frac{5}{12}\sigma_3(n) + \left(\frac{1}{12} - \frac{1}{2}n\right)\sigma(n),$$

by Theorem 12.1. $\qquad\qquad\square$

We now use Theorems 15.4, 15.5 and 15.8 to evaluate $W_{0,2}(n)$ and $W_{1,2}(n)$.

Theorem 15.9. *Let $n \in \mathbb{N}$. If $n \equiv 0 \,(\mathrm{mod}\, 2)$ then*

$$W_{0,2}(n) = \frac{7}{24}\sigma_3(n) + \frac{1}{8}\sigma_3(n/2) + \left(\frac{1}{12} - \frac{1}{2}n\right)\sigma(n),$$

$$W_{1,2}(n) = \frac{1}{8}\sigma_3(n) - \frac{1}{8}\sigma_3(n/2).$$

If $n \equiv 1 \,(\mathrm{mod}\, 2)$ then

$$W_{0,2}(n) = W_{1,2}(n) = \frac{5}{24}\sigma_3(n) + \left(\frac{1}{24} - \frac{1}{4}n\right)\sigma(n).$$

Proof. Suppose first that $n \equiv 0 \,(\mathrm{mod}\, 2)$. By Theorem 15.4 we have

$$W_{0,2}(n) = \frac{1}{24}\left(5\sigma_3(n) + 21\sigma_3(n/2) - 16\sigma_3(n/4)\right)$$
$$+ \frac{1}{24}(1 - 6n)(\sigma(n) + 3\sigma(n/2) - 2\sigma(n/4)).$$

Now as n is even, we have by Theorem 3.1(ii)

$$\sigma(n) - 3\sigma(n/2) + 2\sigma(n/4) = 0$$

and

$$\sigma_3(n) - 9\sigma_3(n/2) + 8\sigma_3(n/4) = 0.$$

Thus

$$W_{0,2}(n) = \frac{7}{24}\sigma_3(n) + \frac{1}{8}\sigma_3(n/2) + \left(\frac{1}{12} - \frac{1}{2}n\right)\sigma(n).$$

The value of $W_{1,2}(n)$ then follows by Theorem 15.8 with $b = 2$.

Now suppose that $n \equiv 1 \pmod 2$. Then, by Theorems 15.7 and 15.8, we have

$$W_{0,2}(n) = W_{1,2}(n)$$

and

$$W_{0,2}(n) + W_{1,2}(n) = \frac{5}{12}\sigma_3(n) + \left(\frac{1}{12} - \frac{1}{2}n\right)\sigma(n)$$

so that

$$W_{0,2}(n) = W_{1,2}(n) = \frac{5}{24}\sigma_3(n) + \left(\frac{1}{24} - \frac{1}{4}n\right)\sigma(n),$$

as required. □

In addition to $\sigma_3(n)$, $\sigma_3(n/3)$ and $\sigma(n)$ the sums $W_{a,3}(n)$ $(a \in \{0, 1, 2\})$ require another function for their evaluation, see the notes at the end of this chapter. Similarly the sums $W_{a,4}(n)$ $(a \in \{0, 1, 2, 3\})$ require for their evaluation a function other than $\sigma_3(n)$, $\sigma_3(n/2)$ and $\sigma(n)$, see the notes at the end of this chapter. For $b \geq 5$ the evaluation of $W_{a,b}(n)$ is unknown.

Example 15.3. Let $n \in \mathbb{N}$. We evaluate the sum $A_9(n)$ for $n \equiv 0 \pmod 3$.

As $n \equiv 0 \pmod 3$ we have

$$A_9(n) = \sum_{\substack{m \in \mathbb{N} \\ m < n/9}} \sigma(m)\sigma(3(n/3 - 3m))$$

$$= \sum_{\substack{m \in \mathbb{N} \\ m < n/9}} \sigma(m)(4\sigma(n/3 - 3m) - 3\sigma(n/9 - m))$$

$$= 4A_3(n/3) - 3A_1(n/9).$$

Appealing to Theorems 12.1 and 15.3 we obtain

$$A_9(n) = \frac{1}{6}\sigma_3(n/3) + \frac{1}{4}\sigma_3(n/9) + \frac{1}{18}(3 - 2n)\sigma(n/3) - \frac{1}{12}(1 + 2n)\sigma(n/9)$$

for $n \equiv 0 \pmod 3$.

Example 15.4. Let $n \in \mathbb{N}$. We use Example 15.3 to evaluate the sum $W_{0,3}(n)$ for $n \equiv 0 \pmod 3$.

We have

$$W_{0,3}(n) = \sum_{\substack{m \in \mathbb{N} \\ m < n \\ m \equiv 0 \,(\mathrm{mod}\ 3)}} \sigma(m)\sigma(n-m)$$

$$= \sum_{\substack{m \in \mathbb{N} \\ 3m < n}} \sigma(3m)\sigma(n-3m)$$

$$= \sum_{\substack{m \in \mathbb{N} \\ m < n/3}} (4\sigma(m) - 3\sigma(m/3))\sigma(n-3m)$$

$$= 4A_3(n) - 3A_9(n).$$

Appealing to Theorem 15.3 and Example 15.3, we obtain

$$W_{0,3}(n) = \frac{1}{6}\sigma_3(n) + \sigma_3(n/3) - \frac{3}{4}\sigma_3(n/9)$$
$$+ \left(\frac{1}{6} - \frac{1}{3}n\right)\sigma(n) - \left(\frac{1}{3} + \frac{2}{3}n\right)\sigma(n/3) + \left(\frac{1}{4} + \frac{1}{2}n\right)\sigma(n/9).$$

Appealing to

$$\sigma_3(n) = 28\sigma_3(n/3) - 27\sigma_3(n/9)$$

and

$$\sigma(n) = 4\sigma(n/3) - 3\sigma(n/9),$$

which are valid for all $n \equiv 0 \,(\mathrm{mod}\ 3)$, we obtain

$$W_{0,3}(n) = \frac{7}{36}\sigma_3(n) + \frac{2}{9}\sigma_3(n/3) + \left(\frac{1}{12} - \frac{1}{2}n\right)\sigma(n).$$

Example 15.5. We use Theorem 15.3 to prove the following result: If q is a prime such that $q \equiv 5, 11 \,(\mathrm{mod}\ 24)$ then there exist positive integers x and y such that

$$q = 2x^2 + 3y^2.$$

Taking $n = q$ in Theorem 15.3, we obtain

$$\sum_{m \in \mathbb{N}, \, m < q/3} \sigma(m)\sigma(q - 3m) = \frac{1}{24}(\sigma_3(q) + (1 - 2q)\sigma(q))$$

$$= \frac{1}{24}((q^3 + 1) + (1 - 2q)(q + 1))$$

$$= \frac{(q^2 - 1)}{8} \frac{(q - 2)}{3}.$$

As $q \equiv 5, 11 \pmod{24}$, both $(q^2 - 1)/8$ and $(q - 2)/3$ are odd integers, and so

$$\sum_{\substack{m \in \mathbb{N} \\ m < q/3}} \sigma(m)\sigma(q - 3m) \equiv 1 \pmod{2}.$$

Hence there exists $i \in \mathbb{N}$ with $i < q/3$ such that $\sigma(q - 3i)\sigma(i) \equiv 1 \pmod{2}$. Thus

$$\sigma(q - 3i) \equiv \sigma(i) \equiv 1 \pmod{2}.$$

Now $\sigma(m) \equiv 1 \pmod{2}$ if and only if $m = t^2$ or $2t^2$ for some $t \in \mathbb{N}$. Thus there exist $x \in \mathbb{N}$ and $y \in \mathbb{N}$ such that

$$q - 3i = x^2 \text{ or } 2x^2, \quad i = y^2 \text{ or } 2y^2.$$

Hence

$$q = x^2 + 3y^2, \ 2x^2 + 3y^2, \ x^2 + 6y^2 \text{ or } 2x^2 + 6y^2.$$

Now

$$x^2 + 3y^2 \equiv 0, 1 \pmod{3}, \ x^2 + 6y^2 \equiv 0, 1 \pmod{3}, \ 2x^2 + 6y^2 \equiv 0 \pmod{2},$$

so

$$q \neq x^2 + 3y^2, \ x^2 + 6y^2, \ 2x^2 + 6y^2.$$

Therefore

$$q = 2x^2 + 3y^2$$

for some $x, y \in \mathbb{N}$ as claimed.

Exercises 15

1. By taking $k = 2$ in Problem 12 of Exercises 13, prove Theorem 15.1.
2. By taking $k = 3$ in Problem 12 of Exercises 13, prove Theorem 15.3.

3. By taking $k = 4$ in Problem 12 of Exercises 13, prove Theorem 15.2.
4. By taking $n = 2M$, where M is odd, in Theorem 15.5 prove that

$$\sum_{m=1}^{M} \sigma(2m - 1)\sigma(2M - 2m + 1) = \sigma_3(M).$$

5. Let $n \in \mathbb{N}$. Set

$$X := \sum_{\substack{m \in \mathbb{N} \\ m < n/2}} \sigma_3(m)\sigma(n - 2m), \quad Y := \sum_{\substack{m \in \mathbb{N} \\ m < n/2}} \sigma(m)\sigma_3(n - 2m).$$

By taking $f(x) = x^4$ in Theorem 14.1 prove that

$$X + Y = \frac{1}{40}\sigma_5(n) + \frac{3}{20}\sigma_5(n/2) + \left(\frac{1}{24} - \frac{1}{16}n\right)\sigma_3(n)$$

$$+ \left(\frac{1}{24} - \frac{1}{8}n\right)\sigma_3(n/2) - \frac{1}{240}\sigma(n) - \frac{1}{240}\sigma(n/2).$$

6. Let $r \in \mathbb{N}$. Let p be a prime. Let $n \in \mathbb{N}$ be such that $n \equiv 0 \pmod{p}$. Prove that

$$A_{rp^2}(n) = (p + 1)A_{rp}(n/p) - pA_r(n/p^2).$$

7. Let $n \in \mathbb{N}$. Use Problem 6, Theorem 15.1 and Theorem 15.2 to prove that

$$A_8(n) = \frac{1}{192}\sigma_3(n) + \frac{1}{64}\sigma_3(n/2) + \frac{1}{16}\sigma_3(n/4) + \frac{1}{3}\sigma_3(n/8)$$

$$+ \left(\frac{1}{24} - \frac{1}{32}n\right)\sigma(n) + \left(\frac{1}{24} - \frac{1}{4}n\right)\sigma(n/8)$$

for $n \equiv 0 \pmod 2$.

8. Let $n \in \mathbb{N}$ be such that $n \equiv 0 \pmod 3$. Use Example 15.3 to prove that

$$A_9(n) = \frac{1}{216}\sigma_3(n) + \frac{1}{27}\sigma_3(n/3) + \frac{3}{8}\sigma_3(n/9)$$

$$+ \left(\frac{1}{24} - \frac{1}{36}n\right)\sigma(n) + \left(\frac{1}{24} - \frac{1}{4}n\right)\sigma(n/9).$$

9. Let $n \in \mathbb{N}$. Let p be a prime. Prove that

$$W_{0,p}(n) = (p + 1)A_p(n) - pA_{p^2}(n).$$

10. Let p be a prime. Let $n \in \mathbb{N}$ be such that $n \equiv 0 \pmod p$. Prove that

$$W_{0,p}(n) = (p + 1)A_p(n) - p(p + 1)A_p(n/p) + p^2A_1(n/p^2).$$

11. Use the ideas of Example 15.5 in conjunction with Theorem 15.1 to prove that if q is a prime with $q \equiv 5 \pmod 8$ then there exist positive integers x and y such that $q = x^2 + y^2$. (This is a special case of the Girard-Fermat theorem (Theorem 7.1).)
12. Use the ideas of Example 15.5 in conjunction with Theorem 15.2 to prove that if q is a prime with $q \equiv 3 \pmod 8$ then there exist positive integers x and y such that $q = x^2 + 2y^2$. (This is a special case of Theorem 7.3.)
13. Evaluate the sum

$$\sum_{\substack{\ell \in \mathbb{N} \\ \ell < n/2}} \sigma^*(\ell)\sigma^*(n - 2\ell)$$

for all $n \in \mathbb{N}$.
14. Let $a \in \mathbb{Z}$ and $b, n \in \mathbb{N}$. Let $d \in \mathbb{N}$ be such that $d \mid b$. Prove that

$$\sum_{j=0}^{(b/d)-1} W_{a+jd,b}(n) = W_{a,d}(n).$$

Notes on Chapter 15

Theorems 15.1 and 15.2 in the case when n is odd are due to Melfi [202]. For general n they are due to Huard, Ou, Spearman and Williams [137, Theorems 2 and 4, pp. 247, 249]. For sums related to $A_4(n)$, see Cheng and Williams [69, Theorem 4.1. p. 570]. For $3 \nmid n$ Theorem 15.3 is due to Melfi [202]. His result was extended to general n by Huard, Ou, Spearman and Williams [137, Theorem 3, p. 248]. Theorems 15.4 and 15.5 are Theorem 5 and Corollary 1 respectively of Huard, Ou, Spearman and Williams [137]. The evaluation of $A_s(n)$ has been carried out for $s = 5$ Lemire and Williams [168], $s = 6$ Alaca and Williams [29], $s = 7$ Lemire and Williams [168], $s = 8$ Williams [270], $s = 9$ Williams [268], $s = 10$ Royer [238], $s = 12$ Alaca, Alaca and Williams [12], $s = 16$ Alaca, Alaca and Williams [18], $s = 18$ Alaca, Alaca and Williams [14], $s = 23$ Chan and Cooper [63] and $s = 24$ Alaca, Alaca and Williams [15]. Functions other than $\sigma_s(n)$ are required for their evaluation. A general approach to the evaluation of $A_s(n)$ has been given by Royer [238] using quasimodular forms. Theorem 15.6 is due to Liouville [171], [172]. It would be interesting to have a proof of Liouville's theorem based on the theory of binary quadratic forms. Theorem 15.9 can be found in Huard, Ou, Spearman and Williams [137, p. 256].

The formulae of Example 15.1 are due to Hahn [121, Theorem 2.7.5, p. 58]. (Note that in equation (2.7.9) of [121] there should be a "2" in front of the term $\tilde{\sigma}_3(n/2)$.)

Example 15.5 is a simple consequence of Gauss's theory of genera. If q is a prime such that $\left(\dfrac{-24}{q}\right) = 1$ then q is represented by exactly one of the two inequivalent positive-definite binary quadratic forms $x^2 + 6y^2$ and $2x^2 + 3y^2$ of discriminant -24. By Gauss's theory of genera we have

$$q = x^2 + 6y^2 \Longleftrightarrow \left(\frac{-3}{q}\right) = \left(\frac{8}{q}\right) = 1 \Longleftrightarrow q \equiv 1, 7 \pmod{24}$$

and

$$q = 2x^2 + 3y^2 \Longleftrightarrow \left(\frac{-3}{q}\right) = \left(\frac{8}{q}\right) = -1 \Longleftrightarrow q \equiv 5, 11 \pmod{24}.$$

The identity of Problem 4 is due to Liouville [170, 1st article, p. 146], see also [93, Vol. 1, p. 287; Vol. 2, p. 329].

The evaluation of $W_{a,3}(n)$ was begun in Melfi [202], Huard, Ou, Spearman and Williams [137], Williams [267] and completed in Williams [268], where it was shown that: if $n \equiv 0 \pmod 3$

$$W_{0,3}(n) = \frac{7}{36}\sigma_3(n) + \frac{2}{9}\sigma_3(n/3) + \left(\frac{1}{12} - \frac{1}{2}n\right)\sigma(n),$$

$$W_{1,3}(n) = \frac{1}{9}\sigma_3(n) - \frac{1}{9}\sigma_3(n/3),$$

$$W_{2,3}(n) = \frac{1}{9}\sigma_3(n) - \frac{1}{9}\sigma_3(n/3);$$

if $n \equiv 1 \pmod 3$

$$W_{0,3}(n) = \frac{11}{72}\sigma_3(n) + \left(\frac{1}{24} - \frac{1}{4}n\right)\sigma(n) + \frac{1}{18}a(n),$$

$$W_{1,3}(n) = \frac{11}{72}\sigma_3(n) + \left(\frac{1}{24} - \frac{1}{4}n\right)\sigma(n) + \frac{1}{18}a(n),$$

$$W_{2,3}(n) = \frac{1}{9}\sigma_3(n) - \frac{1}{9}a(n);$$

if $n \equiv 2 \pmod 3$

$$W_{0,3}(n) = \frac{11}{72}\sigma_3(n) + \left(\frac{1}{24} - \frac{1}{4}n\right)\sigma(n),$$

$$W_{1,3}(n) = \frac{1}{9}\sigma_3(n),$$

$$W_{2,3}(n) = \frac{11}{72}\sigma_3(n) + \left(\frac{1}{24} - \frac{1}{4}n\right)\sigma(n);$$

where the integers $a(n)$ $(n \in \mathbb{N})$ are defined by

$$\sum_{n=1}^{\infty} a(n)q^n = q \prod_{n=1}^{\infty}(1 - q^{3n})^8.$$

The evaluation of $W_{a,4}(n)$ was begun in Melfi [202] and Huard, Ou, Spearman and Williams [137], and completed in Alaca, Alaca and Williams [27], where it was shown that: if $n \equiv 0 \pmod 4$

$$W_{0,4}(n) = \frac{29}{192}\sigma_3(n) + \frac{17}{64}\sigma_3(n/2) + \left(\frac{1}{12} - \frac{1}{2}n\right)\sigma(n),$$

$$W_{1,4}(n) = \frac{1}{16}\sigma_3(n) - \frac{1}{16}\sigma_3(n/2),$$

$$W_{2,4}(n) = \frac{9}{64}\sigma_3(n) - \frac{9}{64}\sigma_3(n/2),$$

$$W_{3,4}(n) = \frac{1}{16}\sigma_3(n) - \frac{1}{16}\sigma_3(n/2);$$

if $n \equiv 1 \pmod 4$

$$W_{0,4}(n) = \frac{11}{96}\sigma_3(n) + \left(\frac{1}{24} - \frac{1}{4}n\right)\sigma(n) + \frac{3}{32}c_8(n),$$

$$W_{1,4}(n) = \frac{11}{96}\sigma_3(n) + \left(\frac{1}{24} - \frac{1}{4}n\right)\sigma(n) + \frac{3}{32}c_8(n),$$

$$W_{2,4}(n) = \frac{3}{32}\sigma_3(n) - \frac{3}{32}c_8(n),$$

$$W_{3,4}(n) = \frac{3}{32}\sigma_3(n) - \frac{3}{32}c_8(n);$$

if $n \equiv 2 \pmod 4$

$$W_{0,4}(n) = \frac{11}{72}\sigma_3(n) + \left(\frac{1}{24} - \frac{1}{4}n\right)\sigma(n),$$

$$W_{1,4}(n) = \frac{1}{18}\sigma_3(n) + \frac{1}{2}c_8(n/2),$$

$$W_{2,4}(n) = \frac{11}{72}\sigma_3(n) + \left(\frac{1}{24} - \frac{1}{4}n\right)\sigma(n),$$

$$W_{3,4}(n) = \frac{1}{18}\sigma_3(n) - \frac{1}{2}c_8(n/2);$$

if $n \equiv 3 \pmod 4$

$$W_{0,4}(n) = \frac{11}{96}\sigma_3(n) + \left(\frac{1}{24} - \frac{1}{4}n\right)\sigma(n) + \frac{3}{32}c_8(n),$$

$$W_{1,4}(n) = \frac{3}{32}\sigma_3(n) - \frac{3}{32}c_8(n),$$

$$W_{2,4}(n) = \frac{3}{32}\sigma_3(n) - \frac{3}{32}c_8(n),$$

$$W_{3,4}(n) = \frac{11}{96}\sigma_3(n) + \left(\frac{1}{24} - \frac{1}{4}n\right)\sigma(n) + \frac{3}{32}c_8(n);$$

where the integers $c_8(n)$ are defined by

$$\sum_{n=1}^{\infty} c_8(n)q^n = q\prod_{n=1}^{\infty}(1 - q^{2n})^4(1 - q^{4n})^4.$$

Recently Alaca, Alaca, Uygul and Williams [10] have determined $W_{a,8}(n)$ for all $a \in \{0, 1, 2, 3, 4, 5, 6, 7\}$ and all $n \in \mathbb{N}$.

16

Sums of Two, Four, Six and Eight Triangular Numbers

The triangular numbers are the nonnegative integers

$$T_k = \frac{1}{2}k(k+1), \ k \in \mathbb{N}_0,$$

so that

$$T_0 = 0, \ T_1 = 1, \ T_2 = 3, \ T_3 = 6, \ T_4 = 10, \ T_5 = 15, \ldots .$$

We set

$$\Delta = \{T_k \mid k \in \mathbb{N}_0\} = \{0, 1, 3, 6, 10, 15, \ldots\}.$$

Let k be a positive integer. We denote the number of representations of $n(\in \mathbb{N}_0)$ as a sum of k triangular numbers by $t_k(n)$, that is,

$$t_k(n) := \text{card } \{(m_1, \ldots, m_k) \in \mathbb{N}_0^k \mid n = \frac{1}{2}m_1(m_1+1) + \cdots + \frac{1}{2}m_k(m_k+1)\}.$$

Clearly $t_k(0) = 1$ for all $k \in \mathbb{N}$. In this chapter we obtain formulae for $t_k(n)$ ($n \in \mathbb{N}_0$) for $k = 2, 4, 6$ and 8.

We begin by using Theorem 9.3 to determine a formula for $t_2(n)$.

Theorem 16.1. *Let $n \in \mathbb{N}_0$. Then*

$$t_2(n) = \sum_{\substack{d \in \mathbb{N} \\ d \mid 4n+1}} \left(\frac{-4}{d}\right).$$

Proof. For $n \in \mathbb{N}_0$ we define the following three sets:

$$X = \left\{ (a, b) \in \mathbb{N}_0^2 \mid n = \frac{1}{2}a(a + 1) + \frac{1}{2}b(b + 1) \right\},$$

$$Y = \left\{ (a, b) \in \mathbb{Z}^2 \mid n = \frac{1}{2}a(a + 1) + \frac{1}{2}b(b + 1) \right\},$$

$$Z = \{(a, b) \in \mathbb{Z}^2 \mid 8n + 2 = a^2 + b^2\}.$$

Clearly

$$t_2(n) = \text{card } X.$$

As $\frac{1}{2}a(a + 1) = \frac{1}{2}(-a - 1)(-a - 1 + 1)$ we have

$$\text{card } Y = 4 \text{ card } X.$$

It is easily checked that the mapping $f : Y \to Z$ given by

$$f((a, b)) = (2a + 1, 2b + 1)$$

is a bijection. Thus

$$\text{card } Y = \text{card } Z = r_2(8n + 2).$$

By Theorem 9.3 we have

$$r_2(8n + 2) = 4 \sum_{\substack{d \in \mathbb{N} \\ d \mid 8n + 2}} \left(\frac{-4}{d} \right).$$

Now the even divisors of $8n + 2$ contribute nothing to the sum so that

$$r_2(8n + 2) = 4 \sum_{\substack{d \in \mathbb{N} \\ d \mid 4n + 1}} \left(\frac{-4}{d} \right).$$

Hence

$$t_2(n) = \text{card } X = \frac{1}{4} \text{ card } Y = \frac{1}{4} \text{ card } Z = \sum_{\substack{d \in \mathbb{N} \\ d \mid 4n + 1}} \left(\frac{-4}{d} \right),$$

completing the proof of the theorem.   □

Example 16.1. We determine all the representations of 748 as the sum of two triangular numbers. The four integers 45, 153, 595 and 703 are triangular numbers as

$$45 = \frac{9 \cdot 10}{2}, \quad 153 = \frac{17 \cdot 18}{2}, \quad 595 = \frac{34 \cdot 35}{2}, \quad 703 = \frac{37 \cdot 38}{2}.$$

We have

$$748 = 45 + 703 = 153 + 595 = 595 + 153 = 703 + 45.$$

By Theorem 16.1 we have

$$t_2(748) = \sum_{\substack{d \in \mathbb{N} \\ d \mid 2993}} \left(\frac{-4}{d}\right)$$

$$= \left(\frac{-4}{1}\right) + \left(\frac{-4}{41}\right) + \left(\frac{-4}{73}\right) + \left(\frac{-4}{2993}\right)$$

$$= 1 + 1 + 1 + 1$$

$$= 4.$$

Thus the four representations of 748 as the sum of two triangular numbers given above comprise all such representations.

Next we determine a formula for the number $t_4(n)$ of representations of a positive integer n as the sum of four triangular numbers.

We begin by observing that

$$\sum_{\substack{(a, b, x, y) \in \mathbb{N}^4 \\ ax + by = n \\ x \equiv 1 \,(\mathrm{mod}\, 2)}} (f(a - b) - f(a + b))$$

$$= \sum_{\substack{(a, b, x, y) \in \mathbb{N}^4 \\ ax + by = n}} (f(a - b) - f(a + b)) - \sum_{\substack{(a, b, x, y) \in \mathbb{N}^4 \\ ax + by = n \\ x \equiv 0 \,(\mathrm{mod}\, 2)}} (f(a - b) - f(a + b))$$

$$= \sum_{\substack{(a, b, x, y) \in \mathbb{N}^4 \\ ax + by = n}} (f(a - b) - f(a + b)) - \sum_{\substack{(a, b, x, y) \in \mathbb{N}^4 \\ 2ax + by = n}} (f(a - b) - f(a + b)).$$

Then, appealing to Theorems 10.1 and 14.1 for the evaluation of the last two sums, we obtain the following theorem.

Theorem 16.2. *Let $n \in \mathbb{N}$. Let $f : \mathbb{Z} \to \mathbb{C}$ be an even function. Then*

$$\sum_{\substack{(a, b, x, y) \in \mathbb{N}^4 \\ ax + by = n \\ x \equiv 1 \,(\mathrm{mod}\, 2)}} (f(a - b) - f(a + b)) = \frac{1}{2} f(0)(\sigma(n) - d(n) + d(n/2))$$

$$+ \frac{1}{2} \sum_{\substack{d \in \mathbb{N} \\ d \mid n}} \left(1 - 2d + \frac{3n}{d}\right) f(d) - \frac{1}{2} \sum_{\substack{d \in \mathbb{N} \\ d \mid n/2}} \left(1 - 2d + \frac{2n}{d}\right) f(d)$$

$$- \sum_{\substack{d \in \mathbb{N} \\ d \mid n}} \left(\sum_{v=1}^{d} f(v)\right) + \sum_{\substack{d \in \mathbb{N} \\ d \mid n/2}} \left(\sum_{v=1}^{d} f(v)\right).$$

We require the following consequence of Theorem 16.2.

Theorem 16.3. *Let $n \in \mathbb{N}_0$. Then*

$$\sum_{\substack{(a, b, x, y) \in \mathbb{N}^4 \\ ax + by = 4n + 2 \\ x \equiv 1 \,(\mathrm{mod}\ 2)}} \left(\frac{-4}{ab}\right) = \sigma(2n + 1).$$

Proof. Replacing n by $4n + 2$ and choosing $f(x) = F_4(x)$ in Theorem 16.2, we obtain using Theorem 3.10 and Example 3.6

$$\sum_{\substack{(a, b, x, y) \in \mathbb{N}^4 \\ ax + by = 4n + 2 \\ x \equiv 1 \,(\mathrm{mod}\ 2)}} \left(\frac{-4}{ab}\right) = \frac{3}{2}\sigma(2n + 1) - \frac{1}{2}d(2n + 1) - \sum_{\substack{d \in \mathbb{N} \\ d \mid 4n + 2}} \left[\frac{d}{4}\right] + \sum_{\substack{d \in \mathbb{N} \\ d \mid 2n + 1}} \left[\frac{d}{4}\right].$$

Appealing to (3.13) we have

$$\sum_{\substack{d \in \mathbb{N} \\ d \mid 4n + 2}} \left[\frac{d}{4}\right] - \sum_{\substack{d \in \mathbb{N} \\ d \mid 2n + 1}} \left[\frac{d}{4}\right] = \sum_{\substack{d \in \mathbb{N} \\ d \mid 4n + 2 \\ 2 \mid d}} \left[\frac{d}{4}\right] = \sum_{\substack{e \in \mathbb{N} \\ e \mid 2n + 1}} \left[\frac{e}{2}\right]$$

$$= \frac{1}{2}\sigma(2n + 1) - \frac{1}{2}d_{1,2}(2n + 1)$$

$$= \frac{1}{2}\sigma(2n + 1) - \frac{1}{2}d(2n + 1).$$

The asserted formula now follows. □

Our next result is a slight reformulation of Theorem 16.3.

Theorem 16.4. *Let $n \in \mathbb{N}_0$. Then*

$$\sum_{\substack{(a, b, x, y) \in \mathbb{N}^4 \\ ax + by = 4n + 2 \\ ax \equiv 1 \,(\mathrm{mod}\ 2)}} \left(\frac{-4}{ab}\right) = \sigma(2n + 1).$$

Proof. As

$$\left(\frac{-4}{ab}\right) = 0, \ \text{ for } a \equiv 0 \,(\mathrm{mod}\ 2),$$

only the terms with $a \equiv 1 \,(\mathrm{mod}\ 2)$ contribute to the sum

$$\sum_{\substack{(a, b, x, y) \in \mathbb{N}^4 \\ ax + by = 4n + 2 \\ x \equiv 1 \,(\mathrm{mod}\ 2)}} \left(\frac{-4}{ab}\right)$$

and the asserted result follows from Theorem 16.3. □

Theorem 16.5. *Let* $n \in \mathbb{N}_0$. *Then*

$$\sum_{\substack{(a,b,x,y) \in \mathbb{N}^4 \\ ax+by=4n+2 \\ ax \equiv 3 \,(\mathrm{mod}\,4)}} \left(\frac{-4}{ab}\right) = 0.$$

Proof. Interchanging the roles of a and x, we obtain

$$\sum_{\substack{(a,b,x,y) \in \mathbb{N}^4 \\ ax+by=4n+2 \\ ax \equiv 3 \,(\mathrm{mod}\,4)}} \left(\frac{-4}{ab}\right) = \sum_{\substack{(a,b,x,y) \in \mathbb{N}^4 \\ ax+by=4n+2 \\ ax \equiv 3 \,(\mathrm{mod}\,4)}} \left(\frac{-4}{xb}\right).$$

As $ax \equiv 3 \,(\mathrm{mod}\,4)$ we have

$$\left(\frac{-4}{xb}\right) = -\left(\frac{-4}{ab}\right).$$

Thus

$$\sum_{\substack{(a,b,x,y) \in \mathbb{N}^4 \\ ax+by=4n+2 \\ ax \equiv 3 \,(\mathrm{mod}\,4)}} \left(\frac{-4}{ab}\right) = -\sum_{\substack{(a,b,x,y) \in \mathbb{N}^4 \\ ax+by=4n+2 \\ ax \equiv 3 \,(\mathrm{mod}\,4)}} \left(\frac{-4}{ab}\right),$$

and the asserted result follows. □

The next theorem gives the result we require for the determination of $t_4(n)$.

Theorem 16.6. *Let* $n \in \mathbb{N}_0$. *Then*

$$\sum_{\substack{(a,b,x,y) \in \mathbb{N}^4 \\ ax+by=4n+2 \\ ax \equiv 1 \,(\mathrm{mod}\,4)}} \left(\frac{-4}{ab}\right) = \sigma(2n+1).$$

Proof. This result follows immediately from Theorems 16.4 and 16.5. □

We are now ready to determine $t_4(n)$. We prove

Theorem 16.7. *Let* $n \in \mathbb{N}_0$. *Then*

$$t_4(n) = \sigma(2n+1).$$

Proof. We have by Theorems 16.1 and 16.6

$$t_4(n) = \sum_{m=0}^{n} t_2(m)t_2(n-m)$$

$$= \sum_{m=0}^{n} \sum_{\substack{a \in \mathbb{N} \\ a \mid 4m+1}} \left(\frac{-4}{a}\right) \sum_{\substack{b \in \mathbb{N} \\ b \mid 4(n-m)+1}} \left(\frac{-4}{b}\right)$$

$$= \sum_{\substack{(a,b,x,y) \in \mathbb{N}^4 \\ ax+by = 4n+2 \\ ax \equiv 1 \,(\mathrm{mod}\,4)}} \left(\frac{-4}{ab}\right)$$

$$= \sigma(2n+1),$$

which is the required result. $\qquad\qquad\qquad\qquad\square$

Example 16.2. By Theorem 16.7 we have $t_4(3) = \sigma(7) = 8$. The eight representations of 3 as the sum of four triangular numbers are

$$3 = 0+0+0+3 = 0+0+3+0 = 0+3+0+0 = 3+0+0+0$$
$$= 1+1+1+0 = 1+1+0+1 = 1+0+1+1 = 0+1+1+1.$$

We now turn to the determination of $t_6(n)$ ($n \in \mathbb{N}_0$). We require some preliminary results.

Theorem 16.8. *For $n \in \mathbb{N}_0$ we have*

$$\sum_{\substack{(a,b,x,y) \in \mathbb{N}^4 \\ ax+by = 4n+3}} \left(\left(\frac{-4}{b-a}\right) - \left(\frac{-4}{a+b}\right)\right)b = -\frac{1}{2}\sum_{\substack{d \in \mathbb{N} \\ d \mid 4n+3}}\left(\frac{-4}{d}\right)d^2.$$

Proof. We extend the definition of the Legendre-Jacobi-Kronecker symbol $\left(\dfrac{-4}{x}\right)$ ($x \in \mathbb{N}$) to all $x \in \mathbb{Z}$ by defining

$$\left(\frac{-4}{-x}\right) = -\left(\frac{-4}{x}\right) \quad (x \in \mathbb{N}), \quad \left(\frac{-4}{0}\right) = 0,$$

so that $\left(\dfrac{-4}{x}\right)$ ($x \in \mathbb{Z}$) is an odd function of x. We choose

$$f(a,b,x,y) = \left(\frac{-4}{a}\right)b, \quad (a,b,x,y) \in \mathbb{Z}^4,$$

in Theorem 13.1. It is easy to check that (13.1) is satisfied. With this choice of f, and with n replaced by $4n+3$ ($n \in \mathbb{N}_0$), the left hand side of

Theorem 13.1 is

$$\sum_{\substack{(a,b,x,y) \in \mathbb{N}^4 \\ ax+by=4n+3}} \left(\left(\frac{-4}{a}\right)b - \left(\frac{-4}{a}\right)(-b) + \left(\frac{-4}{a}\right)(a-b) \right.$$

$$\left. - \left(\frac{-4}{a}\right)(a+b) + \left(\frac{-4}{b-a}\right)b - \left(\frac{-4}{a+b}\right)b \right)$$

$$= \sum_{\substack{(a,b,x,y) \in \mathbb{N}^4 \\ ax+by=4n+3}} \left(\left(\frac{-4}{b-a}\right) - \left(\frac{-4}{a+b}\right) \right)b.$$

The right hand side of Theorem 13.1 is

$$\sum_{\substack{d \in \mathbb{N} \\ d \mid 4n+3}} \sum_{x=1}^{d-1} \left(\left(\frac{-4}{(4n+3)/d}\right)\frac{4n+3}{d} - \left(\frac{-4}{x}\right)(x-d) - \left(\frac{-4}{x}\right)d - \left(\frac{-4}{d}\right)x \right)$$

$$= \sum_{\substack{d \in \mathbb{N} \\ d \mid 4n+3}} \sum_{x=1}^{d-1} \left(\left(\frac{-4}{(4n+3)/d}\right)\frac{4n+3}{d} - \left(\frac{-4}{x}\right)x - \left(\frac{-4}{d}\right)x \right)$$

$$= A_1 - A_2 - A_3,$$

where

$$A_1 := \sum_{\substack{d \in \mathbb{N} \\ d \mid 4n+3}} \sum_{x=1}^{d-1} \left(\frac{-4}{(4n+3)/d}\right)\frac{4n+3}{d},$$

$$A_2 := \sum_{\substack{d \in \mathbb{N} \\ d \mid 4n+3}} \sum_{x=1}^{d-1} \left(\frac{-4}{x}\right)x,$$

$$A_3 := \sum_{\substack{d \in \mathbb{N} \\ d \mid 4n+3}} \sum_{x=1}^{d-1} \left(\frac{-4}{d}\right)x.$$

First we determine A_1. We have

$$A_1 = \sum_{\substack{d \in \mathbb{N} \\ d \mid 4n+3}} (d-1)\left(\frac{-4}{(4n+3)/d}\right)\frac{4n+3}{d}$$

$$= \sum_{\substack{d \in \mathbb{N} \\ d \mid 4n+3}} \left(\frac{4n+3}{d} - 1\right)\left(\frac{-4}{d}\right)d$$

$$= (4n+3)\sum_{\substack{d \in \mathbb{N} \\ d \mid 4n+3}} \left(\frac{-4}{d}\right) - \sum_{\substack{d \in \mathbb{N} \\ d \mid 4n+3}} \left(\frac{-4}{d}\right)d.$$

By Problem 26 of Exercises 3

$$\sum_{\substack{d \in \mathbb{N} \\ d \mid 4n+3}} \left(\frac{-4}{d}\right) = \sum_{\substack{d \in \mathbb{N} \\ d \mid 4n+3}} (-1)^{(d-1)/2} = 0.$$

Thus

$$A_1 = - \sum_{\substack{d \in \mathbb{N} \\ d \mid 4n+3}} \left(\frac{-4}{d}\right) d.$$

Next we determine A_2. It is easy to show that

$$\sum_{x=1}^{d-1} \left(\frac{-4}{x}\right) x = -\left(\frac{-4}{d}\right) \frac{(d-1)}{2}$$

for a positive odd integer d. Hence

$$A_2 = - \sum_{\substack{d \in \mathbb{N} \\ d \mid 4n+3}} \left(\frac{-4}{d}\right) \frac{d-1}{2} = -\frac{1}{2} \sum_{\substack{d \in \mathbb{N} \\ d \mid 4n+3}} \left(\frac{-4}{d}\right) d.$$

Finally we determine A_3. We have

$$A_3 = \sum_{\substack{d \in \mathbb{N} \\ d \mid 4n+3}} \left(\frac{-4}{d}\right) \sum_{x=1}^{d-1} x$$

$$= \sum_{\substack{d \in \mathbb{N} \\ d \mid 4n+3}} \left(\frac{-4}{d}\right) \frac{(d-1)d}{2}$$

$$= \frac{1}{2} \sum_{\substack{d \in \mathbb{N} \\ d \mid 4n+3}} \left(\frac{-4}{d}\right) d^2 - \frac{1}{2} \sum_{\substack{d \in \mathbb{N} \\ d \mid 4n+3}} \left(\frac{-4}{d}\right) d.$$

Equating the left and right hand sides, we obtain the asserted result. □

Theorem 16.9. *For $n \in \mathbb{N}_0$ we have*

$$\sum_{\substack{(a,b,x,y) \in \mathbb{N}^4 \\ ax+2by = 4n+3}} \left(\frac{-4}{a+2b}\right) b = \frac{1}{8} \sum_{\substack{d \in \mathbb{N} \\ d \mid 4n+3}} \left(\frac{-4}{d}\right) d^2.$$

Proof. First we observe that if a is odd then $4b^2 - a^2 \equiv 3 \pmod 4$ so

$$\left(\frac{-4}{2b-a}\right)\left(\frac{-4}{2b+a}\right) = \left(\frac{-4}{4b^2-a^2}\right) = \left(\frac{-4}{3}\right) = -1$$

and thus

$$\left(\frac{-4}{2b-a}\right) = -\left(\frac{-4}{2b+a}\right).$$

Then

$$\sum_{\substack{(a,b,x,y)\,\in\,\mathbb{N}^4 \\ ax+by=4n+3 \\ a\ \text{odd},\ b\ \text{even}}} \left(\left(\frac{-4}{b-a}\right) - \left(\frac{-4}{a+b}\right)\right) b$$

$$= 2 \sum_{\substack{(a,b,x,y)\,\in\,\mathbb{N}^4 \\ ax+2by=4n+3 \\ a\ \text{odd}}} \left(\left(\frac{-4}{2b-a}\right) - \left(\frac{-4}{2b+a}\right)\right) b$$

$$= -4 \sum_{\substack{(a,b,x,y)\,\in\,\mathbb{N}^4 \\ ax+2by=4n+3 \\ a\ \text{odd}}} \left(\frac{-4}{2b+a}\right) b$$

$$= -4 \sum_{\substack{(a,b,x,y)\,\in\,\mathbb{N}^4 \\ ax+2by=4n+3}} \left(\frac{-4}{a+2b}\right) b.$$

Next if b is odd we note that $b^2 - 4a^2 \equiv 1 \pmod 4$ so that

$$\left(\frac{-4}{b-2a}\right)\left(\frac{-4}{b+2a}\right) = \left(\frac{-4}{b^2-4a^2}\right) = \left(\frac{-4}{1}\right) = 1$$

and thus

$$\left(\frac{-4}{b-2a}\right) = \left(\frac{-4}{b+2a}\right).$$

Then

$$\sum_{\substack{(a,b,x,y)\,\in\,\mathbb{N}^4 \\ ax+by=4n+3 \\ a\ \text{even},\ b\ \text{odd}}} \left(\left(\frac{-4}{b-a}\right) - \left(\frac{-4}{a+b}\right)\right) b$$

$$= \sum_{\substack{(a,b,x,y)\,\in\,\mathbb{N}^4 \\ 2ax+by=4n+3 \\ b\ \text{odd}}} \left(\left(\frac{-4}{b-2a}\right) - \left(\frac{-4}{b+2a}\right)\right) b$$

$$= 0.$$

If a and b are both odd then $a + b$ and $a - b$ are both even so

$$\sum_{\substack{(a, b, x, y) \in \mathbb{N}^4 \\ ax + by = 4n + 3 \\ a \text{ odd}, b \text{ odd}}} \left(\left(\frac{-4}{b - a} \right) - \left(\frac{-4}{a + b} \right) \right) b = 0.$$

Thus, as $ax + by = 4n + 3$ implies that a and b are not both even, we have

$$\sum_{\substack{(a, b, x, y) \in \mathbb{N}^4 \\ ax + by = 4n + 3}} \left(\left(\frac{-4}{b - a} \right) - \left(\frac{-4}{a + b} \right) \right) b = -4 \sum_{\substack{(a, b, x, y) \in \mathbb{N}^4 \\ ax + 2by = 4n + 3}} \left(\frac{-4}{a + 2b} \right) b.$$

The assertion of the theorem now follows from Theorem 16.8. $\square$

Theorem 16.10. *For $n \in \mathbb{N}_0$ we have*

$$\sum_{\substack{(a, b, x, y) \in \mathbb{N}^4 \\ ax + 2by = 4n + 3}} \left(\frac{-4}{a} \right) b = -\frac{1}{8} \sum_{\substack{d \in \mathbb{N} \\ d \mid 4n + 3}} \left(\frac{-4}{d} \right) d^2.$$

Proof. We have

$$\left(\frac{-4}{a + 2b} \right) = (-1)^b \left(\frac{-4}{a} \right)$$

so that

$$\sum_{\substack{(a, b, x, y) \in \mathbb{N}^4 \\ ax + 2by = 4n + 3}} \left(\left(\frac{-4}{a} \right) b + \left(\frac{-4}{a + 2b} \right) b \right)$$

$$= 2 \sum_{\substack{(a, b, x, y) \in \mathbb{N}^4 \\ ax + 2by = 4n + 3 \\ b \text{ even}}} \left(\frac{-4}{a} \right) b$$

$$= 4 \sum_{\substack{(a, b, x, y) \in \mathbb{N}^4 \\ ax + 4by = 4n + 3}} \left(\frac{-4}{a} \right) b$$

$$= 4 \sum_{\substack{(a, b, x, y) \in \mathbb{N}^4 \\ ax + 4by = 4n + 3}} \left(\frac{-4}{x} \right) b$$

$$= 2 \sum_{\substack{(a, b, x, y) \in \mathbb{N}^4 \\ ax + 4by = 4n + 3}} \left(\left(\frac{-4}{a} \right) + \left(\frac{-4}{x} \right) \right) b$$

$$= 2 \sum_{\substack{(a,b,x,y) \in \mathbb{N}^4 \\ ax + 4by = 4n+3}} \left(\frac{-4}{a}\right)\left(1 + \left(\frac{-4}{ax}\right)\right)b$$

$$= 2 \sum_{\substack{(a,b,x,y) \in \mathbb{N}^4 \\ ax + 4by = 4n+3}} \left(\frac{-4}{a}\right)(1 + (-1))b$$

$$= 0.$$

The asserted identity now follows from Theorem 16.9. ☐

Theorem 16.11. *For $n \in \mathbb{N}_0$ we have*

$$\sum_{\substack{(a,b,x,y) \in \mathbb{N}^4 \\ ax + 2by = 4n+3 \\ ax \equiv 3 \,(\mathrm{mod}\, 4)}} \left(\frac{-4}{a}\right)b = 0.$$

Proof. We have

$$\sum_{\substack{(a,b,x,y) \in \mathbb{N}^4 \\ ax + 2by = 4n+3 \\ ax \equiv 3 \,(\mathrm{mod}\, 4)}} \left(\frac{-4}{a}\right)b = \sum_{\substack{(a,b,x,y) \in \mathbb{N}^4 \\ ax + 2by = 4n+3 \\ ax \equiv 3 \,(\mathrm{mod}\, 4)}} \left(\frac{-4}{x}\right)b$$

$$= \frac{1}{2} \sum_{\substack{(a,b,x,y) \in \mathbb{N}^4 \\ ax + 2by = 4n+3 \\ ax \equiv 3 \,(\mathrm{mod}\, 4)}} \left(\left(\frac{-4}{a}\right) + \left(\frac{-4}{x}\right)\right)b$$

$$= \frac{1}{2} \sum_{\substack{(a,b,x,y) \in \mathbb{N}^4 \\ ax + 2by = 4n+3 \\ ax \equiv 3 \,(\mathrm{mod}\, 4)}} \left(\frac{-4}{a}\right)\left(1 + \left(\frac{-4}{ax}\right)\right)b$$

$$= \frac{1}{2} \sum_{\substack{(a,b,x,y) \in \mathbb{N}^4 \\ ax + 2by = 4n+3 \\ ax \equiv 3 \,(\mathrm{mod}\, 4)}} \left(\frac{-4}{a}\right)(1 + (-1))b$$

$$= 0,$$

as claimed. ☐

We are now ready to determine the number of representations of a nonnegative integer as a sum of six triangular numbers.

Theorem 16.12. *For $n \in \mathbb{N}_0$ we have*

$$t_6(n) = -\frac{1}{8} \sum_{\substack{d \in \mathbb{N} \\ d \mid 4n+3}} \left(\frac{-4}{d}\right) d^2.$$

Proof. Appealing to Theorems 16.1, 16.7, 16.11 and 16.10, we obtain

$$t_6(n) = \sum_{m=0}^{n} t_2(m) t_4(n-m)$$

$$= \sum_{m=0}^{n} \sum_{\substack{a \in \mathbb{N} \\ a \mid 4m+1}} \left(\frac{-4}{a}\right) \sum_{\substack{b \in \mathbb{N} \\ b \mid 2(n-m)+1}} b$$

$$= \sum_{\substack{(a,b,x,y) \in \mathbb{N}^4 \\ ax + 2by = 4n+3 \\ ax \equiv 1 \,(\mathrm{mod}\ 4)}} \left(\frac{-4}{a}\right) b$$

$$= \sum_{\substack{(a,b,x,y) \in \mathbb{N}^4 \\ ax + 2by = 4n+3}} \left(\frac{-4}{a}\right) b$$

$$= -\frac{1}{8} \sum_{\substack{d \in \mathbb{N} \\ d \mid 4n+3}} \left(\frac{-4}{d}\right) d^2.$$

This completes the proof of the formula for $t_6(n)$. $\qquad\square$

Example 16.3. By Theorem 16.12 we have

$$t_6(3) = -\frac{1}{8} \sum_{\substack{d \in \mathbb{N} \\ d \mid 15}} \left(\frac{-4}{d}\right) d^2 = -\frac{1}{8}(1^2 - 3^2 + 5^2 - 15^2) = -\frac{1}{8}(-208) = 26.$$

Next we determine the number of representations of a nonnegative integer as a sum of eight triangular numbers.

Theorem 16.13. *Let $n \in \mathbb{N}_0$. Then*

$$t_8(n) = \sigma_3(n+1) - \sigma_3((n+1)/2).$$

Proof. Recall from Theorem 16.7 that

$$t_4(n) = \sigma(2n+1), \quad n \in \mathbb{N}_0.$$

Thus

$$t_8(n) = \sum_{m=0}^{n} t_4(m)t_4(n-m)$$

$$= \sum_{m=0}^{n} \sigma(2m+1)\sigma(2n-2m+1)$$

$$= \sum_{m=1}^{n+1} \sigma(2m-1)\sigma((2n+2)-(2m-1)).$$

Appealing to Theorem 15.5, we obtain

$$t_8(n) = \frac{5}{24}\sigma_3(2n+2) - \frac{7}{8}\sigma_3(n+1)$$
$$+ \frac{2}{3}\sigma_3((n+1)/2) - \left(\frac{11}{24} + \frac{1}{2}n\right)\sigma(2n+2)$$
$$+ \left(\frac{11}{8} + \frac{3}{2}n\right)\sigma(n+1) - \left(\frac{11}{12} + n\right)\sigma((n+1)/2).$$

By Theorem 3.1(ii), with $k = 3$, $p = 2$ and n replaced by $n + 1$, we have

$$\sigma_3(2n+2) = 9\sigma_3(n+1) - 8\sigma_3((n+1)/2)$$

and, with $k = 1$, $p = 2$ and n replaced by $n + 1$, we have

$$\sigma(2n+2) = 3\sigma(n+1) - 2\sigma((n+1)/2).$$

Using these we deduce the formula of the theorem. □

Example 16.4. For $n \in \mathbb{N}_0$ we have

$$t_8(2n) = \sigma_3(2n+1)$$

by Theorem 16.13. Thus we have the inequality

$$(2n+1)^3 \leq t_8(2n) \leq (2n+1)^4.$$

We close this chapter by determining the number of representations of $n (\in \mathbb{N}_0)$ in the form $t_1 + t_2 + 2t_3 + 2t_4$, where t_1, t_2, t_3, t_4 are triangular numbers. We set

$$R(n) = \text{card}\{(t_1, t_2, t_3, t_4) \in \Delta^4 \mid n = t_1 + t_2 + 2t_3 + 2t_4\}$$

for $n \in \mathbb{N}_0$. We begin by proving some results that are analogous to Theorems 16.3–16.6.

Theorem 16.14. *For $n \in \mathbb{N}_0$ we have*

$$\sum_{\substack{(a, b, x, y) \in \mathbb{N}^4 \\ 2ax + by = 4n + 3}} \left(\frac{-4}{ab}\right) = \frac{1}{4}\sigma(4n + 3).$$

Proof. From Problem 27 of Exercises 3 and Problem 26 of Exercises 14, we have

$$\sum_{\substack{(a, b, x, y) \in \mathbb{N}^4 \\ 2ax + by = 4n + 3}} (F_4(a - b) - F_4(a + b)) = \frac{1}{4}\sigma(4n + 3).$$

The asserted identity now follows using Theorem 3.10. $\square$

Theorem 16.15. *For $n \in \mathbb{N}_0$ we have*

$$\sum_{\substack{(a, b, x, y) \in \mathbb{N}^4 \\ 2ax + by = 4n + 3 \\ x \equiv 0 \,(\mathrm{mod}\, 2)}} \left(\frac{-4}{ab}\right) = 0.$$

Proof. Interchanging the roles of b and y in the sum in the statement of the theorem, and noting that

$$\left(\frac{-4}{ay}\right) = \left(\frac{-4}{b^2 ay}\right) = \left(\frac{-4}{by}\right)\left(\frac{-4}{ab}\right) = \left(\frac{-4}{3}\right)\left(\frac{-4}{ab}\right) = -\left(\frac{-4}{ab}\right),$$

we obtain

$$\sum_{\substack{(a, b, x, y) \in \mathbb{N}^4 \\ 2ax + by = 4n + 3 \\ x \equiv 0 \,(\mathrm{mod}\, 2)}} \left(\frac{-4}{ab}\right) = \sum_{\substack{(a, b, x, y) \in \mathbb{N}^4 \\ 2ax + by = 4n + 3 \\ x \equiv 0 \,(\mathrm{mod}\, 2)}} \left(\frac{-4}{ay}\right) = -\sum_{\substack{(a, b, x, y) \in \mathbb{N}^4 \\ 2ax + by = 4n + 3 \\ x \equiv 0 \,(\mathrm{mod}\, 2)}} \left(\frac{-4}{ab}\right),$$

from which the assertion of the theorem follows. $\square$

Theorem 16.16. *For $n \in \mathbb{N}_0$ we have*

$$\sum_{\substack{(a, b, x, y) \in \mathbb{N}^4 \\ 2ax + by = 4n + 3 \\ x \equiv 1 \,(\mathrm{mod}\, 2)}} \left(\frac{-4}{ab}\right) = \frac{1}{4}\sigma(4n + 3).$$

Proof. This follows from Theorems 16.14 and 16.15. $\square$

Theorem 16.17. *For $n \in \mathbb{N}_0$ we have*

$$\sum_{\substack{(a, b, x, y) \in \mathbb{N}^4 \\ 2ax + by = 4n + 3 \\ ax \equiv 1 \,(\mathrm{mod}\, 2)}} \left(\frac{-4}{ab}\right) = \frac{1}{4}\sigma(4n + 3).$$

Proof. If a is even then $\left(\dfrac{-4}{ab}\right) = 0$ so

$$\sum_{\substack{(a,b,x,y) \in \mathbb{N}^4 \\ 2ax + by = 4n + 3 \\ x \equiv 1 \,(\mathrm{mod}\, 2)}} \left(\frac{-4}{ab}\right) = \sum_{\substack{(a,b,x,y) \in \mathbb{N}^4 \\ 2ax + by = 4n + 3 \\ ax \equiv 1 \,(\mathrm{mod}\, 2)}} \left(\frac{-4}{ab}\right)$$

and the required result follows from Theorem 16.16. ☐

Theorem 16.18. *For $n \in \mathbb{N}_0$ we have*

$$\sum_{\substack{(a,b,x,y) \in \mathbb{N}^4 \\ 2ax + by = 4n + 3 \\ ax \equiv 3 \,(\mathrm{mod}\, 4)}} \left(\frac{-4}{ab}\right) = 0.$$

Proof. Interchanging the roles of a and x in the sum in the statement of the theorem, and noting that

$$\left(\frac{-4}{xb}\right) = \left(\frac{-4}{a^2 xb}\right) = \left(\frac{-4}{ax}\right)\left(\frac{-4}{ab}\right) = \left(\frac{-4}{3}\right)\left(\frac{-4}{ab}\right) = -\left(\frac{-4}{ab}\right),$$

we obtain

$$\sum_{\substack{(a,b,x,y) \in \mathbb{N}^4 \\ 2ax + by = 4n + 3 \\ ax \equiv 3 \,(\mathrm{mod}\, 4)}} \left(\frac{-4}{ab}\right) = \sum_{\substack{(a,b,x,y) \in \mathbb{N}^4 \\ 2ax + by = 4n + 3 \\ ax \equiv 3 \,(\mathrm{mod}\, 4)}} \left(\frac{-4}{xb}\right) = - \sum_{\substack{(a,b,x,y) \in \mathbb{N}^4 \\ 2ax + by = 4n + 3 \\ ax \equiv 3 \,(\mathrm{mod}\, 4)}} \left(\frac{-4}{ab}\right),$$

from which the asserted result follows. ☐

Theorem 16.19. *For $n \in \mathbb{N}_0$ we have*

$$\sum_{\substack{(a,b,x,y) \in \mathbb{N}^4 \\ 2ax + by = 4n + 3 \\ ax \equiv 1 \,(\mathrm{mod}\, 4)}} \left(\frac{-4}{ab}\right) = \frac{1}{4}\sigma(4n + 3).$$

Proof. This theorem follows from Theorems 16.17 and 16.18. ☐

We are now ready to determine $R(n)$.

Theorem 16.20. *For $n \in \mathbb{N}_0$ the number $R(n)$ of representations of n in the form $t_1 + t_2 + 2t_3 + 2t_4$, where t_1, t_2, t_3 and t_4 are triangular numbers, is given by*

$$R(n) = \frac{1}{4}\sigma(4n + 3).$$

Proof. We have

$$R(n) = \sum_{\substack{m \in \mathbb{N}_0 \\ m \le n/2}} t_2(m) t_2(n - 2m).$$

Hence, by Theorem 16.1, we obtain

$$R(n) = \sum_{\substack{m \in \mathbb{N}_0 \\ m \le n/2}} \left(\sum_{\substack{a \in \mathbb{N} \\ a \mid 4m + 1}} \left(\frac{-4}{a} \right) \right) \left(\sum_{\substack{b \in \mathbb{N} \\ b \mid 4(n - 2m) + 1}} \left(\frac{-4}{b} \right) \right).$$

Now for $a, b \in \mathbb{N}$ we have

$$a \mid 4m + 1, \ b \mid 4(n - 2m) + 1 \text{ for some } m \in \mathbb{N}_0 \text{ with } m \le n/2$$

if and only if

$$4n + 3 = 2ax + by, \ ax \equiv 1 \pmod{4} \text{ for some } x, y \in \mathbb{N}.$$

Hence

$$R(n) = \sum_{\substack{(a, b, x, y) \in \mathbb{N}^4 \\ 2ax + by = 4n + 3 \\ ax \equiv 1 \pmod{4}}} \left(\frac{-4}{ab} \right)$$

and the theorem follows from Theorem 16.19. $\qquad \qquad \square$

An immediate consequence of Theorem 16.20 is that every nonnegative integer is of the form $t_1 + t_2 + 2t_3 + 2t_4$ for some triangular numbers t_1, t_2, t_3, t_4.

Example 16.5. For $n = 6$ Theorem 16.20 gives

$$R(6) = \frac{1}{4} \sigma(27) = 10.$$

The ten representations $(t_1, t_2, t_3, t_4) \in \Delta^4$ in $6 = t_1 + t_2 + 2t_3 + 2t_4$ are

$$(t_1, t_2, t_3, t_4) = (0, 0, 0, 3), \ (0, 0, 3, 0), \ (0, 6, 0, 0), \ (1, 1, 1, 1), \ (1, 3, 0, 1),$$
$$(1, 3, 1, 0), \ (3, 1, 0, 1), \ (3, 1, 1, 0), \ (3, 3, 0, 0), \ (6, 0, 0, 0).$$

Exercises 16

1. Characterize those $n \in \mathbb{N}$ which are not the sum of two triangular numbers.
2. Prove that every positive integer is the sum of four triangular numbers. (Gauss proved that three triangular numbers suffice.)

3. Let d be a positive odd integer. Prove that

$$\sum_{x=1}^{d-1} \left(\frac{-4}{x}\right) x = -\left(\frac{-4}{d}\right) \frac{d-1}{2}.$$

4. Since $R(n) \in \mathbb{N}$ for all $n \in \mathbb{N}_0$ it follows from Theorem 16.20 that

$$\sigma(4n + 3) \equiv 0 \pmod 4$$

for all $n \in \mathbb{N}_0$. Prove this congruence directly from first principles.

5. Let $n \in \mathbb{N}$. Let $f : \mathbb{Z} \to \mathbb{C}$ be an even function. Prove that

$$\sum_{\substack{(a, b, x, y) \in \mathbb{N}^4 \\ ax + by = n \\ x \equiv y \equiv 1 \pmod 2}} (f(a-b) - f(a + b)) = f(0)\sigma(n/2) + \sum_{\substack{d \in \mathbb{N} \\ d \mid n \\ d \nmid n/2}} \left(\frac{n}{d} - d\right) f(d).$$

6. Let $n \in \mathbb{N}$. Deduce from Problem 5 that

$$\sum_{m=1}^{n-1} \sigma^*(m)\sigma^*(n - m) = \frac{1}{4}\sigma_3^*(n) - \frac{n}{4}\sigma^*(n).$$

7. Let $n \in \mathbb{N}$ be even. Let $f : \mathbb{Z} \to \mathbb{C}$ be an even function. Prove that

$$\sum_{\substack{(a, b, x, y) \in \mathbb{N}^4 \\ ax + by = n \\ a \equiv b \equiv x \equiv y \equiv 1 \pmod 2}} (f(a-b) - f(a + b)) = \frac{1}{2} f(0)\sigma^*(n) - \frac{1}{2} \sum_{\substack{d \in \mathbb{N} \\ d \mid n \\ d \nmid n/2}} df(d).$$

8. Let $n \in \mathbb{N}$ be even. Deduce from Problem 7 that

$$\sum_{m=1}^{n/2} \sigma(2m - 1)\sigma(n - (2m - 1)) = \frac{1}{8}\sigma_3(n) - \frac{1}{8}\sigma_3(n/2).$$

9. Deduce Theorem 13.13 from Problem 8.

10. Let $n \in \mathbb{N}$ be odd. Formulate and prove the analogue of Problem 8 for the sum

$$\sum_{m=1}^{(n-1)/2} \sigma(2m - 1)\sigma(n - (2m - 1)).$$

Notes on Chapter 16

Theorem 16.1, which gives a formula for $t_2(n)$, is a classical result. Proofs have been given by Adiga [1], Cooper and Lam [82] and Ono, Robins and Wahl [216, Corollary 1, p. 78]. Theorem 16.2 is of Liouville type but was not stated by him. This theorem is given in Alaca, Alaca, McAfee and Williams [9, Theorem 5.4, p. 16], where many other similar formulae are proved.

Theorems 16.3–16.6 are taken from Huard, Ou, Spearman and Williams [137, pp. 261–2].

Theorem 16.7, which gives a formula for $t_4(n)$, was known to Legendre [163, Vol. 3, pp. 133–4], see Dickson's *History* [93, Vol. 2, p. 19]. A proof of Theorem 16.7 using modular functions has been given by Ono, Robins and Wahl [216]. The proof given in this chapter is due to Huard, Ou, Spearman and Williams [137, Theorem 10, p. 259]. Other proofs have been given by Adiga [1] and Cooper and Lam [82].

Theorems 16.8–16.11 are taken from Huard, Ou, Spearman and Williams [137, pp. 262–5].

Theorem 16.12, which gives a formula for $t_6(n)$, was proved in Ono, Robins and Wahl [216, Theorem 4, p. 81] using modular forms. Our elementary arithmetic proof is based on Huard, Ou, Spearman and Williams [137, Theorem 11, p. 262]. Cooper and Lam [82] have also given a proof.

Theorem 16.13, which gives a formula for $t_8(n)$, was proved in Ono, Robins and Wahl [216, Theorem 5, p. 82] using modular forms. Our elementary arithmetic proof is taken from Huard, Ou, Spearman and Williams [137, Theorem 12, p. 265]. Another proof has been given by Cooper and Lam [82].

Theorems 16.14–16.19 are taken from Williams [266]. Theorem 16.20 is due to Williams [266, p. 235]. Note that in the statement of the theorem in Williams [266] the sum $\displaystyle\sum_{\substack{d \in \mathbb{N} \\ d \mid 4n+3}} (-1)^{(d-1)/2} = 0$.

Ono, Robins and Wahl [216] have given formulae for $t_{10}(n)$, $t_{12}(n)$ and $t_{24}(n)$. We now give their formulae.

The Ono, Robins and Wahl formula for $t_{10}(n)$ ($n \in \mathbb{N}_0$) is

$$t_{10}(n) = \frac{1}{640} \sum_{\substack{d \in \mathbb{N} \\ d \mid 4n+5}} \left(\frac{-4}{d}\right) d^4 - \frac{1}{640} a(4n+5),$$

where $a(n)$ ($n \in \mathbb{N}$) is defined by

$$q \prod_{n=1}^{\infty} (1 - q^n)^4 (1 - q^{2n})^2 (1 - q^{4n})^4 = \sum_{n=1}^{\infty} a(n) q^n.$$

The Ono, Robins and Wahl formula for $t_{12}(n)$ ($n \in \mathbb{N}_0$) is

$$t_{12}(n) = \frac{1}{256} (\sigma_5(2n+3) - b(2n+3)),$$

where $b(n)$ $(n \in \mathbb{N})$ is defined by

$$q \prod_{n=1}^{\infty} (1 - q^{2n})^{12} = \sum_{n=1}^{\infty} b(n) q^n.$$

Cheng and Williams [70, pp. 55–56] have given a simple proof of this result. A different evaluation of $t_{12}(n)$ has been given by McAfee and Williams [199, p. 238].

The Ono, Robins and Wahl formula for $t_{24}(n)$ $(n \in \mathbb{N}_0)$ is

$$t_{24}(n) = \frac{1}{176896} \left(\sigma_{11}^*(n+3) - \tau(n+3) - 2072\tau((n+3)/2) \right),$$

where the Ramanujan tau function $\tau(n)$ $(n \in \mathbb{N})$ was defined in (13.15).

Kac and Wakimoto [151, p. 452] have used the representation theory of affine super-algebras to show that

$$t_{16}(n) = \frac{1}{192} \sum_{\substack{(a, b, x, y) \in \mathbb{N}^4 \\ ax + by = 2n + 4 \\ a \equiv b \equiv x \equiv y \equiv 1 \,(\mathrm{mod}\, 2) \\ a > b}} ab(a^2 - b^2)^2, \quad n \in \mathbb{N}_0,$$

and

$$t_{24}(n) = \frac{1}{72} \sum_{\substack{(a, b, x, y) \in \mathbb{N}^4 \\ ax + by = n + 3 \\ x \equiv y \equiv 1 \,(\mathrm{mod}\, 2) \\ a > b}} ab(a^2 - b^2)^2 \quad n \in \mathbb{N}_0.$$

Elementary proofs of these formulae have been given by Huard and Williams [140].

Other papers on triangular numbers include those of Cooper [75], Liu [184] and Zagier [274].

Problem 5 was stated but not proved by Liouville [170, 1st article, p. 144; 2nd article, p. 194]. Proofs of this result have been given by McAfee [197] and Pepin [219, p. 159], see also Alaca, Alaca, McAfee and Williams [9, Theorem 5.2, p. 16].

Problem 7 was stated but not proved by Liouville [170, 4th article, p. 242]. Proofs have been given by Baskakov [39, p. 344], Bugaev [53, p. 9], Deltour [86, p. 123], Humbert [143], Mathews [195], McAfee [197], Pepin [219, p. 94] and Smith [244, pp. 346–348], [245, Vol. I, p. 348], see also Alaca, Alaca, McAfee and Williams [9, Theorem 5.3, p. 16].

17

Sums of integers of the form $x^2 + xy + y^2$

Let $k \in \mathbb{N}$ and $n \in \mathbb{N}_0$. We recall that

$$r_{2k}(n) = \text{card}\left\{(x_1, x_2, \ldots, x_{2k-1}, x_{2k}) \in \mathbb{Z}^{2k} \mid n = x_1^2 + x_2^2 + \cdots + x_{2k-1}^2 + x_{2k}^2\right\}.$$

Analogously to $r_{2k}(n)$, we define

$$s_{2k}(n) = \text{card}\left\{(x_1, x_2, \ldots, x_{2k-1}, x_{2k}) \in \mathbb{Z}^{2k} \mid n = x_1^2 + x_1 x_2 + x_2^2 \right.$$
$$\left. + \cdots + x_{2k-1}^2 + x_{2k-1} x_{2k} + x_{2k}^2\right\}.$$

We note that $x^2 + xy + y^2 = \left(x + \frac{1}{2}y\right)^2 + \frac{3}{4}y^2 \in \mathbb{N}_0$ for $(x, y) \in \mathbb{Z}^2$ and $x^2 + xy + y^2 = 0$ if and only if $x = y = 0$. Thus $s_{2k}(0) = 1$ for all $k \in \mathbb{N}$. There is a formula, similar to that of Theorem 9.3, for the number $s_2(n)$ of representations of $n(\in \mathbb{N})$ by the form $x^2 + xy + y^2$. This formula is conveniently stated in terms of the Legendre-Jacobi-Kronecker symbol for discriminant -3, which is defined for $d \in \mathbb{N}$ by

$$\left(\frac{-3}{d}\right) = \begin{cases} +1, & \text{if } d \equiv 1 \pmod{3}, \\ -1, & \text{if } d \equiv 2 \pmod{3}, \\ 0, & \text{if } d \equiv 0 \pmod{3}. \end{cases}$$

Theorem 17.1. *Let $n \in \mathbb{N}$. Then the number $s_2(n)$ of $(x_1, x_2) \in \mathbb{Z}^2$ such that*

$$n = x_1^2 + x_1 x_2 + x_2^2$$

is given by

$$s_2(n) = 6 \sum_{\substack{d \in \mathbb{N} \\ d \mid n}} \left(\frac{-3}{d}\right).$$

We refer the reader to the notes section at the end of this chapter for references to proofs of this result. For $n \in \mathbb{N}$ we clearly have

$$s_2(n) = 6d_{1,3}(n) - 6d_{2,3}(n). \tag{17.1}$$

224

Also, by (3.7) and (3.8) with $m = 3$, we have

$$d_{0,3}(n) = d(n/3)$$

and

$$d_{0,3}(n) + d_{1,3}(n) + d_{2,3}(n) = d(n)$$

respectively. Solving these three equations for $d_{0,3}(n)$, $d_{1,3}(n)$ and $d_{2,3}(n)$, we obtain the following result, which is an analogue of Theorem 9.4.

Theorem 17.2. *For $n \in \mathbb{N}$*

$$d_{0,3}(n) = d(n/3),$$
$$d_{1,3}(n) = \frac{1}{2}d(n) - \frac{1}{2}d(n/3) + \frac{1}{12}s_2(n),$$
$$d_{2,3}(n) = \frac{1}{2}d(n) - \frac{1}{2}d(n/3) - \frac{1}{12}s_2(n).$$

Liouville gave a formula for $s_4(n)$ valid for all $n \in \mathbb{N}$. We now prove Liouville's formula in a manner similar to the proof of Theorem 11.1.

Theorem 17.3. *Let $n \in \mathbb{N}$. Then the number $s_4(n)$ of $(x_1, x_2, x_3, x_4) \in \mathbb{Z}^4$ such that*

$$n = x_1^2 + x_1 x_2 + x_2^2 + x_3^2 + x_3 x_4 + x_4^2$$

is given by

$$s_4(n) = 12\sigma(n) - 36\sigma(n/3).$$

Proof. Analogously to (11.1) we have

$$s_4(n) = \sum_{k=0}^{n} s_2(k)s_2(n - k).$$

Since $s_2(0) = 1$, we can write this as

$$s_4(n) - 2s_2(n) = \sum_{k=1}^{n-1} s_2(k)s_2(n - k).$$

Recalling from Theorem 17.1 that

$$s_2(n) = 6 \sum_{\substack{d \in \mathbb{N} \\ d \mid n}} \left(\frac{-3}{d}\right) = 6(d_{1,3}(n) - d_{2,3}(n)),$$

we obtain by the first part of Problem 5 of Exercises 3

$$s_4(n) - 2s_2(n) = 36 \sum_{k=1}^{n-1} \left(\sum_{\substack{a \in \mathbb{N} \\ a \mid k}} \left(\frac{-3}{a} \right) \right) \left(\sum_{\substack{b \in \mathbb{N} \\ b \mid n-k}} \left(\frac{-3}{b} \right) \right)$$

$$= 36 \sum_{\substack{(a,b,x,y) \in \mathbb{N}^4 \\ ax+by=n}} \left(\frac{-3}{ab} \right)$$

$$= 36 \sum_{\substack{(a,b,x,y) \in \mathbb{N}^4 \\ ax+by=n}} (F_3(a-b) - F_3(a+b)).$$

Appealing to Example 10.1, we have

$$\sum_{\substack{(a,b,x,y) \in \mathbb{N}^4 \\ ax+by=n}} (F_3(a-b) - F_3(a+b))$$

$$= \frac{1}{3}\sigma(n) - \sigma(n/3) - \frac{1}{3}(d_{1,3}(n) - d_{2,3}(n))$$

$$= \frac{1}{3}\sigma(n) - \sigma(n/3) - \frac{1}{18}s_2(n),$$

so that

$$s_4(n) = 2s_2(n) + 36 \sum_{\substack{(a,b,x,y) \in \mathbb{N}^4 \\ ax+by=n}} (F_3(a-b) - F_3(a+b))$$

$$= 2s_2(n) + 36 \left(\frac{1}{3}\sigma(n) - \sigma(n/3) - \frac{1}{18}s_2(n) \right)$$

$$= 12\sigma(n) - 36\sigma(n/3),$$

which is Liouville's formula for $s_4(n)$. $\qquad\qquad\square$

We can also give a formula for $s_8(n)$.

Theorem 17.4. *Let $n \in \mathbb{N}$. Then the number $s_8(n)$ of $(x_1, x_2, \ldots, x_8) \in \mathbb{Z}^8$ such that*

$$n = x_1^2 + x_1 x_2 + x_2^2 + x_3^2 + x_3 x_4 + x_4^2 + x_5^2 + x_5 x_6 + x_6^2 + x_7^2 + x_7 x_8 + x_8^2$$

is given by

$$s_8(n) = 24\sigma_3(n) + 216\sigma_3(n/3).$$

Proof. By Theorem 17.3 we have

$$s_4(n) = 12\sigma(n) - 36\sigma(n/3), \quad n \in \mathbb{N}.$$

Thus

$$s_8(n) = \sum_{m=0}^{n} s_4(m)s_4(n-m)$$

$$= 2s_4(n) + \sum_{m=1}^{n-1} s_4(m)s_4(n-m)$$

$$= 24\sigma(n) - 72\sigma(n/3)$$

$$+ \sum_{m=1}^{n-1}(12\sigma(m) - 36\sigma(m/3))(12\sigma(n-m) - 36\sigma((n-m)/3)$$

$$= 24\sigma(n) - 72\sigma(n/3) + 144A_1(n) - 864A_3(n) + 1296A_1(n/3).$$

Appealing to Theorem 12.1 (Besge's formula) for the evaluation of $A_1(n)$ and $A_1(n/3)$, and to Theorem 15.3 for $A_3(n)$, we obtain the asserted formula. $\square$

Next we determine the number of representations of $n \in \mathbb{N}$ by the quaternary quadratic form $x_1^2 + x_1x_2 + x_2^2 + 2(x_3^2 + x_3x_4 + x_4^2)$. Liouville stated without proof a formula for this number in 1863.

Theorem 17.5. *Let $n \in \mathbb{N}$. Then the number of $(x_1, x_2, x_3, x_4) \in \mathbb{Z}^4$ such that*

$$n = x_1^2 + x_1x_2 + x_2^2 + 2(x_3^2 + x_3x_4 + x_4^2)$$

is given by

$$6\sigma(n) - 12\sigma(n/2) + 18\sigma(n/3) - 36\sigma(n/6).$$

Proof. Let $n \in \mathbb{N}$. The number of $(x_1, x_2, x_3, x_4) \in \mathbb{Z}^4$ such that

$$n = x_1^2 + x_1x_2 + x_2^2 + 2(x_3^2 + x_3x_4 + x_4^2)$$

is

$$\sum_{\substack{(k,\ell) \in \mathbb{N}_0^2 \\ k + 2\ell = n}} s_2(k)s_2(\ell) = s_2(n) + s_2(n/2) + \sum_{\substack{(k,\ell) \in \mathbb{N}^2 \\ k + 2\ell = n}} s_2(k)s_2(\ell).$$

Recalling from Theorem 17.1 that

$$s_2(n) = 6 \sum_{\substack{d \in \mathbb{N} \\ d \mid n}} \left(\frac{-3}{d} \right),$$

we obtain (appealing to the first part of Problem 5 of Exercises 3)

$$\sum_{\substack{(k,\,\ell)\,\in\,\mathbb{N}^2 \\ k+2\ell=n}} s_2(k)s_2(\ell) = \sum_{\substack{(k,\,\ell)\,\in\,\mathbb{N}^2 \\ 2k+\ell=n}} s_2(k)s_2(\ell)$$

$$= 36 \sum_{\substack{(a,\,b,\,x,\,y)\,\in\,\mathbb{N}^4 \\ 2ax+by=n}} \left(\frac{-3}{ab}\right)$$

$$= 36 \sum_{\substack{(a,\,b,\,x,\,y)\,\in\,\mathbb{N}^4 \\ 2ax+by=n}} (F_3(a-b) - F_3(a+b)).$$

By Problem 25 of Exercises 14, we have

$$\sum_{\substack{(a,\,b,\,x,\,y)\,\in\,\mathbb{N}^4 \\ 2ax+by=n}} (F_3(a-b) - F_3(a+b))$$

$$= \frac{1}{6}\sigma(n) - \frac{1}{3}\sigma(n/2) + \frac{1}{2}\sigma(n/3) - \sigma(n/6)$$

$$- \frac{1}{6}(d_{1,3}(n) - d_{2,3}(n)) - \frac{1}{6}(d_{1,3}(n/2) - d_{2,3}(n/2))$$

$$= \frac{1}{6}\sigma(n) - \frac{1}{3}\sigma(n/2) + \frac{1}{2}\sigma(n/3) - \sigma(n/6)$$

$$- \frac{1}{36}s_2(n) - \frac{1}{36}s_2(n/2).$$

Thus the required number of representations is

$$s_2(n) + s_2(n/2)$$

$$+ 36\left(\frac{1}{6}\sigma(n) - \frac{1}{3}\sigma(n/2) + \frac{1}{2}\sigma(n/3) - \sigma(n/6) - \frac{s_2(n)}{36} - \frac{s_2(n/2)}{36}\right),$$

which gives the asserted number. $\square$

Our final theorem of this chapter gives Liouville's formula for the number of representations $(x_1, x_2, x_3, x_4) \in \mathbb{Z}^4$ of $n(\in \mathbb{N})$ by the quaternary quadratic form $x_1^2 + x_1x_2 + x_2^2 + 4(x_3^2 + x_3x_4 + x_4^2)$.

Theorem 17.6. *Let $n \in \mathbb{N}$. Then the number of $(x_1, x_2, x_3, x_4) \in \mathbb{Z}^4$ such that*

$$n = x_1^2 + x_1x_2 + x_2^2 + 4(x_3^2 + x_3x_4 + x_4^2)$$

is

$$(-1)^{n+1}(6\sigma(n) - 18\sigma(n/2) - 18\sigma(n/3) + 54\sigma(n/6)).$$

Proof. Suppose first that $n \equiv 2 \pmod 4$. We have

$$x^2 + xy + y^2 \equiv \begin{cases} 0 \pmod 4, & \text{if } x \text{ and } y \text{ are both even,} \\ 1 \pmod 2, & \text{if } x \text{ and } y \text{ are of opposite parity,} \\ 1 \pmod 2, & \text{if } x \text{ and } y \text{ are both odd.} \end{cases}$$

Thus

$$x^2 + xy + y^2 \not\equiv 2 \pmod 4$$

for all integers x and y. Hence

$$n \neq x_1^2 + x_1 x_2 + x_2^2 + 4(x_3^2 + x_3 x_4 + x_4^2)$$

for any integers x_1, x_2, x_3, x_4. For $n \equiv 2 \pmod 4$ we have

$$\sigma(n) = \sigma(2)\sigma(n/2) = 3\sigma(n/2) \text{ and } \sigma(n/3) = 3\sigma(n/6),$$

so that

$$(-1)^{n+1}(6\sigma(n) - 18\sigma(n/2) - 18\sigma(n/3) + 54\sigma(n/6)) = 0,$$

proving the theorem in this case.

Suppose next that $n \equiv 0 \pmod 4$. If

$$n = x_1^2 + x_1 x_2 + x_2^2 + 4(x_3^2 + x_3 x_4 + x_4^2)$$

for some integers x_1, x_2, x_3, x_4 then $x_1 \equiv x_2 \equiv 0 \pmod 2$ and

$$n/4 = (x_1/2)^2 + (x_1/2)(x_2/2) + (x_2/2)^2 + x_3^2 + x_3 x_4 + x_4^2.$$

Conversely if

$$n/4 = x_1^2 + x_1 x_2 + x_2^2 + x_3^2 + x_3 x_4 + x_4^2$$

then

$$n = (2x_1)^2 + (2x_1)(2x_2) + (2x_2)^2 + 4(x_3^2 + x_3 x_4 + x_4^2).$$

Hence there is a one-to-one correspondence between the solutions of

$$n = x_1^2 + x_1 x_2 + x_2^2 + 4(x_3^2 + x_3 x_4 + x_4^2)$$

and those of

$$n/4 = x_1^2 + x_1 x_2 + x_2^2 + x_3^2 + x_3 x_4 + x_4^2.$$

Thus the required number is

$$s_4(n/4) = 12\sigma(n/4) - 36\sigma(n/12)$$

by Theorem 17.3. For $n \equiv 0 \pmod 4$ we have by Theorem 3.1(ii) (with $k = 1$, $p = 2$ and n replaced by $n/2$)

$$\sigma(n) - 3\sigma(n/2) + 2\sigma(n/4) = 0.$$

Thus

$$(-1)^{n+1}(6\sigma(n) - 18\sigma(n/2) - 18\sigma(n/3) + 54\sigma(n/6))$$
$$= -6(\sigma(n) - 3\sigma(n/2)) + 18(\sigma(n/3) - 3\sigma(n/6)) = 12\sigma(n/4) - 36\sigma(n/12),$$

proving the theorem in this case.

Finally we suppose that n is odd. In this case, in order to complete the proof of the theorem, we must show that the required number of representations is $6\sigma(n) - 18\sigma(n/3)$. The number of $(x_1, x_2, x_3, x_4) \in \mathbb{Z}^4$ such that

$$n = x_1^2 + x_1 x_2 + x_2^2 + 4(x_3^2 + x_3 x_4 + x_4^2)$$

is

$$\sum_{\substack{(k,\ell) \in \mathbb{N}_0^2 \\ k + 4\ell = n}} s_2(k)s_2(\ell) = s_2(n) + \sum_{\substack{(k,\ell) \in \mathbb{N}^2 \\ k + 4\ell = n}} s_2(k)s_2(\ell)$$

$$= s_2(n) + 36 \sum_{\substack{(k,\ell) \in \mathbb{N}^2 \\ k + 4\ell = n}} \sum_{\substack{a \in \mathbb{N} \\ a \mid k}} \left(\frac{-3}{a}\right) \sum_{\substack{b \in \mathbb{N} \\ b \mid \ell}} \left(\frac{-3}{b}\right)$$

$$= s_2(n) + 36 \sum_{\substack{(a,b,x,y) \in \mathbb{N}^4 \\ ax + 4by = n}} \left(\frac{-3}{ab}\right)$$

$$= s_2(n) + 36 \sum_{\substack{(a,b,x,y) \in \mathbb{N}^4 \\ 4ax + by = n}} \left(\frac{-3}{ab}\right)$$

$$= s_2(n) + 36 \sum_{\substack{(a,b,x,y) \in \mathbb{N}^4 \\ 4ax + by = n}} (F_3(a - b) - F_3(a + b))$$

$$= s_2(n) - 36 \sum_{\substack{(a,b,x,y) \in \mathbb{N}^4 \\ 4ax + by = n}} (F_3(2a - b) - F_3(2a + b)),$$

as $F_3(a \pm b) = F_3(-a \mp b) = F_3(2a \mp b)$. By Theorem 14.4 with $f = F_3$, we have for n odd

$$\sum_{\substack{(a,b,x,y) \in \mathbb{N}^4 \\ 4ax + by = n}} (F_3(2a - b) - F_3(2a + b))$$

$$= \frac{1}{2} \sum_{\substack{d \in \mathbb{N} \\ d \mid n}} \left(1 + \frac{n}{d}\right) F_3(d) - \sum_{\substack{d \in \mathbb{N} \\ d \mid n}} \sum_{\substack{\ell = 1 \\ \ell \equiv 1 \,(\mathrm{mod}\,2)}}^{d} F_3(\ell)$$

$$= \frac{1}{2} \sum_{\substack{d \in \mathbb{N} \\ d \mid n \\ 3 \mid d}} \left(1 + \frac{n}{d}\right) - \sum_{\substack{d \in \mathbb{N} \\ d \mid n}} \sum_{\substack{\ell = 1 \\ \ell \equiv 1 \,(\mathrm{mod}\,2)}}^{[d/3]} 1$$

$$= \frac{1}{2} \sum_{\substack{d \in \mathbb{N} \\ d \mid n/3}} \left(1 + \frac{n}{3d}\right) - \sum_{\substack{d \in \mathbb{N} \\ d \mid n}} ([d/3] - [d/6])$$

$$= \frac{1}{2} d(n/3) + \frac{1}{2} \sigma(n/3) - \sum_{\substack{d \in \mathbb{N} \\ d \mid n}} ([d/3] - [d/6]) \,.$$

By (3.13) with $m = 3$ we have

$$\sum_{\substack{d \in \mathbb{N} \\ d \mid n}} [d/3] = \frac{1}{3}\sigma(n) - \frac{1}{3}d_{1,3}(n) - \frac{2}{3}d_{2,3}(n).$$

For n odd, we have

$$d_{0,6}(n) = 0,$$
$$d_{1,6}(n) = d_{1,3}(n),$$
$$d_{2,6}(n) = 0,$$
$$d_{3,6}(n) = d(n/3),$$
$$d_{4,6}(n) = 0,$$
$$d_{5,6}(n) = d_{2,3}(n),$$

so that by (3.13) with $m = 6$ we obtain

$$\sum_{\substack{d \in \mathbb{N} \\ d \mid n}} [d/6] = \frac{1}{6}\sigma(n) - \frac{1}{2}d(n/3) - \frac{1}{6}d_{1,3}(n) - \frac{5}{6}d_{2,3}(n).$$

Hence, by (17.1), we obtain

$$\sum_{\substack{d \in \mathbb{N} \\ d \mid n}} ([d/3] - [d/6]) = \frac{1}{6}\sigma(n) + \frac{1}{2}d(n/3) - \frac{1}{36}s_2(n).$$

Thus

$$\sum_{\substack{(a, b, x, y) \in \mathbb{N}^4 \\ 4ax + by = n}} (F_3(2a - b) - F_3(2a + b)) = -\frac{1}{6}\sigma(n) + \frac{1}{2}\sigma(n/3) + \frac{1}{36}s_2(n)$$

and the number of representations is

$$s_2(n) - 36\left(-\frac{1}{6}\sigma(n) + \frac{1}{2}\sigma(n/3) + \frac{1}{36}s_2(n)\right) = 6\sigma(n) - 18\sigma(n/3)$$

as required.   □

Exercises 17

1. Let $n \in \mathbb{N}$. Define $\alpha \in \mathbb{N}_0$ and $N \in \mathbb{N}$ with $3 \nmid N$ by $n = 3^\alpha N$. Prove that
    $$s_4(n) = 12\sigma(N).$$

2. Deduce from Problem 1 that every positive integer n can be written in the form
    $$n = x_1^2 + x_1 x_2 + x_2^2 + x_3^2 + x_3 x_4 + x_4^2$$
    for some integers x_1, x_2, x_3 and x_4.

3. Prove that Theorem 17.3 can be written in the form
    $$s_4(n) = 12 \sum_{\substack{d \in \mathbb{N} \\ d \mid n \\ 3 \nmid d}} d.$$

4. Can every positive integer be written in the form
    $$x_1^2 + x_1 x_2 + x_2^2 + 2(x_3^2 + x_3 x_4 + x_4^2)$$
    for some integers x_1, x_2, x_3 and x_4?

5. Can every positive integer be written in the form
    $$x_1^2 + x_1 x_2 + x_2^2 + 4(x_3^2 + x_3 x_4 + x_4^2)$$
    for some integers x_1, x_2, x_3 and x_4?

6. Prove that the number of representations of the positive integer n by the quadratic form

$$x_1^2 + x_1x_2 + x_2^2 + 4(x_3^2 + x_3x_4 + x_4^2)$$

is

$$\begin{cases} 12\sigma(n/4) - 36\sigma(n/12), & \text{if } n \equiv 0 \pmod 2, \\ 6\sigma(n) - 18\sigma(n/3), & \text{if } n \equiv 1 \pmod 2. \end{cases}$$

7. (i) If x and y are integers prove that

$$x^2 + xy + y^2 \equiv 0 \text{ or } 1 \pmod 3.$$

(ii) If x and y are integers, prove that

$$3(x^2 + xy + y^2) = a^2 + ab + b^2,$$

where the integers a and b are given by

$$a = -x + y, \quad b = 2x + y.$$

(iii) Suppose that x and y are integers such that

$$x^2 + xy + y^2 \equiv 0 \pmod 3.$$

Prove that

$$x \equiv y \pmod 3.$$

Define integers u and v by

$$u = (x - y)/3, \quad v = u + y.$$

Prove that $x = 2u + v$, $y = -u + v$ and

$$\frac{x^2 + xy + y^2}{3} = u^2 + uv + v^2.$$

8. Use Problem 7 to prove that positive integers $n \equiv 2 \pmod 3$ are not represented by the quadratic form

$$x_1^2 + x_1x_2 + x_2^2 + 6(x_3^2 + x_3x_4 + x_4^2).$$

9. Let $n \in \mathbb{N}$. If $n \equiv 0 \pmod 3$ use Problem 7 to prove that the number of representations of n by the quadratic form

$$x_1^2 + x_1x_2 + x_2^2 + 6(x_3^2 + x_3x_4 + x_4^2)$$

is the same as the number of $n/3$ by the form

$$x_1^2 + x_1x_2 + x_2^2 + 2(x_3^2 + x_3x_4 + x_4^2).$$

10. Let $n \in \mathbb{N}$. If $n \equiv 1 \pmod 3$ use Problem 7 to prove that the number of representations of n by the quadratic form

$$x_1^2 + x_1x_2 + x_2^2 + 6(x_3^2 + x_3x_4 + x_4^2)$$

is the same as the number of n by the form

$$x_1^2 + x_1x_2 + x_2^2 + 2(x_3^2 + x_3x_4 + x_4^2).$$

11. Use Problems 8, 9 and 10 and Theorem 17.5 to prove that the number of representations of n by

$$x_1^2 + x_1x_2 + x_2^2 + 6(x_3^2 + x_3x_4 + x_4^2)$$

is

$$\begin{cases} -6\sigma(n) + 12\sigma(n/12) + 30\sigma(n/3) - 60\sigma(n/6), & \text{if } n \equiv 0 \pmod 3, \\ 6\sigma(n) - 12\sigma(n/2), & \text{if } n \equiv 1 \pmod 3, \\ 0, & \text{if } n \equiv 2 \pmod 3. \end{cases}$$

12. Use Problem 7 to prove that positive integers $n \equiv 1 \pmod 3$ are not represented by the quadratic form

$$2(x_1^2 + x_1x_2 + x_2^2) + 3(x_3^2 + x_3x_4 + x_4^2).$$

13. Let $n \in \mathbb{N}$. If $n \equiv 0 \pmod 3$ use Problem 7 to prove that the number of representations of n by the quadratic form

$$2(x_1^2 + x_1x_2 + x_2^2) + 3(x_3^2 + x_3x_4 + x_4^2)$$

is the same as the number of $n/3$ by the form

$$x_1^2 + x_1x_2 + x_2^2 + 2(x_3^2 + x_3x_4 + x_4^2).$$

14. Let $n \in \mathbb{N}$. If $n \equiv 2 \pmod 3$ use Problem 7 to prove that the number of representations of n by the quadratic form

$$2(x_1^2 + x_1x_2 + x_2^2) + 3(x_3^2 + x_3x_4 + x_4^2)$$

is the same as the number of n by the form

$$x_1^2 + x_1x_2 + x_2^2 + 2(x_3^2 + x_3x_4 + x_4^2).$$

15. Use Problems 11, 12 and 13 and Theorem 17.5 to prove that the number of representations of n by

$$2(x_1^2 + x_1x_2 + x_2^2) + 3(x_3^2 + x_3x_4 + x_4^2)$$

is

$$\begin{cases} -6\sigma(n) + 12\sigma(n/2) + 30\sigma(n/3) - 60\sigma(n/6), & \text{if } n \equiv 0 \ (\text{mod } 3), \\ 0, & \text{if } n \equiv 1 \ (\text{mod } 3), \\ 6\sigma(n) - 12\sigma(n/2), & \text{if } n \equiv 2 \ (\text{mod } 3). \end{cases}$$

16. Conjecture and prove a formula for the number of representations of n by the quadratic form

$$x_1^2 + x_1x_2 + x_2^2 + 3(x_3^2 + x_3x_4 + x_4^2).$$

Notes on Chapter 17

Theorem 17.1 is due to Lorenz [191]. In 1840 Dirichlet gave a formula for the number of representations of a positive integer n coprime with d by a set of inequivalent, positive-definite, primitive, integral, binary quadratic forms of discriminant d, see for example Dickson's book [94, Theorem 64, p. 78]. An extension of this formula to all n coprime with the conductor f of the discriminant d has been given by Kaplan and Williams [152] and to all n by Huard, Kaplan and Williams [136, Theorem 9.1, p. 291]. When $d = -3$ we have $f = 1$ and all positive-definite, primitive, integral, binary quadratic forms of discriminant -3 are equivalent to $x^2 + xy + y^2$ and we recover Lorenz's formula from each of these extensions. Lorenz's formula is also given in Dickson's book [94, Problem 2, Exercises XXII, p. 80].

Theorem 17.3 was first stated by Liouville [177] and can be found in Petersson [222] as well as in Lomadze [187, formula (I), p. 12], [188]. The first elementary arithmetic proof of Theorem 17.3 was given by Huard, Ou, Spearman and Williams [137, Theorem 13, p. 266]. This is the proof presented in this chapter. A second elementary proof was given by Chapman [67] based upon ideas of Spearman and Williams [246].

Theorem 17.4 is given in Petersson [222] and in Lomadze [187, formula (III), p. 12], [188]. The proof presented here is the first arithmetic proof of this result. Lomadze [187] gives the formulae

$$s_6(n) = 27 \sum_{\substack{d \in \mathbb{N} \\ d \mid n}} \left(\frac{-3}{n/d}\right) d^2 - 9 \sum_{\substack{d \in \mathbb{N} \\ d \mid n}} \left(\frac{-3}{d}\right) d^2$$

and

$$s_{10}(n) = 27 \sum_{\substack{d \in \mathbb{N} \\ d \mid n}} \left(\frac{-3}{n/d}\right) d^4 + 3 \sum_{\substack{d \in \mathbb{N} \\ d \mid n}} \left(\frac{-3}{d}\right) d^4,$$

as well as more complicated formulae for $s_{12}(n), \ldots, s_{34}(n)$.

A result equivalent to Theorem 17.5 was stated by Liouville [178] but not proved by him. A proof was given by Alaca, Alaca and Williams [13, Theorem 13, p. 180] using properties of the two-dimensional theta functions defined by J. M. Borwein and P. B. Borwein [50]. The proof given here is the first arithmetic proof of this result.

Theorem 17.6 is due to Alaca, Alaca and Williams [13, Theorem 15, p. 181]. The proof presented here is the first arithmetic proof of this result.

The number of representations of $n \in \mathbb{N}$ by the form

$$k(x_1^2 + x_1 x_2 + x_2^2) + \ell(x_3^2 + x_3 x_4 + x_4^2)$$

was conjectured by Liouville [178], [179], [180], [181] and determined by Walfisz [257] in the cases $(k, \ell) = (1, 2), (1, 3), (1, 4), (1, 6)$ and $(2, 3)$.

The number of representations of n by the quadratic form

$$x_1^2 + x_1 x_2 + x_2^2 + x_3^2 + x_3 x_4 + x_4^2 + k(x_5^2 + x_5 x_6 + x_6^2 + x_7^2 + x_7 x_8 + x_8^2)$$

is known for $k = 2$ Alaca and Williams [29], $k = 3$ Williams [268], $k = 4$ Alaca, Alaca and Williams [12], $k = 6$ Alaca, Alaca and Williams [14] and $k = 8$ Alaca, Alaca and Williams [15].

Following Borwein, Borwein and Garvan [51, p. 36], we set

$$a(q) := \sum_{x,y=-\infty}^{\infty} q^{x^2+xy+y^2}, \quad q \in \mathbb{C}, \quad |q| < 1.$$

By Theorem 17.3 we have

$$a(q) = \sum_{n=0}^{\infty} s_2(n)q^n = 1 + \sum_{n=1}^{\infty} s_2(n)q^n$$

$$= 1 + 6 \sum_{n=1}^{\infty} \sum_{\substack{d \in \mathbb{N} \\ d \mid n}} \left(\frac{-3}{d}\right) q^n = 1 + 6 \sum_{d,e=1}^{\infty} \left(\frac{-3}{d}\right) q^{de}$$

so that

$$a(q) = 1 + 6\sum_{d=1}^{\infty}\left(\frac{-3}{d}\right)\frac{q^d}{1-q^d}.$$

This Lambert series expansion of $a(q)$ is due to Lorenz [191, p. 111], see also Borwein, Borwein and Garvan [51, p. 43].

Appealing to Theorem 17.3, we obtain

$$a^2(q) = \sum_{n=0}^{\infty} s_4(n)q^n$$

$$= 1 + \sum_{n=1}^{\infty} s_4(n)q^n$$

$$= 1 + \sum_{n=1}^{\infty}(12\sigma(n) - 36\sigma(n/3))q^n$$

$$= 1 + 12\sum_{n=1}^{\infty}\sigma(n)q^n - 36\sum_{n=1}^{\infty}\sigma(n)q^{3n}$$

$$= 1 + 12\sum_{n=1}^{\infty}\sum_{\substack{d \in \mathbb{N} \\ d \mid n}} dq^n - 36\sum_{n=1}^{\infty}\sum_{\substack{d \in \mathbb{N} \\ d \mid n}} dq^{3n}$$

$$= 1 + 12\sum_{d,e=1}^{\infty} dq^{de} - 36\sum_{d,e=1}^{\infty} dq^{3de}$$

$$= 1 + 12\sum_{d=1}^{\infty}\frac{dq^d}{1-q^d} - 36\sum_{d=1}^{\infty}\frac{dq^{3d}}{1-q^{3d}}$$

so that

$$a^2(q) = 1 + 12\sum_{\substack{n=1 \\ 3 \nmid n}}^{\infty}\frac{nq^n}{1-q^n}.$$

This Lambert series expansion of $a^2(q)$ is due to Ramanujan, see Andrews and Berndt [30, Part I, p. 402].

The formula

$$a^3(q) = 1 - 9\sum_{n=1}^{\infty}\left(\frac{-3}{n}\right)\frac{n^2q^n}{1-q^n} + 27\sum_{n=1}^{\infty}\frac{n^2q^n}{1+q^n+q^{2n}}$$

was given by Ramanujan, see Andrews and Berndt [30, Part I, p. 402]. A different (but equivalent) formula was given by Alaca, Alaca and Williams [23, p. 97].

In a similar manner from Theorem 17.4 we obtain the formula

$$a^4(q) = 1 + 24 \sum_{n=1}^{\infty} \frac{n^3 q^n}{1 - q^n} + 216 \sum_{n=1}^{\infty} \frac{n^3 q^{3n}}{1 - q^{3n}},$$

which is due to Ramanujan, see Andrews and Berndt [30, Part I, p. 403]. For more formulae of this type, see Lomadze [186].

From Theorem 17.5 we obtain

$$a(q)a(q^2) = 1 + \sum_{n=1}^{\infty} (6\sigma(n) - 12\sigma(n/2) + 18\sigma(n/3) - 36\sigma(n/6))q^n.$$

A simple proof of this formula has been given by Alaca, Alaca and Williams [13, p. 191]. From this formula we obtain the Lambert series expansion

$$a(q)a(q^2) = 1 + 6 \sum_{\substack{n=1 \\ n \equiv 1 \,(\mathrm{mod}\,2)}}^{\infty} \frac{nq^n}{1-q^n} + 18 \sum_{\substack{n=1 \\ n \equiv 1 \,(\mathrm{mod}\,2)}}^{\infty} \frac{nq^{3n}}{1-q^{3n}}.$$

In a similar manner from Theorem 17.6 we obtain

$$a(q)a(q^4) = 1 + 6 \sum_{\substack{n=1 \\ n \equiv 1 \,(\mathrm{mod}\,2) \\ n \not\equiv 0 \,(\mathrm{mod}\,3)}}^{\infty} \frac{nq^n}{1-q^{2n}} + 12 \sum_{\substack{n=1 \\ n \not\equiv 0 \,(\mathrm{mod}\,3)}}^{\infty} \frac{nq^{4n}}{1-q^{4n}}.$$

The Lambert series expansions of $a(q)a(q^3)$, $a(q)a(q^4)$, $a(q)a(q^6)$ and $a(q^2)a(q^3)$ follow from the results of Alaca, Alaca and Williams [13, Theorems 14, 15, 16 and 17].

18

Representations by $x^2 + y^2 + z^2 + 2t^2$, $x^2 + y^2 + 2z^2 + 2t^2$ and $x^2 + 2y^2 + 2z^2 + 2t^2$

For $n \in \mathbb{N}_0$ we define

$$R_1(n) := \operatorname{card}\{(x, y, z, t) \in \mathbb{Z}^4 \mid n = x^2 + y^2 + z^2 + 2t^2\},$$
$$R_2(n) := \operatorname{card}\{(x, y, z, t) \in \mathbb{Z}^4 \mid n = x^2 + y^2 + 2z^2 + 2t^2\},$$
$$R_3(n) := \operatorname{card}\{(x, y, z, t) \in \mathbb{Z}^4 \mid n = x^2 + 2y^2 + 2z^2 + 2t^2\}.$$

Clearly $R_1(0) = R_2(0) = R_3(0) = 1$.

In 1860–61 Liouville asserted without proof formulae for $R_1(n)$, $R_2(n)$ and $R_3(n)$. In this chapter we show that Liouville's formulae for $R_1(n)$, $R_2(n)$ and $R_3(n)$ are simple consequences of the elementary arithmetic theorems proved in Chapter 14. We begin with $R_2(n)$.

Theorem 18.1. *Let $n \in \mathbb{N}$. Then the number $R_2(n)$ of solutions $(x, y, z, t) \in \mathbb{Z}^4$ of*

$$n = x^2 + y^2 + 2z^2 + 2t^2$$

is given by

$$R_2(n) = 4\sigma(n) - 4\sigma(n/2) + 8\sigma(n/4) - 32\sigma(n/8).$$

Proof. We have

$$R_2(n) = \sum_{\substack{(\ell, m) \in \mathbb{N}_0^2 \\ \ell + 2m = n}} \sum_{\substack{(x, y) \in \mathbb{Z}^2 \\ x^2 + y^2 = \ell}} 1 \sum_{\substack{(z, t) \in \mathbb{N}_0^2 \\ z^2 + t^2 = m}} 1$$

$$= \sum_{\substack{(\ell, m) \in \mathbb{N}_0^2 \\ \ell + 2m = n}} r_2(\ell) r_2(m)$$

$$= r_2(n) + r_2(n/2) + \sum_{\substack{(\ell, m) \in \mathbb{N}^2 \\ \ell + 2m = n}} r_2(\ell) r_2(m).$$

239

By Theorem 9.3, Theorem 3.10 and Problem 26 of Exercises 14, we obtain

$$\sum_{\substack{(\ell, m) \in \mathbb{N}^2 \\ \ell + 2m = n}} r_2(\ell)r_2(m) = \sum_{\substack{(\ell, m) \in \mathbb{N}^2 \\ 2\ell + m = n}} r_2(\ell)r_2(m)$$

$$= 16 \sum_{\substack{(\ell, m) \in \mathbb{N}^2 \\ 2\ell + m = n}} \sum_{\substack{a \in \mathbb{N} \\ a \mid \ell}} \left(\frac{-4}{a}\right) \sum_{\substack{b \in \mathbb{N} \\ b \mid m}} \left(\frac{-4}{b}\right)$$

$$= 16 \sum_{\substack{(a, b, x, y) \in \mathbb{N}^4 \\ 2ax + by = n}} \left(\frac{-4}{ab}\right)$$

$$= 16 \sum_{\substack{(a, b, x, y) \in \mathbb{N}^4 \\ 2ax + by = n}} (F_4(a - b) - F_4(a + b))$$

$$= 4\sigma(n) - 4\sigma(n/2) + 8\sigma(n/4) - 32\sigma(n/8)$$
$$\quad - 4(d_{1,4}(n) - d_{3,4}(n)) - 4(d_{1,4}(n/2) - d_{3,4}(n/2))$$

$$= 4\sigma(n) - 4\sigma(n/2) + 8\sigma(n/4) - 32\sigma(n/8)$$
$$\quad - r_2(n) - r_2(n/2),$$

so that

$$R_2(n) = 4\sigma(n) - 4\sigma(n/2) + 8\sigma(n/4) - 32\sigma(n/8)$$

as asserted.   □

Before proving Liouville's formulae for $R_1(n)$ and $R_3(n)$ we recall that the Legendre-Jacobi-Kronecker symbol for discriminant 8 is defined for $d \in \mathbb{N}$ by

$$\left(\frac{8}{d}\right) = \begin{cases} 0, & \text{if } d \equiv 0 \pmod{2}, \\ 1, & \text{if } d \equiv 1, 7 \pmod{8}, \\ -1, & \text{if } d \equiv 3, 5 \pmod{8}. \end{cases}$$

We extend the definition of $\left(\frac{8}{d}\right)$ $(d \in \mathbb{N})$ to all $d \in \mathbb{Z}$ by

$$\left(\frac{8}{-d}\right) = \left(\frac{8}{d}\right) \ (d \in \mathbb{N}), \quad \left(\frac{8}{0}\right) = 0,$$

so that $\left(\frac{8}{d}\right)$ is an even function of d. The following two results are easily proved: for $a, b \in \mathbb{N}$

$$\left(\frac{8}{2a - b}\right) - \left(\frac{8}{2a + b}\right) = 2\left(\frac{-4}{a}\right)\left(\frac{-8}{b}\right), \tag{18.1}$$

and for $m \in \mathbb{N}$

$$\sum_{k=1}^{m} \left(\frac{8}{k}\right) = \begin{cases} \dfrac{1}{2}\left(\dfrac{-8}{m}\right) + \dfrac{1}{2}\left(\dfrac{8}{m}\right), & \text{if } 2 \nmid m, \\[2mm] \left(\dfrac{-4}{m/2}\right), & \text{if } 2 \mid m. \end{cases} \tag{18.2}$$

We are now ready to prove Liouville's formulae for $R_1(n)$ and $R_3(n)$.

Theorem 18.2. *Let $n \in \mathbb{N}$. Then the number $R_1(n)$ of solutions $(x, y, z, t) \in \mathbb{Z}^4$ of*

$$n = x^2 + y^2 + z^2 + 2t^2$$

is given by

$$R_1(n) = 8 \sum_{\substack{d \in \mathbb{N} \\ d \mid n}} \frac{n}{d}\left(\frac{8}{d}\right) - 2 \sum_{\substack{d \in \mathbb{N} \\ d \mid n}} d\left(\frac{8}{d}\right)$$

and the number $R_3(n)$ of solutions $(x, y, z, t) \in \mathbb{Z}^4$ of

$$n = x^2 + 2y^2 + 2z^2 + 2t^2$$

is given by

$$R_3(n) = 4 \sum_{\substack{d \in \mathbb{N} \\ d \mid n}} \frac{n}{d}\left(\frac{8}{d}\right) - 2 \sum_{\substack{d \in \mathbb{N} \\ d \mid n}} d\left(\frac{8}{d}\right).$$

Proof. First we prove the formula for $R_1(n)$. We choose

$$f(x) = \left(\frac{8}{x}\right), \quad x \in \mathbb{Z},$$

in Theorem 14.2, and after a short calculation using (18.1) and (18.2), we obtain

$$\sum_{\substack{(a, b, x, y) \in \mathbb{N}^4 \\ ax + by = n}} \left(\frac{-4}{a}\right)\left(\frac{-8}{b}\right) = -\frac{1}{2}\sum_{\substack{d \in \mathbb{N} \\ d \mid n}} \left(\frac{-4}{d}\right) - \frac{1}{4}\sum_{\substack{d \in \mathbb{N} \\ d \mid n}} \left(\frac{-8}{d}\right) \tag{18.3}$$

$$- \frac{1}{4}\sum_{\substack{d \in \mathbb{N} \\ d \mid n}} d\left(\frac{8}{d}\right) + \sum_{\substack{d \in \mathbb{N} \\ d \mid n}} \frac{n}{d}\left(\frac{8}{d}\right).$$

For $n \in \mathbb{N}_0$ we set for convenience

$$a(n) := \text{card}\{(x, y) \in \mathbb{Z}^2 \mid n = x^2 + y^2\},$$

and

$$b(n) := \text{card}\{(x, y) \in \mathbb{Z}^2 \mid n = x^2 + 2y^2\},$$

so that $a(n) = r_2(n)$. Clearly $a(0) = b(0) = 1$. We recall that

$$a(n) = 4 \sum_{\substack{d \in \mathbb{N} \\ d \mid n}} \left(\frac{-4}{d}\right), \quad b(n) = 2 \sum_{\substack{d \in \mathbb{N} \\ d \mid n}} \left(\frac{-8}{d}\right), \quad n \in \mathbb{N}.$$

Then, for $n \in \mathbb{N}$, we have

$$R_1(n) = \sum_{k=0}^{n} a(k)b(n-k)$$

$$= a(n) + b(n) + \sum_{k=1}^{n-1} a(k)b(n-k)$$

$$= 4 \sum_{\substack{d \in \mathbb{N} \\ d \mid n}} \left(\frac{-4}{d}\right) + 2 \sum_{\substack{d \in \mathbb{N} \\ d \mid n}} \left(\frac{-8}{d}\right) + 8 \sum_{\substack{(a,b,x,y) \in \mathbb{N}^4 \\ ax+by=n}} \left(\frac{-4}{a}\right)\left(\frac{-8}{b}\right),$$

that is by (18.3)

$$R_1(n) = 8 \sum_{\substack{d \in \mathbb{N} \\ d \mid n}} \frac{n}{d}\left(\frac{8}{d}\right) - 2 \sum_{\substack{d \in \mathbb{N} \\ d \mid n}} d\left(\frac{8}{d}\right),$$

as asserted.

Next we prove the formula for $R_3(n)$. Let $n \in \mathbb{N}$. Choosing

$$f(x) = \left(\frac{8}{x}\right), \quad x \in \mathbb{Z},$$

in Theorem 14.3, we obtain after a short calculation using (18.1) and (18.2)

$$\sum_{\substack{(a,b,x,y) \in \mathbb{N}^4 \\ 2ax+by=n}} \left(\frac{-4}{a}\right)\left(\frac{-8}{b}\right) = -\frac{1}{2} \sum_{\substack{d \in \mathbb{N} \\ d \mid n/2}} \left(\frac{-4}{d}\right) - \frac{1}{4} \sum_{\substack{d \in \mathbb{N} \\ d \mid n}} \left(\frac{-8}{d}\right) \qquad (18.4)$$

$$- \frac{1}{4} \sum_{\substack{d \in \mathbb{N} \\ d \mid n}} d\left(\frac{8}{d}\right) + \frac{1}{2} \sum_{\substack{d \in \mathbb{N} \\ d \mid n}} \frac{n}{d}\left(\frac{8}{d}\right).$$

Then

$$R_3(n) = \sum_{\substack{k \in \mathbb{N}_0 \\ k \le n/2}} a(k)b(n-2k)$$

$$= a(n/2) + b(n) + \sum_{\substack{k \in \mathbb{N} \\ k < n/2}} a(k)b(n-2k)$$

$$= 4 \sum_{\substack{d \in \mathbb{N} \\ d \mid n/2}} \left(\frac{-4}{d}\right) + 2 \sum_{\substack{d \in \mathbb{N} \\ d \mid n}} \left(\frac{-8}{d}\right)$$

$$+ 8 \sum_{\substack{k \in \mathbb{N} \\ 1 \le k < n/2}} \sum_{\substack{a \in \mathbb{N} \\ a \mid k}} \left(\frac{-4}{a}\right) \sum_{\substack{b \in \mathbb{N} \\ b \mid n - 2k}} \left(\frac{-8}{b}\right)$$

$$= 4 \sum_{\substack{d \in \mathbb{N} \\ d \mid n/2}} \left(\frac{-4}{d}\right) + 2 \sum_{\substack{d \in \mathbb{N} \\ d \mid n}} \left(\frac{-8}{d}\right) + 8 \sum_{\substack{(a, b, x, y) \in \mathbb{N}^4 \\ 2ax + by = n}} \left(\frac{-4}{a}\right) \left(\frac{-8}{b}\right),$$

that is by (18.4)

$$R_3(n) = 4 \sum_{\substack{d \in \mathbb{N} \\ d \mid n}} \frac{n}{d} \left(\frac{8}{d}\right) - 2 \sum_{\substack{d \in \mathbb{N} \\ d \mid n}} d \left(\frac{8}{d}\right),$$

as claimed. $\qquad\square$

Example 18.1. We use the formula

$$b(n) = 2 \sum_{\substack{d \in \mathbb{N} \\ d \mid n}} \left(\frac{-8}{d}\right), \quad n \in \mathbb{N},$$

to determine the number of representations of a nonnegative integer as the sum of a triangular number and a square.

We begin by noting that if u and v are integers then

$$u^2 + 2v^2 \equiv 1 \pmod{8} \iff u \equiv 1 \pmod{2}, v \equiv 0 \pmod{2}.$$

Then, for $n \in \mathbb{N}_0$, we have

$$\text{card} \left\{ (x, y) \in \mathbb{N}_0 \times \mathbb{Z} \mid n = \frac{1}{2} x(x + 1) + y^2 \right\}$$

$$= \frac{1}{2} \text{card} \left\{ (x, y) \in \mathbb{Z}^2 \mid n = \frac{1}{2} x(x + 1) + y^2 \right\}$$

$$= \frac{1}{2} \text{card} \{ (x, y) \in \mathbb{Z}^2 \mid 8n + 1 = (2x + 1)^2 + 8y^2 \}$$

$$= \frac{1}{2} \text{card} \{ (u, v) \in \mathbb{Z}^2 \mid 8n + 1 = u^2 + 2v^2, u \equiv 1 \pmod{2}, v \equiv 0 \pmod{2} \}$$

$$= \frac{1}{2} \text{card} \{ (u, v) \in \mathbb{Z}^2 \mid 8n + 1 = u^2 + 2v^2 \}$$

$$= \frac{1}{2} b(8n + 1),$$

so the number of representations of $n \in \mathbb{N}_0$ as the sum of a triangular number and a square is

$$\sum_{\substack{d \in \mathbb{N} \\ d \,|\, 8n+1}} \left(\frac{-8}{d} \right).$$

Example 18.2. The number of representations of 7 as the sum of a triangular number and a square is

$$\sum_{\substack{d \in \mathbb{N} \\ d \,|\, 57}} \left(\frac{-8}{d} \right) = \left(\frac{-8}{1} \right) + \left(\frac{-8}{3} \right) + \left(\frac{-8}{19} \right) + \left(\frac{-8}{57} \right) = 1 + 1 + 1 + 1 = 4.$$

The four representations are

$$7 = 3 + (\pm 2)^2 = 6 + (\pm 1)^2.$$

Example 18.3. We use Theorem 18.2 to determine the number of representations of a nonnegative integer n as the sum of three squares and a triangular number.

Let $n \in \mathbb{N}_0$. Suppose that

$$8n + 1 = x^2 + y^2 + z^2 + 2t^2$$

for some integers x, y, z and t. If $t \equiv 1 \pmod 2$ then

$$x^2 + y^2 + z^2 = 8n + 1 - 2t^2 \equiv 7 \pmod 8,$$

which is impossible as, for any integers x, y and z, we have

$$x^2 + y^2 + z^2 \equiv 0, 1, 2, 3, 4, 5, 6 \pmod 8.$$

Hence $t \equiv 0 \pmod 2$. Then

$$x^2 + y^2 + z^2 = 8n + 1 - 2t^2 \equiv 1 \pmod 8$$

so

$$x \equiv y \equiv 0 \pmod 2, \quad x \equiv y \pmod 4, \quad z \equiv 1 \pmod 2$$

or

$$x \equiv z \equiv 0 \pmod 2, \quad x \equiv z \pmod 4, \quad y \equiv 1 \pmod 2$$

or

$$y \equiv z \equiv 0 \pmod 2, \quad y \equiv z \pmod 4, \quad x \equiv 1 \pmod 2.$$

Thus

$$R_1(8n + 1) = 3 \operatorname{card} \left\{ (x, y, z, t) \in \mathbb{Z}^4 \mid 8n + 1 = x^2 + y^2 + z^2 + 2t^2, \right.$$
$$x \equiv y \equiv 0 \pmod 2, \ x \equiv y \pmod 4,$$
$$\left. z \equiv 1 \pmod 2, \ t \equiv 0 \pmod 2 \right\}.$$

If x and y are integers such that

$$x \equiv y \equiv 0 \pmod 2, \quad x \equiv y \pmod 4,$$

then

$$a = \frac{1}{4}(x + y), \quad b = \frac{1}{4}(x - y),$$

are integers such that

$$x^2 + y^2 = 8a^2 + 8b^2.$$

Conversely, if a and b are integers, then

$$x = 2a + 2b, \quad y = 2a - 2b,$$

are integers such that

$$x^2 + y^2 = 8a^2 + 8b^2$$

and

$$x \equiv y \equiv 0 \pmod 2, \quad x \equiv y \pmod 4.$$

Hence

$$R_1(8n+1) = 3 \operatorname{card} \left\{ (a, b, c, d) \in \mathbb{Z}^4 \mid 8n+1 = 8a^2 + 8b^2 + 8c^2 + (2d + 1)^2 \right\}$$
$$= 3 \operatorname{card} \left\{ (a, b, c, d) \in \mathbb{Z}^4 \mid n = a^2 + b^2 + c^2 + d(d + 1)/2 \right\}$$
$$= 6 \operatorname{card} \left\{ (a, b, c, d) \in \mathbb{Z}^3 \times \mathbb{N}_0 \mid n = a^2 + b^2 + c^2 + d(d+1)/2 \right\}.$$

By Theorem 18.2 we have

$$R_1(8n + 1) = 8 \sum_{\substack{d \in \mathbb{N} \\ d \mid 8n + 1}} \frac{8n + 1}{d} \left(\frac{8}{d} \right) - 2 \sum_{\substack{d \in \mathbb{N} \\ d \mid 8n + 1}} d \left(\frac{8}{d} \right)$$

$$= 8 \sum_{\substack{d \in \mathbb{N} \\ d \mid 8n + 1}} d \left(\frac{8}{(8n + 1)/d} \right) - 2 \sum_{\substack{d \in \mathbb{N} \\ d \mid 8n + 1}} d \left(\frac{8}{d} \right)$$

$$= 6 \sum_{\substack{d \in \mathbb{N} \\ d \mid 8n + 1}} d \left(\frac{8}{d} \right).$$

Thus, equating the two expressions for $R_1(8n + 1)$, we obtain

$$\text{card}\left\{(a, b, c, d) \in \mathbb{Z}^3 \times \mathbb{N}_0 \mid n = a^2 + b^2 + c^2 + d(d + 1)/2\right\} = \sum_{\substack{d \in \mathbb{N} \\ d \mid 8n + 1}} d\left(\frac{8}{d}\right),$$

which is the required formula.

Example 18.4. The number of representations of 7 as the sum of three squares and a triangular number is

$$\sum_{\substack{d \in \mathbb{N} \\ d \mid 57}} d\left(\frac{8}{d}\right) = 1\left(\frac{8}{1}\right) + 3\left(\frac{8}{3}\right) + 19\left(\frac{8}{19}\right) + 57\left(\frac{8}{57}\right) = 1 - 3 - 19 + 57 = 36.$$

The thirty-six representations arise as follows: six from $0 + 0 + 1 + 6$, six from $0 + 0 + 4 + 3$, and twenty-four from $1 + 1 + 4 + 1$.

Exercises 18

1. Prove (18.1).
2. Prove (18.2).
3. Complete the details of the proof of (18.3).
4. Complete the details of the proof of (18.4).
5. Prove that

$$\sum_{\substack{d \in \mathbb{N} \\ d \mid n}} d\left(\frac{8}{d}\right)$$

 is a multiplicative function of n.
6. Prove that

$$\sum_{\substack{d \in \mathbb{N} \\ d \mid n}} \frac{n}{d}\left(\frac{8}{d}\right)$$

 is a multiplicative function of n.
7. Prove from first principles that

$$R_3(2n) = R_1(n), \quad n \in \mathbb{N}.$$

8. Let $n \in \mathbb{N}$ be odd. Prove from first principles that

$$R_1(n) = 3R_3(n), \quad \text{if } n \equiv 1, 7 \pmod 8$$

and

$$R_1(n) = \frac{5}{3} R_3(n), \quad \text{if } n \equiv 3, 5 \pmod 8.$$

Hint: Consider the identity

$$x^2 + 2y^2 + 2z^2 + 2t^2 = x^2 + (y+z)^2 + (y-z)^2 + 2t^2.$$

9. Let $n \in \mathbb{N}$. As $R_3(n) \geq 0$ it follows from Theorem 18.2 that

$$\sum_{\substack{d \in \mathbb{N} \\ d \mid n}} d\left(\frac{8}{d}\right) \leq 2 \sum_{\substack{d \in \mathbb{N} \\ d \mid n}} \frac{n}{d} \left(\frac{8}{d}\right).$$

Prove the stronger inequality

$$\sum_{\substack{d \in \mathbb{N} \\ d \mid n}} d\left(\frac{8}{d}\right) \leq \sum_{\substack{d \in \mathbb{N} \\ d \mid n}} \frac{n}{d} \left(\frac{8}{d}\right).$$

10. Let $n \in \mathbb{N}$. Define $\alpha \in \mathbb{N}_0$ and $N \in \mathbb{N}$ with N odd by $n = 2^\alpha N$. Use Theorem 18.1 to prove that

$$R_2(n) = \begin{cases} 4\sigma(N), & \text{if } \alpha = 0, \\ 8\sigma(N), & \text{if } \alpha = 1, \\ 24\sigma(N), & \text{if } \alpha \geq 2. \end{cases}$$

11. Let $n \in \mathbb{N}$. Define $\alpha \in \mathbb{N}_0$ and $N \in \mathbb{N}$ with N odd by $n = 2^\alpha N$. Prove that

$$\sum_{\substack{d \in \mathbb{N} \\ d \mid n}} d\left(\frac{8}{d}\right) = \sum_{\substack{d \in \mathbb{N} \\ d \mid N}} d\left(\frac{8}{d}\right).$$

12. Let $n \in \mathbb{N}$. Define $\alpha \in \mathbb{N}_0$ and $N \in \mathbb{N}$ with N odd by $n = 2^\alpha N$. Prove that

$$\sum_{\substack{d \in \mathbb{N} \\ d \mid n}} \frac{n}{d}\left(\frac{8}{d}\right) = 2^\alpha \left(\frac{8}{N}\right) \sum_{\substack{d \in \mathbb{N} \\ d \mid N}} d\left(\frac{8}{d}\right).$$

13. Let $n \in \mathbb{N}$. Define $\alpha \in \mathbb{N}_0$ and $N \in \mathbb{N}$ with N odd by $n = 2^\alpha N$. Prove that

$$R_1(n) = 2\left(2^{\alpha+2}\left(\frac{8}{N}\right) - 1\right) \sum_{\substack{d \in \mathbb{N} \\ d \mid N}} d\left(\frac{8}{d}\right).$$

14. Let $n \in \mathbb{N}$. Define $\alpha \in \mathbb{N}_0$ and $N \in \mathbb{N}$ with N odd by $n = 2^\alpha N$. Prove that

$$R_3(n) = 2\left(2^{\alpha+1}\left(\frac{8}{N}\right) - 1\right) \sum_{\substack{d \in \mathbb{N} \\ d \mid N}} d\left(\frac{8}{d}\right).$$

15. Deduce the result of Problem 7 directly from Problems 13 and 14.

16. Deduce the results of Problem 8 directly from Problems 13 and 14.

17. Let $n \in \mathbb{N}$. Define $\alpha \in \mathbb{N}_0$ and $N \in \mathbb{N}$ with N odd by $n = 2^\alpha N$. Use Problems 13 and 14 to prove that

$$R_1(n) = \frac{((2^{2\alpha+3} - 1) + 2^{\alpha+1}\left(\frac{8}{N}\right))}{2^{2\alpha+2} - 1} R_3(n).$$

18. Let $n \in \mathbb{N}$. Define $\alpha \in \mathbb{N}_0$ and $N \in \mathbb{N}$ with N odd by $n = 2^\alpha N$. For each (necessarily odd) prime p dividing N, let $\alpha_p(N)$ be the largest positive integer such that $p^{\alpha_p(n)} \mid N$. Prove that

$$R_1(n) = 2\left(2^{\alpha+2}\left(\frac{8}{N}\right) - 1\right) \prod_{p \mid N} \frac{1 - p^{\alpha_p(N)+1}\left(\frac{8}{p}\right)^{\alpha_p(N)+1}}{1 - p\left(\frac{8}{p}\right)}.$$

19. Let $n \in \mathbb{N}$. Define $\alpha \in \mathbb{N}_0$ and $N \in \mathbb{N}$ with N odd by $n = 2^\alpha N$. For each (necessarily odd) prime p dividing N, let $\alpha_p(N)$ be the largest positive integer such that $p^{\alpha_p(n)} \mid N$. Prove that

$$R_3(n) = 2\left(2^{\alpha+1}\left(\frac{8}{N}\right) - 1\right) \prod_{p \mid N} \frac{1 - p^{\alpha_p(N)+1}\left(\frac{8}{p}\right)^{\alpha_p(N)+1}}{1 - p\left(\frac{8}{p}\right)}.$$

20. Let $n \in \mathbb{N}$ be odd. Prove that

$$\sum_{\substack{d \in \mathbb{N} \\ d \mid n}} d\left(\frac{8}{d}\right) > 0 \quad \text{if } n \equiv 1, 7 \pmod 8$$

and

$$\sum_{\substack{d \in \mathbb{N} \\ d \mid n}} d\left(\frac{8}{d}\right) < 0 \quad \text{if } n \equiv 3, 5 \pmod 8.$$

21. Prove that every positive integer is of the form $x^2 + y^2 + z^2 + 2t^2$ for some integers x, y, z and t.

22. Prove that every positive integer is of the form $x^2 + y^2 + 2z^2 + 2t^2$ for some integers x, y, z and t.

23. Prove that every positive integer is of the form $x^2 + 2y^2 + 2z^2 + 2t^2$ for some integers x, y, z and t.

24. Prove or disprove

$$R_1(n) \geq R_3(n), \quad n \in \mathbb{N}.$$

25. Characterize those positive integers which cannot be expressed as the sum of a triangular number and a square.

Notes on Chapter 18

Theorem 18.1 was stated without proof by Liouville [174] and proved by Pepin [219, pp. 185, 188], [221, p. 40], Bachmann [34], Chan [66, p. 68], Deutsch [88], Fine [106, p. 74] and Alaca, Alaca, Lemire and Williams [5, Theorem 1.8, p. 284].

Theorem 18.2 was stated without proof by Liouville [175]. A search of the literature revealed eight proofs of the formula for $R_1(n)$ and four of the formula for $R_3(n)$. In 1884 Pepin [219, pp. 189–196] gave long proofs of the formulae of Theorem 18.2 using Liouville's elementary methods and recurrence relations between $R_1(n)$ and $R_3(n)$ as well as between $R_1(2^{\alpha} N)$ and $R_1(N)$. In 1901 Petr [223, p. 8] gave some theta function identities from which a proof of the formula for $R_1(n)$ can be deduced. In 1964 Benz [42, pp. 168–175] gave proofs of the formulae of Theorem 18.2 using theta functions and recurrence relations such as the easily proved relation $R_3(2n) = R_1(n)$ ($n \in \mathbb{N}$). In 1968 Demuth [87, pp. 241–243] used Siegel's mass formula to prove the formula for $R_1(n)$. In 1974 Wild [259] used modular forms to prove the formula for $R_1(n)$. In his book on hypergeometric series published in 1988, Fine [106, p. 75] gave an analytic proof of the formula for $R_1(n)$. Alaca, Alaca, Lemire and Williams [8] have given proofs of the formulae for $R_1(n)$ and $R_3(n)$ using theta function identities. Williams [271] gave an elementary arithmetic proof of the formulae for $R_1(n)$ and $R_3(n)$. The proof of Theorem 8.2 is taken from Williams [271].

From Problems 21–23 we know that every positive integer is represented by each of the forms $x^2 + y^2 + z^2 + 2t^2$, $x^2 + y^2 + 2z^2 + 2t^2$ and $x^2 + 2y^2 + 2z^2 + 2t^2$. Ramanujan [231], [232, p. 170] listed 55 such forms with this property. However some of Ramanujan's arguments relied upon unproved results about several ternary forms. These results were later proved by Dickson [90], [95, Vol. V, pp. 255–262], [92, Chapter 5, pp. 86–114]. Dickson [89], [95, Vol. III, pp. 443–444], [91], [95, Vol. I, pp. 461–478] noted that the form $x^2 + 2y^2 + 5z^2 + 5t^2$ in Ramanujan's list does not represent the integer

15 and he completed the proof that the remaining 54 forms given by Ramanujan represent every positive integer. In this connection see also Kloosterman [154], [155].

Problem 7 is taken from Benz [42, p. 171]. Problem 8 is taken from Benz [42, pp. 173, 174]. The formulation of $R_1(n)$ and $R_3(n)$ given in Problems 13 and 14 respectively appears in Benz [42, p. 174].

For results on representing positive integers as mixed sums of squares and triangular numbers, the interested reader should consult Adiga, Cooper and Han [2], Barrucand, Cooper and Hirschhorn [38], Guo, Pan and Sun [120], Lam [161], Oh and Sun [213], and Sun [250]. Recently Berkovich and Jagy [43] have proved certain positivity results for mixed sums of squares and triangular numbers such as

$$\text{card}\{(x, y) \in \mathbb{Z}^2 \mid n = x(x + 1)/2 + y^2 + z^2\}$$
$$\geq \text{card}\{(x, y) \in \mathbb{Z}^2 \mid n = x(x + 1)/2 + 3y^2 + 3z^2\}$$

for all $n \in \mathbb{N}$.

The number of representations of a positive integer n by the quaternary quadratic form $ax^2 + by^2 + cz^2 + dt^2$ has been determined for forms other than $x^2 + y^2 + z^2 + 2t^2$, $x^2 + y^2 + 2z^2 + 2t^2$ and $x^2 + 2y^2 + 2z^2 + 2t^2$. A historical summary of results is given in Cooper [79]. We just mention two such results. The number of representations of n by the form $x^2 + y^2 + 3z^2 + 3t^2$ is

$$4\sigma(n) - 8\sigma(n/2) - 12\sigma(n/3) + 16\sigma(n/4) + 24\sigma(n/6) - 48\sigma(n/12),$$

see for example Alaca, Alaca, Lemire and Williams [5, p. 297], and the number of representations of n by the form $x^2 + y^2 + 4z^2 + 4t^2$ is

$$\left(2 + 2\left(\frac{-4}{n}\right)\right)\sigma(n) - 2\sigma(n/2) + 8\sigma(n/8) - 32\sigma(n/16),$$

see for example Alaca, Alaca, Lemire and Williams [5, p. 298]. The interested reader should consult Alaca [3], Alaca, Alaca, Lemire and Williams [4], [5], [6], [7], [8], Alaca, Alaca and Williams [17], Bachmann [34], Cooper [78], [79], Griffiths [116], Kloosterman [154], [155], Köhler [156], Pepin [221] and Petersson [222].

For representations by a sextenary quadratic form $a_1x_1^2 + \cdots + a_6x_6^2$, see Alaca, Alaca, Uygul and Williams [11], Alaca, Alaca and Williams [19], [20], [23] and [26] and Nazimoff [210, pp. 24–27].

19

Sums of Eight and Twelve Squares

In this chapter we prove Jacobi's formula for the number $r_8(n)$ of representations of a positive integer n as a sum of eight squares as well as Liouville's formula for the number $r_{12}(n)$ of representations of an even positive integer as a sum of twelve squares.

We make use of Theorems 11.1, 12.1 and 15.2 to determine $r_8(n)$.

Theorem 19.1. *Let n be a positive integer. Then*

$$r_8(n) = 16(-1)^n \sum_{\substack{d \in \mathbb{N} \\ d \mid n}} (-1)^d d^3.$$

Proof. Let n and s denote positive integers. Recall from (15.1)

$$A_s(n) := \sum_{\substack{k \in \mathbb{N} \\ k < n/s}} \sigma(k)\sigma(n - sk).$$

From Besge's formula (Theorem 12.1) we have

$$A_1(n) = \frac{5}{12}\sigma_3(n) + \left(\frac{1}{12} - \frac{1}{2}n\right)\sigma(n), \qquad (19.1)$$

and from Theorem 15.2 that

$$A_4(n) = \frac{1}{48}\sigma_3(n) + \frac{1}{16}\sigma_3(n/2) + \frac{1}{3}\sigma_3(n/4)$$
$$+ \left(\frac{1}{24} - \frac{1}{16}n\right)\sigma(n) + \left(\frac{1}{24} - \frac{1}{4}n\right)\sigma(n/4). \qquad (19.2)$$

We have

$$r_8(n) = \sum_{k=0}^{n} r_4(k)r_4(n - k) = 2r_4(n) + \sum_{k=1}^{n-1} r_4(k)r_4(n - k), \qquad (19.3)$$

251

as $r_4(0) = 1$. Appealing to Jacobi's formula for $r_4(n)$ (Theorem 11.1), namely,

$$r_4(n) = 8\sigma(n) - 32\sigma(n/4), \tag{19.4}$$

we obtain

$$\sum_{k=1}^{n-1} r_4(k)r_4(n-k) = 64S_1 - 256S_2 - 256S_3 + 1024S_4, \tag{19.5}$$

where

$$S_1 = \sum_{k=1}^{n-1} \sigma(k)\sigma(n-k), \tag{19.6}$$

$$S_2 = \sum_{k=1}^{n-1} \sigma(k/4)\sigma(n-k), \tag{19.7}$$

$$S_3 = \sum_{k=1}^{n-1} \sigma(k)\sigma((n-k)/4), \tag{19.8}$$

$$S_4 = \sum_{k=1}^{n-1} \sigma(k/4)\sigma((n-k)/4). \tag{19.9}$$

Clearly $S_1 = A_1(n)$ and changing the summation variable in (19.8) from k to $n - k$ shows that $S_3 = S_2$. Since the only terms in S_2 and S_4 which do not vanish are those for which $4 \mid k$, replacing k by $4k$ in (19.7) and (19.9), we find that $S_2 = A_4(n)$ and $S_4 = A_1(n/4)$. Appealing to (19.1) and (19.2) for the values of $A_1(n)$ and $A_4(n)$, and to (19.4) for the value of $r_4(n)$, we obtain from (19.3) and (19.5)–(19.9)

$$r_8(n) = 16\sigma_3(n) - 32\sigma_3(n/2) + 256\sigma_3(n/4). \tag{19.10}$$

Examining the three possibilities $2 \nmid n$, $2 \parallel n$ and $4 \mid n$ individually, we find that the right hand side of (19.10) is the same as the right hand side of the formula in the theorem. $\qquad\square$

In 1864 Liouville stated a formula for $r_{12}(n)$ valid for all even positive integers. We make use of Theorems 3.1, 9.5, 13.11 and 19.1 to give an elementary proof of Liouville's formula. Recall from (13.13)

$$S_{e,f}(n) := \sum_{m=1}^{n-1} \sigma_e(m)\sigma_f(n-m), \quad e, f, n \in \mathbb{N}. \tag{19.11}$$

We require the value of $S_{1,3}(n)$. From Example 12.2 we have

$$S_{1,3}(n) = \frac{7}{80}\sigma_5(n) + \left(\frac{1}{24} - \frac{1}{8}n\right)\sigma_3(n) - \frac{1}{240}\sigma(n). \qquad (19.12)$$

It is also convenient to define the twisted convolution sum

$$A_{e,f}(n) := T_{e,f,2}(n) = \sum_{\substack{m \in \mathbb{N} \\ m < n/2}} \sigma_e(m)\sigma_f(n - 2m), \quad e, f, n \in \mathbb{N}, \qquad (19.13)$$

see (13.14). We just need the values of $A_{1,3}(n)$ and $A_{3,1}(n)$, which were given in Theorem 13.11. We have

$$A_{1,3}(n) = \frac{1}{48}\sigma_5(n) + \frac{1}{15}\sigma_5(n/2) + \left(\frac{1}{24} - \frac{1}{16}n\right)\sigma_3(n) - \frac{1}{240}\sigma(n/2) \qquad (19.14)$$

and

$$A_{3,1}(n) = \frac{1}{240}\sigma_5(n) + \frac{1}{12}\sigma_5(n/2) + \left(\frac{1}{24} - \frac{1}{8}n\right)\sigma_3(n/2) - \frac{1}{240}\sigma(n). \qquad (19.15)$$

Theorem 19.2. *Let n be an even positive integer. Then*

$$r_{12}(n) = 8\sigma_5(n) - 512\sigma_5(n/4).$$

Proof. Let n be an even positive integer. Set $n = 2N$, where $N \in \mathbb{N}$. By Theorem 3.1(ii) we have

$$\sigma(2N) - 3\sigma(N) + 2\sigma(N/2) = 0. \qquad (19.16)$$

Then, by Theorem 9.5, we deduce that

$$r_4(n) = r_4(2N) = 8\sigma(2N) - 32\sigma(N/2) = 24\sigma(N) - 48\sigma(N/2). \qquad (19.17)$$

Again by Theorem 3.1(ii) we have

$$\sigma_3(2N) - 9\sigma_3(N) + 8\sigma_3(N/2) = 0. \qquad (19.18)$$

Then, by Theorem 19.1, we obtain

$$\begin{aligned}
r_8(n) &= r_8(2N) \\
&= 16 \sum_{\substack{d \in \mathbb{N} \\ d \mid 2N}} (-1)^d d^3 \\
&= 16 \sum_{\substack{d \in \mathbb{N} \\ d \mid 2N}} (1 + (-1)^d)d^3 - 16 \sum_{\substack{d \in \mathbb{N} \\ d \mid 2N}} d^3 \\
&= 256\sigma_3(N) - 16\sigma_3(2N),
\end{aligned}$$

and thus by (19.18)

$$r_8(n) = r_8(2N) = 112\sigma_3(N) + 128\sigma_3(N/2). \tag{19.19}$$

Now

$$r_{12}(n) = \sum_{k=0}^{n} r_4(n-k)r_8(k) \tag{19.20}$$

so that

$$r_{12}(n) = r_4(n) + r_8(n) + \sum_{k=1}^{n-1} r_4(n-k)r_8(k). \tag{19.21}$$

Appealing to (19.17), (19.19) and (19.21), we obtain

$$r_{12}(n) = 112\sigma_3(N) + 128\sigma_3(N/2) + 24\sigma(N) - 48\sigma(N/2) + T_0 + T_1, \tag{19.22}$$

where

$$T_i := \sum_{\substack{k=1 \\ k \equiv i \,(\mathrm{mod}\, 2)}}^{2N-1} r_4(2N-k)r_8(k), \quad i = 0, 1. \tag{19.23}$$

It remains to evaluate T_0 and T_1.

We first evaluate T_0. Appealing to (19.23), (19.17) and (19.19), we obtain

$$T_0 = \sum_{\substack{k=1 \\ 2\mid k}}^{2N-1} r_4(2N-k)r_8(k) = \sum_{k=1}^{N-1} r_4(2N-2k)r_8(2k)$$

$$= \sum_{k=1}^{N-1}(24\sigma(N-k) - 48\sigma((N-k)/2)(112\sigma_3(k) + 128\sigma_3(k/2)),$$

that is

$$T_0 = 2688U_1 - 5376U_2 + 3072U_3 - 6144U_4, \tag{19.24}$$

where

$$U_1 := \sum_{k=1}^{N-1} \sigma(N-k)\sigma_3(k) = S_{1,3}(N),$$

$$U_2 := \sum_{k=1}^{N-1} \sigma((N-k)/2)\sigma_3(k) = \sum_{\substack{k \in \mathbb{N} \\ k < N/2}} \sigma(k)\sigma_3(N-2k) = A_{1,3}(N),$$

$$U_3 := \sum_{k=1}^{N-1} \sigma(N-k)\sigma_3(k/2) = \sum_{\substack{k \in \mathbb{N} \\ k < N/2}} \sigma(N-2k)\sigma_3(k) = A_{3,1}(N),$$

$$U_4 := \sum_{k=1}^{N-1} \sigma((N-k)/2)\sigma_3(k/2) = \sum_{\substack{k \in \mathbb{N} \\ k < N/2}} \sigma((N/2)-k)\sigma_3(k) = S_{1,3}(N/2),$$

so that from (19.12), (19.14) and (19.15), we obtain

$$T_0 = 136\sigma_5(N) - 640\sigma_5(N/2) - 112\sigma_3(N) - 128\sigma_3(N/2)$$
$$- 24\sigma(N) + 48\sigma(N/2). \tag{19.25}$$

Now we turn to the evaluation of T_1. By (19.23) and Theorems 9.5 and 19.1, we obtain

$$T_1 = \sum_{\substack{k=1 \\ 2 \nmid k}}^{2N-1} r_4(2N-k)r_8(k)$$

$$= \sum_{k=1}^{N} r_4(2N-(2k-1))r_8(2k-1)$$

$$= 128 \sum_{k=1}^{N} \sigma(2N-(2k-1))\sigma_3(2k-1),$$

that is

$$T_1 = 128V_1 - 128V_2, \tag{19.26}$$

where

$$V_1 := \sum_{k=1}^{2N-1} \sigma(2N-k)\sigma_3(k), \quad V_2 := \sum_{k=1}^{N-1} \sigma(2N-2k)\sigma_3(2k). \tag{19.27}$$

First we evaluate V_1. By Theorem 3.1(ii) we have

$$\sigma_5(2N) - 33\sigma_5(N) + 32\sigma_5(N/2) = 0. \tag{19.28}$$

Then, by (19.27), (19.11), (19.12), (19.16), (19.18) and (19.28), we obtain

$$240V_1 = 240S_{1,3}(2N)$$
$$= 21\sigma_5(2N) + (10-60N)\sigma_3(2N) - \sigma(2N)$$
$$= 693\sigma_5(N) + (90-540N)\sigma_3(N) - 3\sigma(N)$$
$$- 672\sigma_5(N/2) - (80-480N)\sigma_3(N/2) + 2\sigma(N/2).$$

Next we evaluate V_2. By (19.27), (19.16), (19.18), (19.11) and (19.13), we obtain

$$V_2 = \sum_{k=1}^{N-1} (3\sigma(N-k) - 2\sigma((N-k)/2)(9\sigma_3(k) - 8\sigma_3(k/2))$$
$$= 27S_{1,3}(N) - 18A_{1,3}(N) - 24A_{3,1}(N) + 16S_{1,3}(N/2).$$

Then, by (19.12), (19.14) and (19.15), we deduce

$$240V_2 = 453\sigma_5(N) + (90 - 540N)\sigma_3(N) - 3\sigma(N)$$
$$- 432\sigma_5(N/2) + (-80 + 480N)\sigma_3(N/2) + 2\sigma(N/2).$$

Hence, by (19.26), we have

$$T_1 = 128\sigma_5(N) - 128\sigma_5(N/2). \tag{19.29}$$

Thus, by (19.22), (19.25) and (19.29), we obtain

$$r_{12}(2N) = 264\sigma_5(N) - 768\sigma_5(N/2).$$

Therefore for $n \equiv 0 \pmod 2$ we have by (19.28)

$$r_{12}(n) = 264\sigma_5(n/2) - 768\sigma_5(n/4) = 8\sigma_5(n) - 512\sigma_5(n/4),$$

which is Liouville's formula. $\square$

Glaisher [112, p. 480], [113] gave a formula for $r_{12}(n)$ valid for all $n \in \mathbb{N}$ in 1907. A simple proof of a formula equivalent to Glaisher's formula is given in Williams [272]. Define the integers $b(n)$ ($n \in \mathbb{N}$) by

$$\sum_{n=1}^{\infty} b(n)q^n := q \prod_{n=1}^{\infty} (1 - q^{2n})^{12}, \quad q \in \mathbb{C}, \ |q| < 1. \tag{19.30}$$

Then

$$r_{12}(n) = 8\sigma_5(n) - 512\sigma_5(n/4) + 16b(n), \quad n \in \mathbb{N}.$$

In the expansion of the right hand side of (19.30) in powers of q only odd powers of q occur so we have

$$b(n) = 0, \ \text{if } n \equiv 0 \pmod 2,$$

and we recover Theorem 19.2.

Theorem 19.1 gives the number of representations $(x_1, \ldots, x_8) \in \mathbb{Z}^8$ of a positive integer n by the octonary quadratic form $x_1^2 + \cdots + x_8^2$. We conclude this chapter by briefly discussing the determination of the number $N(n)$ of representations of n by the similar quadratic form $x_1^2 + x_2^2 + x_3^2 + x_4^2 + 2x_5^2 + 2x_6^2 + 2x_7^2 + 2x_8^2$. As this form can be written as $x_1^2 + x_2^2 + x_3^2 + x_4^2 +$

$2(x_5^2 + x_6^2 + x_7^2 + x_8^2)$, $N(n)$ is given as a convolution sum analogous to that for $r_8(n)$ in (19.3), namely,

$$N(n) = \sum_{k=0}^{[n/2]} r_4(k)r_4(n - 2k),$$

so that

$$N(n) = r_4(n) + r_4(n/2) + \sum_{\substack{k \in \mathbb{N} \\ k < n/2}} r_4(k)r_4(n - 2k).$$

Appealing to (19.4), we obtain

$$N(n) = 8\sigma(n) - 32\sigma(n/4) + 8\sigma(n/2) - 32\sigma(n/8)$$
$$+ 64W_1 - 256W_2 - 256W_3 + 1024W_4,$$

where

$$W_1 := \sum_{\substack{k \in \mathbb{N} \\ k < n/2}} \sigma(k)\sigma(n - 2k),$$

$$W_2 := \sum_{\substack{k \in \mathbb{N} \\ k < n/2}} \sigma(k/4)\sigma(n - 2k),$$

$$W_3 := \sum_{\substack{k \in \mathbb{N} \\ k < n/2}} \sigma(k)\sigma((n - 2k)/4),$$

$$W_4 := \sum_{\substack{k \in \mathbb{N} \\ k < n/2}} \sigma(k/4)\sigma((n - 2k)/4).$$

Clearly

$$W_1 = A_2(n),$$
$$W_3 = \sum_{\substack{\ell \in \mathbb{N} \\ \ell < n/4}} \sigma(n/2 - 2\ell)\sigma(\ell) = A_2(n/2),$$
$$W_4 = \sum_{\substack{\ell \in \mathbb{N} \\ \ell < n/8}} \sigma(\ell)\sigma(n/4 - 2\ell) = A_2(n/4).$$

Thus the values of W_1, W_3 and W_4 follow from Theorem 15.1, which gives the value of $A_2(n)$. However

$$W_2 = \sum_{\substack{\ell \in \mathbb{N} \\ \ell < n/8}} \sigma(\ell)\sigma(n - 8\ell) = A_8(n),$$

and we have not determined the value of the twisted convolution sum $A_8(n)$ in this book. In this case the values of $\sigma_3(n)$, $\sigma_3(n/2)$, $\sigma_3(n/4)$, $\sigma_3(n/8)$, $\sigma(n)$, $\sigma(n/2)$, $\sigma(n/4)$ and $\sigma(n/8)$ do not suffice to evaluate the sum. We require in addition the integers $k(n)$ $(n \in \mathbb{N})$ defined by

$$\sum_{n=1}^{\infty} k(n)q^n := q \prod_{n=1}^{\infty} (1 - q^{2n})^4 (1 - q^{4n})^4, \quad q \in \mathbb{C}, \ |q| < 1. \qquad (19.31)$$

Only odd powers of q occur in the expansion of the right hand side of (19.31) as a power series in q, so

$$k(n) = 0, \quad \text{if } n \equiv 0 \pmod 2.$$

Expanding the first few terms of the product as a power series in q, we find

$$k(1) = 1, \ k(3) = -4, \ k(5) = -2, \ k(7) = 24, \ k(9) = -11.$$

It was shown in Williams [270] that

$$A_8(n) = \frac{1}{192}\sigma_3(n) + \frac{1}{64}\sigma_3(n/2) + \frac{1}{16}\sigma_3(n/4) + \frac{1}{3}\sigma_3(n/8)$$
$$+ \left(\frac{1}{24} - \frac{1}{32}n \right) \sigma(n) + \left(\frac{1}{24} - \frac{1}{4}n \right) \sigma(n/8) - \frac{1}{64}k(n).$$

Using the values of W_1, W_2, W_3 and W_4, we obtain

$$N(n) = 4\sigma_3(n) - 4\sigma_3(n/2) - 16\sigma_3(n/4) + 256\sigma_3(n/8) + 4k(n).$$

This is the result we were seeking. In particular, for an even positive integer n, we see that the number of representations of n by the octonary quadratic form

$$x_1^2 + x_2^2 + x_3^2 + x_4^2 + 2x_5^2 + 2x_6^2 + 2x_7^2 + 2x_8^2$$

is given by

$$4\sigma_3(n) - 4\sigma_3(n/2) - 16\sigma_3(n/4) + 256\sigma_3(n/8).$$

Exercises 19

1. Let $n \in \mathbb{N}$. Prove that

$$(-1)^n \sum_{\substack{d \in \mathbb{N} \\ d \mid n}} (-1)^d d^3 = \sigma_3(n) - 2\sigma_3(n/2) + 16\sigma_3(n/4).$$

2. Let $n \in \mathbb{N}$. Use Theorem 3.1(ii) to prove that

$$\sigma_3(n/4) = \frac{9}{8}\sigma_3(n/2) - \frac{(1 + (-1)^n)}{16}\sigma_3(n).$$

3. Use Problems 1 and 2 and Theorem 19.1 to show that

$$r_8(n) = 16(-1)^{n-1}(\sigma_3(n) - 16\sigma_3(n/2)).$$

4. Deduce from Problem 3 that $r_8(n) > 0$ for all $n \in \mathbb{N}$.

5. Prove that $k(n)$ cannot be given linearly in terms of $\sigma_3(n)$ and $\sigma(n)$ for all positive odd integers n.

6. Let n be an even positive integer. Use Theorems 3.1(ii), 15.1 and 15.2 to prove that

$$A_8(n) = \frac{1}{192}\sigma_3(n) + \frac{1}{64}\sigma_3(n/2) + \frac{1}{16}\sigma_3(n/4) + \frac{1}{3}\sigma_3(n/8)$$
$$+ \left(\frac{1}{24} - \frac{1}{32}n\right)\sigma(n) + \left(\frac{1}{24} - \frac{1}{4}n\right)\sigma(n/8).$$

Notes on Chapter 19

The formula

$$r_8(n) = 16(-1)^n \sum_{\substack{d \in \mathbb{N} \\ d \mid n}} (-1)^d d^3, \quad n \in \mathbb{N},$$

first appeared implicitly in the work of Jacobi [146], [148, Vol. I, §§40–42, pp. 159–170] and explicitly for odd n in the work of Eisenstein [98], [101, Vol. 1, p. 501]. The standard arithmetic proof of Theorem 19.1 uses an elementary identity due to Liouville, see Nathanson [209, p. 402], to show that the function on the right hand side of Theorem 19.1 satisfies the same recurrence relation as $r_8(n)$ with the same initial conditions so that the two functions are the same, see for example Nathanson [209, pp. 441–445]. The proof in this chapter was given in Williams [265].

Liouville's formula for $r_{12}(n)$ for $n \equiv 2 \pmod 4$ appeared in [173] and for all $n \equiv 0 \pmod 2$ in [176]. The proof of Theorem 19.2 presented here is due to Huard and Williams [139]. An historical account of formulae for $r_{12}(n)$ is also given in [139].

The evaluation of $A_8(n)$ given in this chapter was derived in Williams [270, Theorem 1, p. 388] and the formula for $N(n)$ was obtained as a corollary [270, Theorem 2, p. 388]. The analogous formulae for $A_{16}(n)$ and

$$\text{card}\{(x_1, \ldots, x_8) \in \mathbb{Z}^8 \mid n = x_1^2 + x_2^2 + x_3^2 + x_4^2 + 4x_5^2 + 4x_6^2 + 4x_7^2 + 4x_8^2\}$$

were proved in Alaca, Alaca and Williams [18]. As a consequence of these formulae the number of representations of a positive integer $n \equiv 0 \pmod 4$ by the octonary quadratic form

$$x_1^2 + x_2^2 + x_3^2 + x_4^2 + 4x_5^2 + 4x_6^2 + 4x_7^2 + 4x_8^2$$

is

$$\sigma_3(n) + 3\sigma_3(n/2) - 68\sigma_3(n/4) + 48\sigma_3(n/8) + 256\sigma_3(n/16).$$

Formulae for the number of representations by other octonary quadratic forms can be found in Alaca, Alaca and Williams [12], [14], [15], [21], [25], Alaca and Williams [29], Petersson [222], and Williams [268], [269].

Liouville [182] gave a formula for $r_{10}(n)$ valid for all $n \in \mathbb{N}$ in 1865 as did Glaisher [112] in 1907. Alaca, Alaca and Williams [22] gave a proof of the formula

$$r_{10}(n) = \frac{4}{5} \sum_{\substack{d \in \mathbb{N} \\ d \mid n}} \left(\frac{-4}{d} \right) d^4 + \frac{64}{5} \sum_{\substack{d \in \mathbb{N} \\ d \mid n}} \left(\frac{-4}{n/d} \right) d^4 + \frac{32}{5} w(n),$$

where the integers $w(n)(n \in \mathbb{N})$ are defined by

$$\sum_{n=1}^{\infty} w(n)q^n := q \prod_{n=1}^{\infty} (1 - q^n)^4 (1 - q^{2n})^2 (1 - q^{4n})^4.$$

There is a vast literature on the representations of integers as sums of squares. The interested reader is referred to the books by Grosswald [119], Hardy [125, Chap. IX], Moreno and Wagstaff [207], and Nathanson [209, Chap. 14]. Some relevant papers are Alaca, Alaca and Williams [24] ($4s$ squares), Bell [41] ($2s$ squares), Bhargava and Adiga [49] (2 and 4 squares), Carlitz [57] (24 squares), Carlitz [58] (4 and 6 squares), Carlitz [59] (4 squares), Chan [65] (6 squares), Chan and Chua [62] (32 squares), Chan and Krattenthaler [64] ($4s^2$ and $4s(s + 1)$ squares), Cooper [75] ($2s$ squares), Cooper [76] (5, 7 and 9 squares), Cooper [77] (11 and 13 squares), Cooper and Lam [82] (2, 4, 6 and 8 squares), Eisenstein [99], [101, Vol. II, p. 505] (5 squares), Ewell [104] (16 squares), Glaisher [112] (2, 4, 6, 8, 10, 12, 14, 16 and 18 squares), Glaisher [114] (14, 16 squares), Glaisher [115] (18 squares), Hardy [122], [123], [124] (5, 7 squares), Hirschhorn [132] (4 squares), Hirschhorn [133] (2 squares), Hirschhorn [134] (4 squares), Huard and Williams [138] (16 squares), Huard and Williams [139] (12 squares), Krätzel [157] ($4s$ squares), Krätzel [158] ($4s + 2$ squares), Liouville [182] (10 squares), Liouville [173] (12 squares), Liu [183] (12, 16, 20 and 24 squares), Long and Yang [190] ($4s^2$ and $4s(s + 1)$ squares), McAfee and Williams [198] (6 squares), Milne [203], [204] ($4s^2$ and $4s(s + 1)$ squares), Mordell [205] ($2s$ squares), Mordell [206] ($2s + 1$ squares), Ono [215] ($4s^2$ and $4s(s + 1)$ squares), Petr [224] (10, 12 squares), van der

Pol [227] (8, 16 and 24 squares), Ramanujan [230, p. 184], [232, p. 162] ($2s$ squares), Rankin [234] (20 squares), Rosengren [237] (18 squares, $2s^2$ squares), Smith [245, Vol. II, pp. 623–680] (5 squares), Stanley [248] (7 squares) and Williams [272] (12 squares).

Other relevant papers are Bulygin [54], [55], [56], Carlitz [60], Cooper and Hirschhorn [81], Estermann [103], Lomadze [185], Mathews [196], Rankin [233], [235] and Uspensky [252], [253].

A connection between squares and triangular numbers has been given by Cooper and Hirschhorn [80].

20

Concluding Remarks

In this book we have used Liouville's ideas to give elementary proofs of many arithmetic formulae. Using more advanced mathematics, such as the theory of modular forms, we can prove these and other formulae. In this chapter we sketch how the theory of modular forms can be used to prove arithmetic identities. To keep the discussion concrete, we focus on using modular forms to prove the following three identities, which are valid for all $n \in \mathbb{N}$, namely,

$$\sum_{m=1}^{n-1} \sigma_3(m)\sigma_3(n-m) = \frac{1}{120}\sigma_7(n) - \frac{1}{120}\sigma_3(n), \tag{20.1}$$

$$\sum_{m=1}^{n-1} \sigma_5(m)\sigma_5(n-m) = \frac{65}{174132}\sigma_{11}(n) + \frac{1}{252}\sigma_5(n) - \frac{3}{691}\tau(n), \tag{20.2}$$

$$r_4(n) = 8\sigma(n) - 32\sigma(n/4). \tag{20.3}$$

The first of these is Theorem 13.5. The second identity cannot be proved by Liouville's method as it contains the non-elementary function $\tau(n)$ (the Ramanujan tau function, see (13.15)). The third identity is Jacobi's formula for $r_4(n)$ (Theorem 11.1).

The set of 2×2 matrices with integer entries and determinant 1 is a group with respect to matrix multiplication. This group is denoted by $SL_2(\mathbb{Z})$ ("SL" stands for Special Linear Group) and is called the modular group, that is

$$SL_2(\mathbb{Z}) = \left\{ \begin{bmatrix} a & b \\ c & d \end{bmatrix} \mid a, b, c, d \in \mathbb{Z}, \quad ad - bc = 1 \right\}.$$

Let $\mathbb{H}$ denote the upper half of the complex plane, that is

$$\mathbb{H} := \{z \in \mathbb{C} \mid \mathrm{Im}(z) > 0\}.$$

The formula

$$\mathrm{Im}\left(\frac{az+b}{cz+d}\right) = \frac{\mathrm{Im}(z)}{|cz+d|^2},$$

shows that if $z \in \mathbb{H}$ then $\dfrac{az+b}{cz+d} \in \mathbb{H}$. Let $k \in \mathbb{N}$. If $f : \mathbb{H} \to \mathbb{C}$ is a meromorphic function, which satisfies the transformation formula

$$f\left(\frac{az+b}{cz+d}\right) = (cz+d)^k f(z)$$

for $\begin{bmatrix} a & b \\ c & d \end{bmatrix} \in SL_2(\mathbb{Z})$ and $z \in \mathbb{H}$, then f is said to be weakly modular of weight k for $SL_2(\mathbb{Z})$. If f is weakly modular of weight k for $SL_2(\mathbb{Z})$ then it satisfies

$$f(z+1) = f(z), \quad z \in \mathbb{H},$$

as $\begin{bmatrix} 1 & 1 \\ 0 & 1 \end{bmatrix} \in SL_2(\mathbb{Z})$. Thus f has a Fourier expansion

$$f(q) = \sum_{n \in \mathbb{Z}} a_n q^n, \quad q = e^{2\pi i z}, \quad z \in \mathbb{H},$$

in a suitable neighborhood of the origin. As f is meromorphic only finitely many of the a_n with $n < 0$ are nonzero. In particular if $a_n = 0$ for $n < 0$ then we say that f is holomorphic at ∞. A function $f : \mathbb{H} \to \mathbb{C}$, which is

    (i)   weakly modular of weight k for $SL_2(\mathbb{Z})$,
    (ii) holomorphic on $\mathbb{H}$,
    (iii) holomorphic at ∞,

is called a modular form of weight k for $SL_2(\mathbb{Z})$. The set of all modular forms of weight k for $SL_2(\mathbb{Z})$ is denoted by $M_k(SL_2(\mathbb{Z}))$. It is known that $M_k(SL_2(\mathbb{Z}))$ is a finite-dimensional vector space over $\mathbb{C}$. When k is an even positive integer the dimension of $M_k(SL_2(\mathbb{Z}))$ is given by

$$\dim(M_k(SL_2(\mathbb{Z}))) = \left[\frac{k}{12}\right] + \alpha_k,$$

where

$$\alpha_k = \begin{cases} 1, & \text{if } k \not\equiv 2 \pmod{12}, \\ 0, & \text{if } k \equiv 2 \pmod{12}. \end{cases}$$

Let k be an even integer with $k \geq 2$. The Eisenstein series $E_k(q)$ is defined by

$$E_k(q) := 1 - \frac{2k}{B_k} \sum_{n=1}^{\infty} \sigma_{k-1}(n) q^n, \quad q = e^{2\pi i z}, \quad z \in \mathbb{H},$$

where the Bernoulli numbers B_m ($m \in \mathbb{N}_0$) were defined in (3.17) (or (3.18)). As $B_2 = 1/6$, $B_4 = -1/30$, $B_6 = 1/42$, $B_8 = -1/30$, $B_{10} = 5/66$ and $B_{12} = -691/2730$, we have

$$E_2(q) := 1 - 24 \sum_{n=1}^{\infty} \sigma(n)q^n,$$

$$E_4(q) := 1 + 240 \sum_{n=1}^{\infty} \sigma_3(n)q^n,$$

$$E_6(q) := 1 - 504 \sum_{n=1}^{\infty} \sigma_5(n)q^n,$$

$$E_8(q) := 1 + 480 \sum_{n=1}^{\infty} \sigma_7(n)q^n,$$

$$E_{10}(q) := 1 - 264 \sum_{n=1}^{\infty} \sigma_9(n)q^n,$$

$$E_{12}(q) := 1 + \frac{65520}{691} \sum_{n=1}^{\infty} \sigma_{11}(n)q^n.$$

The Eisenstein series $E_2(q)$ is not a modular form. However, it is known that for $k \geq 4$ the Eisenstein series $E_k(q)$ is a modular form of weight k for the modular group, that is

$$E_k \in M_k(SL_2(\mathbb{Z})), \quad k \in \mathbb{N}, \quad k \text{ (even)} \geq 4.$$

If f_1 is a modular form of weight k_1 for $SL_2(\mathbb{Z})$ and f_2 is a modular form of weight k_2 for $SL_2(\mathbb{Z})$ then $f_1 f_2$ is a modular form of weight $k_1 + k_2$ for $SL_2(\mathbb{Z})$. Thus we have

$$E_\ell^{k/\ell} \in M_k(SL_2(\mathbb{Z})), \quad k \text{ (even)} \geq 4, \quad \ell \text{ (even)} \geq 4, \quad \ell \mid k.$$

We now prove (20.1). We have

$$E_4^2, E_8 \in M_8(SL_2(\mathbb{Z})).$$

By (20.4) we have

$$\dim(M_8(SL_2(\mathbb{Z}))) = \left[\frac{8}{12}\right] + \alpha_8 = 0 + 1 = 1.$$

Hence there exists $c \in \mathbb{C}$ such that

$$E_8 = cE_4^2.$$

Thus

$$1 + 480 \sum_{n=1}^{\infty} \sigma_7(n)q^n = c \left(1 + 240 \sum_{n=1}^{\infty} \sigma_3(n)q^n \right)^2.$$

Equating constant terms, we deduce that $c = 1$. Then, equating coefficients of q^n ($n \in \mathbb{N}$), we deduce

$$480\sigma_7(n) = 480\sigma_3(n) + 240^2 \sum_{m=1}^{n-1} \sigma_3(m)\sigma_3(n-m)$$

from which (20.1) follows.

Next we prove (20.2). We have

$$E_4^3, \ E_6^2, \ E_{12} \in M_{12}(SL_2(\mathbb{Z})).$$

Thus

$$\Delta := \frac{1}{1728}(E_4^3 - E_6^2) \in M_{12}(SL_2(\mathbb{Z})).$$

Ramanujan has shown that

$$\Delta(q) = q \prod_{n=1}^{\infty}(1 - q^n)^{24}$$

so that by (13.15) we have

$$\Delta(q) = \sum_{n=1}^{\infty} \tau(n)q^n,$$

where $\tau(n)$ is the Ramanujan tau function. We show that E_{12} and Δ are linearly independent over $\mathbb{C}$. For suppose

$$aE_{12} + b\Delta = 0$$

for some $a, b \in \mathbb{C}$. Taking $q = 0$ we obtain

$$aE_{12}(0) + b\Delta(0) = 0$$

so that (as $E_{12}(0) = 1$, $\Delta(0) = 0$) we have $a = 0$ and thus $b = 0$. Now

$$\dim M_{12}(SL_2(\mathbb{Z})) = \left[\frac{12}{12}\right] + \alpha_{12} = 1 + 1 = 2$$

so that $\{E_{12}, \Delta\}$ is a basis for the vector space $M_{12}(SL_2(\mathbb{Z}))$. Hence there exist $c, d \in \mathbb{C}$ such that

$$E_6^2 = cE_{12} + d\Delta.$$

Hence

$$\left(1 - 504\sum_{n=1}^{\infty}\sigma_5(n)q^n\right)^2 = c\left(1 + \frac{65520}{691}\sum_{n=1}^{\infty}\sigma_{11}(n)q^n\right) + d\sum_{n=1}^{\infty}\tau(n)q^n.$$

Taking $q = 0$ we obtain $c = 1$. Equating coefficients of q, we deduce

$$-1008 = \frac{65520}{691} + d$$

so that

$$d = \frac{-762048}{691}.$$

Equating coefficients of q^n ($n \in \mathbb{N}$), we have

$$-1008\sigma_5(n) + 504^2\sum_{m=1}^{n-1}\sigma_5(m)\sigma_5(n-m) = \frac{65520}{691}\sigma_{11}(n) - \frac{762048}{691}\tau(n),$$

from which (20.2) follows.

Finally we prove (20.3). We require the subgroup

$$\Gamma_0(4) := \left\{\begin{bmatrix} a & b \\ c & d \end{bmatrix} \in SL_2(\mathbb{Z}) \mid c \equiv 0 \pmod 4\right\}$$

of $SL_2(\mathbb{Z})$. We note that $\begin{bmatrix} 1 & 1 \\ 0 & 1 \end{bmatrix} \in \Gamma_0(4)$ and $[SL_2(\mathbb{Z}) : \Gamma_0(4)] = 6$. Ramanujan's theta function $\varphi(q)$ is defined by

$$\varphi(q) := \sum_{n\in\mathbb{Z}} q^{n^2}.$$

Clearly

$$\varphi^4(q) = \sum_{n=0}^{\infty} r_4(n)q^n.$$

The function $\varphi^4(q)$ is not weakly modular of weight 2 for $SL_2(\mathbb{Z})$. However it satisfies

$$\varphi^4\left(\frac{az+b}{cz+d}\right) = (cz+d)^2\varphi^4(z)$$

for $\begin{bmatrix} a & b \\ c & d \end{bmatrix} \in \Gamma_0(4)$ and $z \in \mathbb{H}$ so we say that $\varphi^4(q)$ is weakly modular of weight 2 for $\Gamma_0(4)$. Let $k \in \mathbb{N}$. A function $f : \mathbb{H} \to \mathbb{C}$, which is

(i)   weakly modular of weight k for $\Gamma_0(4)$,
(ii) holomorphic on $\mathbb{H}$,
(iii) and such that $(cz+d)^{-k} f \left(\frac{az+b}{cz+d} \right)$ is holomorphic at ∞

$\qquad$ for all $\begin{bmatrix} a & b \\ c & d \end{bmatrix} \in SL_2(\mathbb{Z})$,

is called a modular form of weight k for $\Gamma_0(4)$. The set of all modular forms of weight k for $\Gamma_0(4)$ is denoted by $M_k(\Gamma_0(4))$. It is known that $M_k(\Gamma_0(4))$ is a finite-dimensional vector space over $\mathbb{C}$. In particular we have $\varphi^4(q) \in M_2(\Gamma_0(4))$ and $\dim M_2(\Gamma_0(4)) = 2$. The Eisenstein series

$$E_2(q) = 1 - 24 \sum_{n=1}^{\infty} \sigma(n)q^n$$

is such that

$$E_2(q) - 2E_2(q^2) \in M_2(\Gamma_0(4))$$

and

$$E_2(q^2) - 2E_2(q^4) \in M_2(\Gamma_0(4)).$$

Since

$$E_2(q) - 2E_2(q^2) = -1 - 24q - 24q^2 - \cdots$$

and

$$E_2(q^2) - 2E_2(q^4) = -1 - 24q^2 - 24q^4 - \cdots$$

it is clear that $E_2(q^2) - 2E_2(q^4)$ and $E_2(q^2) - 2E_2(q^4)$ are linearly independent over $\mathbb{C}$ and thus form a basis for $M_2(\Gamma_0(4))$. As $\varphi(q)^4 \in M_2(\Gamma_0(4))$ there exist $e, f \in \mathbb{C}$ such that

$$\varphi(q)^4 = e(E_2(q) - 2E_2(q^2)) + f(E_2(q^2) - 2E_2(q^4)).$$

Thus

$$1 + 8q + 24q^2 + \cdots = e(-1 - 24q - 24q^2 - \cdots)$$
$$+ f(-1 - 24q^2 - 24q^4 - \cdots).$$

Equating constant terms and the coefficients of q, we obtain

$$-e - f = 1, \quad -24e = 8,$$

so that

$$e = -1/3, \quad f = -2/3.$$

Therefore

$$\varphi^4(q) = -\frac{1}{3}(E_2(q) - 2E_2(q^2)) - \frac{2}{3}(E_2(q^2) - 2E_2(q^4)).$$

Hence

$$\sum_{n=0}^{\infty} r_4(n)q^n = -\frac{1}{3}E_2(q) + \frac{4}{3}E_2(q^4)$$

$$= -\frac{1}{3}\left(1 - 24\sum_{n=1}^{\infty}\sigma(n)q^n\right) + \frac{4}{3}\left(1 - 24\sum_{n=1}^{\infty}\sigma(n)q^{4n}\right)$$

$$= 1 + 8\sum_{n=1}^{\infty}\sigma(n)q^n - 32\sum_{n=1}^{\infty}\sigma(n)q^{4n}$$

$$= 1 + 8\sum_{n=1}^{\infty}\sigma(n)q^n - 32\sum_{n=1}^{\infty}\sigma(n/4)q^n.$$

Equating coefficients of q^n ($n \in \mathbb{N}$), we deduce

$$r_4(n) = 8\sigma(n) - 32\sigma(n/4),$$

which is Jacobi's formula for $r_4(n)$.

The reader will find proofs or references to proofs of the facts used in these remarks in Kilford's book on modular forms [153].

References

[1] C. Adiga, *On the representations of an integer as a sum of two or four triangular numbers*, Nihonkai Math. J. **3** (1992), 125–131.

[2] C. Adiga, S. Cooper and J. H. Han, *A general relation between sums of squares and sums of triangular numbers*, Int. J. Number Theory **1** (2005), 175–182.

[3] A. Alaca, *Representations by quaternary quadratic forms whose coefficients are 1, 3 and 9*, Acta Arith. **136** (2009), 151–166.

[4] A. Alaca, S. Alaca, M. F. Lemire and K. S. Williams, *Jacobi's identity and representations of integers by certain quaternary quadratic forms*, Int. J. Modern Math. **2** (2007), 143–176.

[5] A. Alaca, S. Alaca, M. F. Lemire and K. S. Williams, *Nineteen quaternary quadratic forms*, Acta Arith. **130** (2007), 277–310.

[6] A. Alaca, S. Alaca, M. F. Lemire and K. S. Williams, *Theta function identities and representations by certain quaternary quadratic forms*, Int. J. Number Theory **4** (2008), 218–239.

[7] A. Alaca, S. Alaca, M. F. Lemire and K. S. Williams, *Theta function identities and representations by certain quaternary quadratic forms II*, Int. Math. Forum **3** (2008), 539–579.

[8] A. Alaca, S. Alaca, M. F. Lemire and K. S. Williams, *The number of representations of a positive integer by certain quaternary quadratic forms*, Int. J. Number Theory **5** (2009), 13–40.

[9] A. Alaca, S. Alaca, E. McAfee and K. S. Williams, *Lambert series and Liouville's identities*, Dissertationes Math. **445** (2007), 1–72.

[10] A. Alaca, S. Alaca, F. Uygul and K. S. Williams, *Evaluation of the sums*
$$\sum_{\substack{m=1 \\ m \equiv a \,(\mathrm{mod}\, 8)}}^{n-1} \sigma(m)\sigma(n-m),$$ *preprint, 2009.*

[11] A. Alaca, S. Alaca, F. Uygul and K. S. Williams, *Representations by sextenary quadratic forms whose coefficients are 1, 2 and 4*, Acta Arith. **141** (2010), 289–309.

[12] A. Alaca, S. Alaca and K. S. Williams, *Evaluation of the convolution sums* $\sum_{\ell+12m=n} \sigma(\ell)\sigma(m)$ *and* $\sum_{3\ell+4m=n} \sigma(\ell)\sigma(m)$, Adv. Theor. Appl. Math. **1** (2006), 27–48.

[13] A. Alaca, S. Alaca and K. S. Williams, *On the two-dimensional theta functions of the Borweins*, Acta Arith. **124** (2006), 177–195.

[14] A. Alaca, S. Alaca and K. S. Williams, *Evaluation of the convolution sums* $\sum_{\ell+18m=n} \sigma(\ell)\sigma(m)$ *and* $\sum_{2\ell+9m=n} \sigma(\ell)\sigma(m)$, Int. Math. Forum **2** (2007), 45–68.

[15] A. Alaca, S. Alaca and K. S. Williams, *The convolution sums* $\sum_{\ell+24m=n} \sigma(\ell)\sigma(m)$ *and* $\sum_{3\ell+8m=n} \sigma(\ell)\sigma(m)$, Math. J. Okayama Univ. **49** (2007), 93–111.

[16] A. Alaca, S. Alaca and K. S. Williams, *The simplest proof of Jacobi's six squares theorem*, Far East J. Math. Sci. **27** (2007), 187–192.

[17] A. Alaca, S. Alaca and K. S. Williams, *On the quaternary forms* $x^2 + y^2 + z^2 + 5t^2$, $x^2 + y^2 + 5z^2 + 5t^2$ *and* $x^2 + 5y^2 + 5z^2 + 5t^2$, JP J. Algebra Number Theory Appl. **9** (2007), 37–53.

[18] A. Alaca, S. Alaca and K. S. Williams, *The convolution sum* $\sum_{m<n/16} \sigma(m)\sigma(n - 16m)$, Canad. Math. Bull. **51** (2008), 3–14.

[19] A. Alaca, S. Alaca and K. S. Williams, *Berndt's curious formula*, Int. J. Number Theory **4** (2008), 677–689.

[20] A. Alaca, S. Alaca and K. S. Williams, *Liouville's sextenary quadratic forms* $x^2 + y^2 + z^2 + t^2 + 2u^2 + 2v^2$, $x^2 + y^2 + 2z^2 + 2t^2 + 2u^2 + 2v^2$ *and* $x^2 + 2y^2 + 2z^2 + t^2 + 2u^2 + 4v^2$, Far East J. Math. Sci. **30** (2008), 547–556.

[21] A. Alaca, S. Alaca and K. S. Williams, *Seven octonary quadratic forms*, Acta Arith. **135** (2008), 339–350.

[22] A. Alaca, S. Alaca and K. S. Williams, *Some identities involving theta functions*, J. Number Theory **129** (2009), 1404–1431.

[23] A. Alaca, S. Alaca and K. S. Williams, *Some new theta function identities with applications to sextenary quadratic forms*, J. Combinatorics and Number Theory **1**(1) (2009), 89–98.

[24] A. Alaca, S. Alaca and K. S. Williams, *Sums of 4k squares: a polynomial approach*, J. Combinatorics and Number Theory **1**(2) (2009), 133–152.

[25] A. Alaca, S. Alaca and K. S. Williams, *Fourteen octonary quadratic forms*, Int. J. Number Theory **6** (2010), 37–50.

[26] A. Alaca, S. Alaca and K. S. Williams, *Sextenary quadratic forms and an identity of Klein and Fricke*, Int. J. Number Theory **6** (2010), 169–183.

[27] A. Alaca, S. Alaca and K. S. Williams, *Evaluation of the sums*
$$\sum_{\substack{m=1 \\ m \equiv a \,(\mathrm{mod}\,4)}}^{n-1} \sigma(m)\sigma(n - m),$$ Czech. Math. J. **59** (2009), 847–859.

[28] S. Alaca and K. S. Williams, *Introductory Algebraic Number Theory*, Cambridge University Press, 2004.

[29] S. Alaca and K. S. Williams, *Evaluation of the convolution sums* $\sum_{\ell+6m=n} \sigma(\ell)\sigma(m)$ *and* $\sum_{2\ell+3m=n} \sigma(\ell)\sigma(m)$, J. Number Theory **124** (2007), 491–510.

[30] G. E. Andrews and B. C. Berndt, *Ramanujan's Lost Notebook*, Springer-Verlag, New York, Part I, 2005; Part II, 2009.

[31] G. E. Andrews, S. B. Ekhad and D. Zeilberger, *A short proof of Jacobi's formula for the number of representations of an integer as a sum of four squares*, Amer. Math. Monthly **100** (1993), 274–276.

[32] N. C. Ankeny, *Sums of three squares*, Proc. Amer. Math. Soc. **8** (1957), 316–319.

[33] L. Aubry, Sphinx-Oedipe, numéro spécial, March 1914, 1–14; errata, 39.

[34] P. Bachmann, *Niedere Zahlentheorie*, Chelsea Publ. Co., New York, 1968.

[35] R. P. Bambah and S. Chowla, *The residue of Ramanujan's tau function $\tau(n)$ to the modulus 2^8*, J. London Math. Soc. **22** (1947), 140–147.

[36] C. W. Barnes, *The representation of primes of the form $4n + 1$ as the sum of two squares*, Enseign. Math. **18** (1972), 289–299.

[37] C. W. Barnes, *A construction of Gauss*, Enseign. Math. **20** (1974), 1–7.

[38] P. Barrucand, S. Cooper and M. Hirschhorn, *Relation between squares and triangles*, Discrete Math. **248** (2002), 245–247.

[39] S. I. Baskakov, *A method for finding arithmetic identities and its application to the theory of arithmetic functions*, Mat. Sb. **10** (1882–83), 313–380. (in Russian)

[40] P. T. Bateman, *On the representations of a number as the sum of three squares*, Trans. Amer. Math. Soc. **71** (1951), 70–101.

[41] E. T. Bell, *On the number of representations of $2n$ as a sum of $2r$ squares*, Bull. Amer. Math. Soc. **26** (1919), 19–25.

[42] E. Benz, *Über die Anzahl Darstellungen einer Zahl n durch gewisse quaternäre quadratische Formen: Beweise, welche auf Identitäten aus dem Gebiete der Thetafunktionen basieren*. Dissertation, Zürich 1964, in "Studien zur Theorie der quadratischen Formen", editors B. L. van der Waerden und H. Gross, Stuttgart, 1968, pp. 165–198.

[43] A. Berkovich and W. J. Jagy, *Ternary quadratic forms, modular equations and certain positivity conjectures*, preprint, 2009.

[44] B. C. Berndt, *Ramanujan's Notebooks*, Springer-Verlag, New York, Part I, 1985; Part II, 1989; Part III, 1991; Part IV, 1994; Part V, 1998.

[45] B. C. Berndt, *Number Theory in the Spirit of Ramanujan*, Amer. Math. Soc., Providence, Rhode Island, USA, 2006.

[46] B. C. Berndt and R. J. Evans, *Chapter 15 of Ramanujan's second notebook: Part 2, Modular forms*, Acta Arith. **47** (1986), 123–142.

[47] B. C. Berndt, R. J. Evans and K. S. Williams, *Gauss and Jacobi sums*, Wiley, New York, 1998.

[48] M. Besge, *Extrait d'une lettre de M. Besge à M. Liouville*, J. Math. Pures Appl. **7** (1862), 256.

[49] S. Bhargava and C. Adiga, *Simple proofs of Jacobi's two and four square theorems*, Int. J. Math. Educ. Sci. Technol. (U.K.) **1–3** (1988), 779–782.

[50] J. M. Borwein and P. B. Borwein, *A cubic counterpart of Jacobi's identity and the AGM*, Trans. Amer. Math. Soc. **323** (1991), 691–701.

[51] J. M. Borwein, P. B. Borwein and F. G. Garvan, *Some cubic modular identities of Ramanujan*, Trans. Amer. Math. Soc. **343** (1994), 35–47.

[52] J. Brillhart, *Note on representing a prime as a sum of two squares*, Math. Comp. **26** (1972), 1011–1012.

[53] N. V. Bugaev, *Some applications of the theory of elliptic functions to the theory of discontinuous functions*, Mat. Sb. **12** (1885), 1–21. (in Russian)

[54] V. Bulygin, *Sur une application des fonctions elliptiques au problème de représentation des nombres entiers par une somme de carrés*, Bull. Acad. Imp. Sci. St. Petersburg (Sér. VI) **8** (1914), 389–404.

[55] V. Bulygin, *Sur la représentation d'un nombre entier par une somme de carrés*, Comptes Rendus Paris **158** (1914), 328–330.

[56] V. Bulygin, *Sur la représentation d'un nombre entier par une somme de carrés*, Comptes Rendus Paris **161** (1915), 28–30.

[57] L. Carlitz, *On the representation of an integer as a sum of twenty-four squares*, Indag. Math. **17** (1955), 504–506.

[58] L. Carlitz, *Note on sums of four and six squares*, Proc. Amer. Math. Soc. **8** (1957), 120–124.

[59] L. Carlitz, *A short proof of Jacobi's four square theorem*, Proc. Amer. Math. Soc. **17** (1966), 768–769.

[60] L. Carlitz, *Bulygin's method for sums of squares*, J. Number Theory **5** (1973), 405–412.

[61] A. Cauchy, *Sur les formes quadratiques de certaines puissances des nombres premiers ou du quadruple de ces puissances*, Oeuvres complètes 1er, tome 3, pp. 390–437.

[62] H. H. Chan and K. S. Chua, *Representations of integers as sums of 32 squares*, Ramanujan J. **7** (2003), 79–89.

[63] H. H. Chan and S. Cooper, *Powers of theta functions*, Pacific J. Math. (to appear).

[64] H. H. Chan and C. S. Krattenthaler, *Recent progress in the study of representations of integers as sums of squares*, Bull. London Math. Soc. **37** (2005), 818–826.

[65] S. H. Chan, *An elementary proof of Jacobi's six squares theorem*, Amer. Math. Monthly **111** (2004), 806–811.

[66] S. H. Chan, *On cranks of partitions, generalized Lambert series, and basic hypergeometric series*, Ph. D. thesis, University of Illinois at Urbana-Champaign, 2005, 101pp.

[67] R. Chapman, *Representations of integers by the form $x^2 + xy + y^2 + z^2 + zt + t^2$*, Int. J. Number Theory **4** (2008), 709–714.

[68] N. Cheng, *Convolution sums involving divisor functions*, M. Sc. thesis, Carleton University, Ottawa, Ontario, Canada, 2003.

[69] N. Cheng and K. S. Williams, *Convolution sums involving the divisor function*, Proc. Edinburgh Math. Soc. **47** (2004), 561–572.

[70] N. Cheng and K. S. Williams, *Evaluation of some convolution sums involving the sum of divisors functions*, Yokohama Math. J. **52** (2005), 39–57.

[71] S. Chowla, *Note on a certain arithmetical sum*, Proc. Nat. Inst. Sci. India **13** (1947), 233.

[72] S. Chowla, *The Collected Papers of Sarvadaman Chowla*, Vol. I (1925–1935), Vol. II (1936–1961), Vol. III (1962–1986), edited by J. G. Huard and K. S. Williams, Centre de Recherches Mathématiques, Montréal, Canada, 1999.

[73] S. Chowla, B. Dwork and R. J. Evans, *On the mod p^2 determination of $\binom{(p-1)/2}{(p-1)/4}$*, J. Number Theory **24** (1986), 188–196.

[74] G. Chrystal, *Joseph Liouville*, Proc. Royal Soc. Edinburgh **14** (1888), 83–91.

[75] S. Cooper, *On sums of an even number of squares, and an even number of triangular numbers: an elementary approach based on Ramanujan's $_1\psi_1$ summation formula*, Contemporary Math. **291** (2001), 115–137.

[76] S. Cooper, *Sums of five, seven, and nine squares*, Ramanujan J. **6** (2002), 469–490.

[77] S. Cooper, *On the number of representations of certain integers as sums of eleven or thirteen squares*, J. Number Theory **103** (2003), 135–162.

[78] S. Cooper, *On the number of representations of integers by certain quadratic forms*, Bull. Austral. Math. Soc. **78** (2008), 129–140.

[79] S. Cooper, *On the number of representations of integers by certain quadratic forms, II*, J. Combinatorics and Number Theory **1**(2) (2009), 153–182.

[80] S. Cooper and M. Hirschhorn, *A combinatorial proof of a result from number theory*, Integers: Electronic J. Combinatorial Number Theory **4** (2004), 1–5.

[81] S. Cooper and M. Hirschhorn, *On the number of primitive representations of integers as sums of squares*, Ramanujan J. **13** (2007), 9–27.

[82] S. Cooper and H. Y. Lam, *Sums of two, four, six and eight squares and triangular numbers: An elementary approach*, Indian J. Math. **44** (2002), 21–40.

[83] J. B. Cosgrave and K. Dilcher, *Mod p^3 analogues of theorems of Gauss and Jacobi on binomial coefficients*, preprint, 2009.

[84] H. Davenport, *The geometry of numbers*, Math. Gazette **31** (1947), 206–210.

[85] H. Davenport, *The Higher Arithmetic*, sixth edition, Cambridge University Press, 1992.

[86] A. Deltour, *Continuants: Applications à la théorie des nombres: troisème partie*, Nouv. Ann. Math. **11** (1911), 116–129.

[87] P. Demuth, *Die Zahl der Darstellungen einer natürlichen Zahl durch spezielle quaternäre quadratische Formen aufgrund der Siegelschen Massformel*, in "Studien zur Theorie der quadratischen Formen", editors B. L. van der Waerden und H. Gross, Stuttgart, 1968, pp. 224–254.

[88] J. I. Deutsch, *A quaternionic proof of the representation formula of a quaternary quadratic formula*, J. Number Theory **113** (2005), 149–174.

[89] L. E. Dickson, *Quadratic forms which represent all integers*, Proc. Nat. Acad. Sci. (USA) **12** (1926), 756–757.

[90] L. E. Dickson, *Integers represented by positive ternary quadratic forms*, Bull. Amer. Math. Soc. **33** (1927), 63–70.

[91] L. E. Dickson, *Quaternary quadratic forms representing all integers*, Amer. J. Math. **49** (1927), 39–56.

[92] L. E. Dickson, *Modern Elementary Theory of Numbers*, University of Chicago Press, Chicago, 1939.

[93] L. E. Dickson, *History of the Theory of Numbers*, Vols. I–III, Chelsea Publ. Co., New York, 1952.

[94] L. E. Dickson, *Introduction to the Theory of Numbers*, Dover Publications Inc., New York, 1957.

[95] L. E. Dickson, *The Collected Mathematical Papers of Leonard Eugene Dickson*, Chelsea Publ. Co., New York, Vols. I–V, 1975; Vol. VI, 1983.

[96] J. D. Dixon, *Another proof of Lagrange's theorem*, Amer. Math. Monthly **71** (1964), 286–288.

[97] C.-E. Dumont, *Histoire de la ville et des seigneurs de Commercy, Vol. 3*, Bar-le-Duc, France, 1843.

[98] G. Eisenstein, *Neue Theoreme der höheren Arithmetik*, J. Reine Angew. Math. **35** (1847), 117–136.

[99] G. Eisenstein, *Note sur la représentation d'un nombre par la somme de cinq carrés*, J. Reine Angew. Math. **35** (1847), 368.

[100] G. Eisenstein, *Zur Theorie der quadratischen Zerfällung der Primzahlen* 8n + 3, 7n + 2 *and* 7n + 4, J. Reine Angew. Math. **37** (1848), 97–126.

[101] G. Eisenstein, *Mathematische Werke*, Vols. I, II, Chelsea Publishing Co., New York, 1989.

[102] N. D. Elkies, *Lattices, linear codes, and invariants, Part I*, Notices Amer. Math. Soc. **47** (2000), 1238–1245.

[103] T. Estermann, *On the representations of a number as a sum of squares*, Acta Arith. **2** (1936), 47–79.

[104] J. A. Ewell, *On sums of sixteen squares*, Rocky Mountain J. Math. **17** (1987), 295–299.

[105] W. L. Ferrar, *Higher Algebra*, Oxford University Press, 1956.

[106] N. J. Fine, *Basic Hypergeometric Series and Applications*, American Mathematical Society, Providence, RI, 1988.

[107] C. F. Gauss, *Werke*, Vol. I (1863), Vol. II (1863), Vol. III (1866), Vol. IV (1873), Vol. V (1867), Vol. VI (1874), Vol. VII (1906), Vol. VIII (1900), Vol. IX (1903), Vol. X (first part 1917, second part 1922–1933), Vol. XI (first part 1927, second part 1924–1929), Vol. XII (1929), Gesellschaft der Wissenschaften zu Göttingen.

[108] O. Giraud, *Periodic orbits and semiclassical form factor in barrier billiards*, Comm. Math. Physics **260** (2005), 183–201.

[109] J. W. L. Glaisher, *On the square of the series in which the coefficients are the sums of the divisors of the exponents*, Mess. Math. **14** (1884), 156–163.

[110] J. W. L. Glaisher, *On certain sums of products of quantities depending upon the divisors of a number*, Mess. Math. **15** (1885), 1–20.

[111] J. W. L. Glaisher, *Expressions for the first five powers of the series in which the coefficients are the sums of the divisors of the exponents*, Mess. Math. **15** (1885), 33–36.

[112] J. W. L. Glaisher, *On the numbers of representations of a number as a sum of* 2r *squares, where* 2r *does not exceed eighteen*, Proc. London Math. Soc. **5** (1907), 479–490.

[113] J. W. L. Glaisher, *On the representations of a number as the sum of two, four, six, eight, ten and twelve squares*, Quart. J. Pure and Appl. Math. **38** (1907), 1–62.

[114] J. W. L. Glaisher, *On the representation of a number as the sum of fourteen and sixteen squares*, Quart. J. Pure and Appl. Math. **38** (1907), 178–236.

[115] J. W. L. Glaisher, *On the representations of a number as the sum of eighteen squares*, Quart. J. Pure and Appl. Math. **38** (1907), 289–351.

[116] L. W. Griffiths, *Representation of integers in the form* $x^2 + 2y^2 + 3z^2 + 6w^2$, Amer. J. Math. **51** (1929), 61–66.

[117] C. C. Grosjean, *An infinite set of recurrence formulae for the divisor sums, Part I*, Bull. Soc. Math. Belgique **29** (1977), 3–49.

[118] C. C. Grosjean, *An infinite set of recurrence formulae for the divisor sums, Part II*, Bull. Soc. Math. Belgique **29** (1977), 95–138.

[119] E. Grosswald, *Representations of integers as sums of squares*, Springer-Verlag, Berlin, 1984.

[120] S. Guo, H. Pan and Z.-W. Sun, *Mixed sums of squares and triangular numbers (II)*, preprint, 2007.

[121] H. Hahn, *Eisenstein series, analogues of the Roger-Ramanujan functions, and partition identities*, Ph. D. thesis, University of Illinois at Urbana-Champaign, 2004, 136pp.

[122] G. H. Hardy, *On the representation of a number as the sum of any number of squares, and in particular of five or seven*, Proc. London Math. Soc. **17** (1918), xxii–xxiv.

[123] G. H. Hardy, *On the representation of a number as the sum of any number of squares, and in particular of five or seven*, Proc. Nat. Acad. Sci. USA **4** (1918), 189–193.

[124] G. H. Hardy, *On the representation of a number as the sum of any number of squares, and in particular of five*, Trans. Amer. Math. Soc. **21** (1920), 255–284. (Errata **29** (1927), 845–847.)

[125] G. H. Hardy, *Ramanujan*, Chelsea Publ. Co., N.Y., Third Edition, 1978.

[126] G. H. Hardy and E. M. Wright, *An Introduction to the Theory of Numbers*, Fourth edition, Oxford University Press, 1960.

[127] K. Hardy, J. B. Muskat and K. S. Williams, *A deterministic algorithm for solving* $n = fu^2 + gv^2$ *in coprime integers u and v*, Math. Comp. **55** (1990), 327–343.

[128] K. Hardy, J. B. Muskat and K. S. Williams, *Solving* $n = au^2 + buv + cv^2$ *using the Euclidean algorithm*, Utilitas Math. **38** (1990), 225–236.

[129] D. R. Heath-Brown, *Fermat's two-squares theorem*, Invariant (1984), 3–5.

[130] C. Hermite, *Note au sujet de l'article précédent*, J. Math. Pures Appl. **13** (1848), 15.

[131] C. Hermite, *Oeuvres*, Vol. I (1905), Vol. II (1908), Gauthier-Villars, Paris.

[132] M. D. Hirschhorn, *A simple proof of Jacobi's four-square theorem*, J. Austral. Math. Soc. Ser. A **32** (1982), 61–67.

[133] M. D. Hirschhorn, *A simple proof of Jacobi's two-square theorem*, Amer. Math. Monthly **92** (1985), 579–580.

[134] M. D. Hirschhorn, *A simple proof of Jacobi's four-square theorem*, Proc. Amer. Math. Soc. **101** (1987), 436–438.

[135] M. D. Hirschhorn and J. A. Sellers, *On representations of a number as a sum of three squares*, Discrete Math. **199** (1999), 85–101.

[136] J. G. Huard, P. Kaplan and K. S. Williams, *The Chowla-Selberg formula for genera*, Acta Arith. **73** (1995), 271–301.

[137] J. G. Huard, Z. M. Ou, B. K. Spearman and K. S. Williams, *Elementary evaluation of certain convolution sums involving divisor functions*, Number Theory for the Millennium II, edited by M. A. Bennett, B. C. Berndt, N. Boston, H. G. Diamond, A. J. Hildebrand, and W. Philipp, A. K. Peters, Natick, Massachusetts, 2002, pp. 229–274.

[138] J. G. Huard and K. S. Williams, *Sums of sixteen squares*, Far East J. Math. **7** (2002), 147–164.

[139] J. G. Huard and K. S. Williams, *Sums of twelve squares*, Acta Arith. **109** (2003), 195–204.

[140] J. G. Huard and K. S. Williams, *Sums of sixteen and twenty-four triangular numbers*, Rocky Mountain J. Math. **35** (2005), 857–868.

[141] T. Huber, *Zeros of generalized Rogers-Ramanujan series and topics from Ramanujan's theory of elliptic functions*, Ph. D. thesis, University of Illinois, 2007.

[142] R. H. Hudson and K. S. Williams, *Binomial coefficients and Jacobi sums*, Trans. Amer. Math. Soc. **281** (1984), 431–505.

[143] G. Humbert, *Démonstration analytique d'une formule de Liouville*, Bull. Sci. Math. **34** (1910), 29–31.

[144] A. E. Ingham, *Some asymptotic formulae in the theory of numbers*, J. London Math. Soc. **2** (1927), 202–208.

[145] T. Jackson, *A short proof that every prime $p \equiv 3 \pmod 8$ is of the form $x^2 + 2y^2$*, Amer. Math. Monthly **107** (2000), 447.

[146] C. G. J. Jacobi, *Fundamenta Nova Theoriae Functionum Ellipticarum*, Sumtibus Fratrum Borntraeger, Regiomonti, 1829.

[147] C. G. J. Jacobi, *Über die Kreistheilung und ihre Anwendung auf die Zahlentheorie*, J. Reine Angew. Math. **30** (1846), 166–182.

[148] C. G. J. Jacobi, *Gesammelte Werke*, Vols. I–VIII, Chelsea Publishing Company, New York, 1969.

[149] E. Jacobsthal, *Anwedungen einer Formel aus der Theorie der quadratischen Reste*, Dissertation, Berlin, 1906.

[150] E. Jacobsthal, *Über die Darstellung der Primzahlen der Form $4n + 1$ als Summe Zweier Quadrate*, J. Reine Angew. Math. **132** (1907), 238–245.

[151] V. G. Kac and M. Wakimoto, *Integrable highest weight modules over affine superalgebras and number theory, Lie theory and geometry,* 415–456, *Progr. Math.,* 123, *Birkhäuser Boston, Boston, MA,* 1994.

[152] P. Kaplan and K. S. Williams, *On a formula of Dirichlet*, Far East J. Math. Sci. **5** (1997), 153–157.

[153] L. J. P. Kilford, *Modular Forms: A Classical and Computational Introduction*, Imperial College Press, London, 2008.

[154] H. D. Kloosterman, *On the representation of numbers in the form $ax^2 + by^2 + cz^2 + dt^2$*, Acta Math. **49** (1926), 407–464.

[155] H. D. Kloosterman, *On the representation of numbers in the form $ax^2 + by^2 + cz^2 + dt^2$*, Proc. London Math. Soc. **25** (1926), 143–173.

[156] G. Köhler, *On two of Liouville's quaternary forms*, Arch. Math. **54** (1990), 465–473.

[157] E. Krätzel, *Über die Anzahl der Darstellungen von natürlichen Zahlen als Summe von $4k$ Quadraten*, Wiss. Z. Friedrich-Schiller-Univ. Jena **10** (1960/61), 33–37.

[158] E. Krätzel, *Über die Anzahl der Darstellungen von natürlichen Zahlen als Summe von $4k + 2$ Quadraten*, Wiss. Z. Friedrich-Schiller-Univ. Jena **11** (1962), 115–120.

[159] D. B. Lahiri, *On Ramanujan's function $\tau(n)$ and the divisor function $\sigma(n)$-I*, Bull. Calcutta Math. Soc. **38** (1946), 193–206.

[160] D. B. Lahiri, *On Ramanujan's function $\tau(n)$ and the divisor function $\sigma_k(n)$-II*, Bull. Calcutta Math. Soc. **39** (1947), 33–52.

[161] H. Y. Lam, *The number of representations by sums of squares and triangular numbers*, Electronic J. Comb. Number Theory **7** (2007) A28, 14pp.

[162] E. Landau, *Elementary Number Theory*, Chelsea Publ. Co., New York, N. Y., 1958.

[163] A. M. Legendre, *Traité des Fonctions Elliptiques*, Vol. I (1825), Vol. II (1826), Vol. III (1828), Huzard-Courcier, Paris.

[164] A. M. Legendre, *Théorie des Nombres*, Vols. 1, 2, quatrième édition, Blanchard, Paris, 1955.

[165] D. H. Lehmer, *Some functions of Ramanujan*, Math. Student **27** (1959), 105–116.

[166] D. H. Lehmer, *Computer technology applied to the theory of numbers* in Studies in Number Theory, Math. Assoc. Amer., 1969, pp. 117–151.

[167] D. H. Lehmer, *Selected Papers*, Vols. I–III, Charles Babbage Research Centre, St. Pierre, Manitoba, Canada, 1981.

[168] M. Lemire and K. S. Williams, *Evaluation of two convolution sums involving the sum of divisors function*, Bull. Austral. Math. Soc. **73** (2006), 107–115.

[169] J. Levitt, *On a problem of Ramanujan*, M. Phil. thesis, University of Nottingham, 1978.

[170] J. Liouville, *Sur quelques formules générales qui peuvent être utiles dans la théorie des nombres,*
(*premier article*), J. Math. Pures Appl. **3** (1858), 143–152.
(*deuxième article*), J. Math. Pures Appl. **3** (1858), 193–200.
(*troisième article*), J. Math. Pures Appl. **3** (1858), 201–208.
(*quatrième article*), J. Math. Pures Appl. **3** (1858), 241–250.
(*cinqième article*), J. Math. Pures Appl. **3** (1858), 273–288.
(*sixième article*), J. Math. Pures Appl. **3** (1858), 325–336.
(*septième article*), J. Math. Pures Appl. **4** (1859), 1–8.
(*huitième article*), J. Math. Pures Appl. **4** (1859), 73–80.
(*neuvième article*), J. Math. Pures Appl. **4** (1859), 111–120.
(*dixième article*), J. Math. Pures Appl. **4** (1859), 195–204.
(*onzième article*), J. Math. Pures Appl. **4** (1859), 281–304.
(*douzième article*), J. Math. Pures Appl. **5** (1860), 1–8.
(*treizième article*), J. Math. Pures Appl. **9** (1864), 249–256.
(*quatorzième article*), J. Math. Pures Appl. **9** (1864), 281–288.
(*quinzième article*), J. Math. Pures Appl. **9** (1864), 321–336.
(*seizième article*), J. Math. Pures Appl. **9** (1864), 389–400.
(*dix-septième article*), J. Math. Pures Appl. **10** (1865), 135–144.
(*dix-huitième article*), J. Math. Pures Appl. **10** (1865), 169–176.

[171] J. Liouville, *Note à l'occasion d'un mémoire de M. Bouniakowsky*, J. Math. Pures Appl. **2** (1857), 424.

[172] J. Liouville, *Démonstration d'un théorème sur les nombres premiers de la forme* $8\mu + 3$, J. Math. Pures Appl. **3** (1858), 84–88.

[173] J. Liouville, *Nombre des représentatations du double d'un entier impair sous la forme d'une somme de douze carrés*, J. Math. Pures Appl. **5** (1860), 143–146.

[174] J. Liouville, *Sur la forme* $x^2 + y^2 + 2(z^2 + t^2)$, J. Math. Pures Appl. **5** (1860), 269–272.

[175] J. Liouville, *Sur les deux formes* $x^2 + y^2 + z^2 + 2t^2$, $x^2 + 2(y^2 + z^2 + t^2)$, J. Math. Pures Appl. **6** (1861), 225–230.

[176] J. Liouville, *Extrait d'une lettre adressée à M. Besge*, J. Math. Pures Appl. **9** (1864), 296–298.

[177] J. Liouville, *Sur la forme* $x^2 + xy + y^2 + z^2 + zt + t^2$, J. Math. Pures Appl. **8** (1863), 141–144.

[178] J. Liouville, *Sur la forme* $x^2 + xy + y^2 + 2z^2 + 2zt + 2t^2$, J. Math. Pures Appl. **8** (1863), 308–310.

[179] J. Liouville, *Sur la forme* $x^2 + xy + y^2 + 6z^2 + 6zt + 6t^2$, J. Math. Pures Appl. **9** (1864), 181–182.

[180] J. Liouville, *Sur la forme* $2x^2 + 2xy + 2y^2 + 3z^2 + 3zt + 3t^2$, J. Math. Pures Appl. **9** (1864), 183–184.

[181] J. Liouville, *Sur la forme* $x^2 + xy + y^2 + 3z^2 + 3zt + 3t^2$, J. Math. Pures Appl. **9** (1864), 223–224.

[182] J. Liouville, *Nombre des représentatations d'un entier quelconque sous la forme d'une somme de dix carrés*, J. Math. Pures Appl. **11** (1865), 1–8.

[183] Z.-G. Liu, *On the representation of integers as sums of squares*, Contemporary Math. **291** (2001), 163–176.

[184] Z.-G. Liu, *An identity of Ramanujan and the representation of integers as sums of triangular numbers*, Ramanujan J. **7** (2003), 407–434.

[185] G. A. Lomadze, *On the representation of numbers by sums of squares*, Akad. Nauk. Gruzin. SSR Trudy Tbiliss. Mat. Inst. Razmadze **16** (1948), 231–275. (in Russian)

[186] G. A. Lomadze, *Über die Darstellung der Zahlen durch einige quaternäre Formen*, Acta Arith. **5** (1959), 125–170.

[187] G. A. Lomadze, *Representation of numbers by sums of the quadratic forms* $x_1^2 + x_1 x_2 + x_2^2$, Acta Arith. **54** (1989), 9–36. (in Russian)

[188] G. A. Lomadze, *On the representations of numbers by a direct sum of the quadratic forms* $x_1^2 + x_1 x_2 + x_2^2$, Trudy Tbiliss. Univ. **288** (1989), 5–21.

[189] G. A. Lomadze, *On the representations of numbers by a direct sum of the quadratic forms* $x_1^2 + x_1 x_2 + 2x_2^2$, Trudy Tbiliss. Univ. **299** (1990), 12–58.

[190] L. Long and Y. Yang, *A short proof of Milne's formulas for sums of integer squares*, Int. J. Number Theory **1** (2005), 533–551.

[191] L. Lorenz, *Bidrag til tallenes theori*, Tidsskrift for Mathematik **1** (1871), 97–114.

[192] J. Lützen, *Joseph Liouville 1809–1882: Master of Pure and Applied Mathematics*, Springer-Verlag, 1990.

[193] P. A. MacMahon, *Divisors of numbers and their continuations in the theory of partitions*, Proc. London Math. Soc. (2) **19** (1920), 75–113.

[194] P. A. MacMahon, *Collected Papers*, Vol. I (1978), Vol. II (1986), MIT Press, Cambridge, MA.

[195] G. B. Mathews, *On a theorem of Liouville's*, Proc. London Math. Soc. **25** (1893–94), 85–92.

[196] G. B. Mathews, *On the representation of a number as a sum of squares*, Proc. London Math. Soc. **27** (1895–96), 55–60.

[197] E. McAfee, *A three term arithmetic formula of Liouville type with application to sums of six squares*, M. Sc. thesis, Carleton University, Ottawa, Canada, 2004.

[198] E. McAfee and K. S. Williams, *Sums of six squares*, Far East J. Math. Sci. **16** (2005), 17–41.

[199] E. McAfee and K. S. Williams, *An arithmetic formula of Liouville*, J. Théor. Nombres Bordeaux **18** (2006), 223–239.

[200] E. McAfee and K. S. Williams, *An arithmetic formula of Liouville type and an extension of an identity of Ramanujan*, JP J. Algebra Number Theory Appl. **6** (2006), 33–56.

[201] E. Meissner, *Über die zahlenthoretischen Formeln Liouville's*, Vierteljahrsschrift Naturforschende Gesellschaft in Zurich **52** (1907), 156–216.

[202] G. Melfi, *On some modular identities*, Number Theory (K. Györy, A. Pethö, and V. Sós, eds.), de Gruyter, Berlin, 1998, 371–382.

[203] S. Milne, *New infinite families of exact sums of squares formulas, Jacobi elliptic functions and Ramanujan's tau function*, Proc. Nat. Acad. Sci. USA **93** (1996), 15004–15008.

[204] S. Milne, *Infinite families of exact sums of squares formulas, Jacobi elliptic functions, continued fractions, and Schur functions*, Ramanujan J. **6** (2002), 7–149.

[205] L. J. Mordell, *On the representations of numbers as a sum of 2r squares*, Quart. J. Pure Appl. Math. **48** (1917), 93–104.

[206] L. J. Mordell, *On the representations of a number as a sum of an odd number of squares*, Trans. Cambridge Philos. Soc. **22** (1918), 259–276.

[207] C. J. Moreno and S. S. Wagstaff, *Sums of squares of integers*, Chapman and Hall/CRC, 2006.

[208] J. B. Muskat, *A refinement of the Hardy-Muskat-Williams algorithm for solving $n = fu^2 + gv^2$*, Utilitas Math. **41** (1992), 109–117.

[209] M. B. Nathanson, *Elementary Methods in Number Theory*, Springer New York, 2000.

[210] P. S. Nazimoff, *Applications of the Theory of Elliptic Functions to the Theory of Numbers*, Moscow, 1884. (in Russian)(American Edition by A. E. Ross, University of Chicago, 1928.)

[211] I. Niven, H. S. Zuckerman and H. L. Montgomery, *An Introduction to the Theory of Numbers, Fifth Edition*, John Wiley and Sons, Inc., New York, 1991.

[212] J. Nunemacher and R. M. Young, *On the sum of consecutive kth powers*, Math. Mag. **60** (1987), 237–238.

[213] B.-K. Oh and Z.-W. Sun, *Mixed sums of squares and triangular numbers (III)*, J. Number Theory **129** (2009), 964–969.

[214] C. D. Olds, *Continued Fractions*, Random House, New York, 1963.

[215] K. Ono, *Representations of integers as sums of suares*, J. Number Theory **95** (2002), 253–258.

[216] K. Ono, S. Robins and P. T. Wahl, *On the representation of integers as sums of triangular numbers*, Aequationes Math. **50** (1995), 73–94.

[217] C. O'Sullivan, *Identities from the holomorphic projection of modular forms*, Number Theory for the Millennium III, edited by M. A. Bennett, B. C. Berndt, N. Boston, H. G. Diamond, A. J. Hildebrand, and W. Philipp, A. K. Peters, Natick, Massachusetts, 2002, pp. 87–106.

[218] Z. M. Ou, *Unpublished research note*, 2000.

[219] T. Pepin, *Étude sur quelques formules d'analyse utiles dans la théorie des nombres*, Atti della Accademia Pontifica de Nuovi Lincei **38** (1884–85), 139–196.

[220] T. Pepin, *Sur quelques formules d'analyse utiles dans les théorie des nombres*, J. Math. Pures Appl. **4** (1888), 83–127.

[221] T. Pepin, *Sur quelques formes quadratiques quaternaires*, J. Math. Pures Appl. **6** (1890), 5–67.

[222] H. Petersson, *Modulfunktionen und quadratische Formen*, Springer Berlin-Heidelberg-New York, 1982.

[223] K. Petr, *O počtu tříd forem kvadratických záporného diskriminantu*, Rozpravy České Akademie Cisare Frantiska Josefa 1 **10** (1901), 1–22.

[224] K. Petr, *Über die Anzahl der Darstellungen einer Zahl als Summe von zehn und zwölf Quadraten*, Archiv Math. Phys. **11** (1907), 83–85.

[225] C. M. Piuma, *Dimostrazione di alcune formole del Sig. Liouville*, Giornale di Mat. **4** (1866), 1–14, 65–75, 193–201.

[226] B. van der Pol, *On a non-linear partial differential equation satisfied by the logarithm of the Jacobian theta-functions, with arithmetical applications.* Parts I, II, Indag. Math. **13** (1951), 261–271, 272–284.

[227] B. van der Pol, *The representation of numbers as sums of eight, sixteen and twenty-four squares*, Indag. Math. **16** (1954), 349–361.

[228] C. Radoux, *Une nouvelle formule de récurrence pour les sommes de Ramanujan*, Bull. Soc. Math. Belgique **27** (1975), 59–65.

[229] V. Ramamani, *On some identities conjectured by Srinivasa Ramanujan found in his lithographed notes connected with partition theory and elliptic modular functions*, Ph. D. thesis, University of Mysore, 1970.

[230] S. Ramanujan, *On certain arithmetical functions*, Trans. Cambridge Philos. Soc. **22** (1916), 159–184.

[231] S. Ramanujan, *On the expression of a number in the form $ax^2 + by^2 + cz^2 + du^2$*, Proc. Cambridge Philos. Soc. **19** (1917), 11–21.

[232] S. Ramanujan, *Collected Papers*, AMS Chelsea Publishing, Providence, RI, USA, 2000.

[233] R. A. Rankin, *On the representations of a number as a sum of squares and certain related identities*, Proc. Cambridge Philos. Soc. **41** (1945), 1–11.

[234] R. A. Rankin, *On the representation of a number as the sum of any number of squares, and in particular of twenty*, Acta Arith. **7** (1962), 399–407.

[235] R. A. Rankin, *Sums of squares and cusp forms*, Amer. J. Math. **87** (1965), 857–860.

[236] R. A. Rankin, *Elementary proofs of relations between Eisenstein series*, Proc. Roy. Soc. Edinburgh **76A** (1976), 107–117.

[237] H. Rosengren, *Sums of squares from elliptic pfaffians*, Int. J. Number Theory **4** (2008), 873–902.

[238] E. Royer, *Evaluating convolution sums of the divisor function by quasimodular forms*, Int. J. Number Theory **3** (2007), 231–261.

[239] O. Schreier and E. Sperner, *Introduction to Modern Algebra and Matrix Theory*, Chelsea Publ. Co., New York, 1955.

[240] J. A. Serret, *Sur un théorème relatif aux nombres entiers*, J. Math. Pures Appl. **13** (1848), 12–14.

[241] K. Shavgulidze, *On the number of representations of integers by the sums of quadratic forms $x_1^2 + x_1 x_2 + 3x_2^2$*, Int. J. Number Theory **5** (2009), 515–525.

[242] N.-P. Skoruppa, *A quick combinatorial proof of Eisenstein series identities*, J. Number Theory **43** (1993), 68–73.

[243] H. J. S. Smith, *De composition numerorum primorum formae $4\lambda + 1$ ex duobus quadratis*, J. Reine Angew. Math. **50** (1855), 91–92.

[244] H. J. S. Smith, *Report on the Theory of Numbers*, originally published as a Report of the British Association in six parts, I (1859), II (1860), III (1861), IV (1862), V (1863), VI (1865), reprinted in book form, Chelsea, NY, 1964.

[245] H. J. S. Smith, *The Collected Mathematical Papers of Henry John Stephen Smith*, Vol. I (1894), Vol. II (1894); reprinted in two volumes, Chelsea, New York, 1965.

[246] B. K. Spearman and K. S. Williams, *The simplest arithmetic proof of Jacobi's four squares theorem*, Far East J. Math. Sci. **2** (2000), 433–439.

[247] D. Stander, *Makers of modern mathematics: Joseph Liouville*, Bull. Inst. Math. Appl. **24** (1988), 59–60.

[248] G. K. Stanley, *On the representations of a number as the sum of seven squares*, J. London Math. Soc. **2** (1927), 91–96.

[249] M. Stern, *Eine Bermerkung zur Zahlentheorie*, J. Reine Angew. Math. **32** (1846), 89–90.

[250] Z.-W. Sun, *Mixed sums of squares and triangular numbers*, Acta Arith. **127** (2007), 103–113.

[251] R. Taton, *Biography of Joseph Liouville*, Encyclopaedia Britannica.

[252] J. V. Uspensky, *Sur la représentation des nombres par les sommes des carrés*, Communications de la Société mathématique de Kharkow (Sér. 2) **14** (1913), 31–64. (in Russian)

[253] J. V. Uspensky, *Note sur le nombre de représentations des nombres par une somme d'un nombre pair de carrés*, Bulletin de l'Académie des Sciences de l'URSS, Leningrad (Sér. 6) **19** (1925), 647–662.

[254] J. V. Uspensky and M. A. Heaslet, *Elementary Number Theory*, McGraw-Hill Book Company, New York and London, 1939.

[255] Y. Varouchas, *Une démonstration élémentaire du théorème des deux carrés*, LaCaverne, I. R. E. M. de Lorraine, France, Bulletin No. 6, février 1984, pp. 31–39.

[256] B. A. Venkov, *Elementary Number Theory*, Wolters-Noordhoff Publishing, Groningen, The Netherlands, 1970.

[257] A. Walfisz, *Zur additiven Zahlentheorie. VI*, Trav. Inst. Math. Tbilissi (Trudy Tbiliss. Mat. Inst.) **5** (1938), 197–254.

[258] E. T. Whittaker and G. N. Watson, *A Course of Modern Analysis*, Cambridge University Press, 1963.

[259] H. Wild, *Die Anzahl der Darstellungen einer natürlichen Zahl durch die Form* $x^2 + y^2 + z^2 + 2t^2$, Abh. Math. Sem. Univ. Hamburg **40** (1974), 132–135.

[260] K. S. Williams, *Heath-Brown's elementary proof of the Girard-Fermat theorem*, Carleton Coordinates, Department of Mathematics and Statistics, Carleton University, Ottawa, Ontario, Canada, January 1985, pp. 4–5.

[261] K. S. Williams, *An identity of Liouville*, Carleton Coordinates, Department of Mathematics and Statistics, Carleton University, Ottawa, Ontario, Canada, January 1992, pp. 3–4.

[262] K. S. Williams, *On finding the solutions of* $n = au^2 + buv + cv^2$ *in integers u and v*, Utilitas Math. **46** (1994), 3–19.

[263] K. S. Williams, *Bernoulli's identity without calculus*, Math. Mag. **70** (1997), 47–50.

[264] K. S. Williams, *On the relative sizes of A and B in* $p = A^2 + B^2$, *where p is a prime* $\equiv 1 \pmod 4$, Far East J. Math. Sci. **3** (2001), 129–132.

[265] K. S. Williams, *An arithmetic proof of Jacobi's eight squares theorem*, Far East J. Math. Sci. **3** (2001), 1001–1005.

[266] K. S. Williams, $n = \Delta + \Delta + 2(\Delta + \Delta)$, Far East J. Math. Sci. **11** (2003), 233–244.

[267] K. S. Williams, *A cubic transformation formula for* $_2F_1(1/3, 2/3; 1; z)$ *and some arithmetic convolution formulae*, Math. Proc. Cambridge Philos. Soc. **137** (2004), 519–539.

[268] K. S. Williams, *The convolution sum* $\sum_{m<n/9} \sigma(m)\sigma(n-9m)$, Int. J. Number Theory **1** (2005), 193–205.

[269] K. S. Williams, *On a double series of Chan and Ong*, Georgian Math. J. **13** (2006), 793–805.

[270] K. S. Williams, *The convolution sum* $\sum_{m<n/8} \sigma(m)\sigma(n-8m)$, Pacific J. Math. **228** (2006), 387–396.

[271] K. S. Williams, *On the representations of a positive integer by the forms* $x^2 + y^2 + z^2 + 2t^2$ *and* $x^2 + 2y^2 + 2z^2 + 2t^2$, Int. J. Modern Math. **3** (2008), 225–230.

[272] K. S. Williams, *On Liouville's twelve squares theorem*, Far East J. Math. Sci. **29** (2008), 239–242.

[273] D. Zagier, *A one sentence proof that every prime* $p \equiv 1$ (mod 4) *is a sum of two squares*, Amer. Math. Monthly **97** (1990), 144.

[274] D. Zagier, *A proof of the Kac-Wakimoto affine denominator formula for the strange series*, Math. Res. Letters Monthly **7** (2000), 597–604.

Index

283